T0318383

Sensory Panel Management

Related titles

Integrating the Packaging and Product Experience in Food and Beverages
(ISBN: 978-0-08-100356-5)

Developing Food Products for Consumers with Special Dietary Needs
(ISBN: 978-0-08-100329-9)

Discrimination Testing in Sensory Science
(ISBN: 978-0-08-101009-9)

Individual Differences in Sensory and Consumer Science
(ISBN: 978-0-08-101000-6)

Woodhead Publishing Series in Food Science, Technology and Nutrition

Sensory Panel Management

A Practical Handbook for Recruitment, Training and Performance

Lauren Rogers

ELSEVIER

WP
WOODHEAD
PUBLISHING
An imprint of Elsevier

Woodhead Publishing is an imprint of Elsevier
The Officers' Mess Business Centre, Royston Road, Duxford, CB22 4QH, United Kingdom
50 Hampshire Street, 5th Floor, Cambridge, MA 02139, United States
The Boulevard, Langford Lane, Kidlington, OX5 1GB, United Kingdom

Notices
Knowledge and best practice in this field are constantly changing. As new research and experience
broaden our understanding, changes in research methods, professional practices, or medical treatment
may become necessary.

Practitioners and researchers must always rely on their own experience and knowledge in evaluating and
using any information, methods, compounds, or experiments described herein. In using such information
or methods they should be mindful of their own safety and the safety of others, including parties for
whom they have a professional responsibility.

To the fullest extent of the law, neither the Publisher nor the authors, contributors, or editors, assume any
liability for any injury and/or damage to persons or property as a matter of products liability, negligence
or otherwise, or from any use or operation of any methods, products, instructions, or ideas contained in
the material herein.

Library of Congress Cataloging-in-Publication Data
A catalog record for this book is available from the Library of Congress

British Library Cataloguing-in-Publication Data
A catalogue record for this book is available from the British Library

ISBN: 978-0-08-101001-3 (print)
ISBN: 978-0-08-101115-7 (online)

For information on all Woodhead Publishing publications visit our website at
https://www.elsevier.com/books-and-journals

Working together
to grow libraries in
developing countries

www.elsevier.com • www.bookaid.org

Publisher: Andre Gerharc Wolff
Acquisition Editor: Megan Ball
Editorial Project Manager: Karen Miller
Production Project Manager: Swapna Srinivasan
Designer: Miles Hitchen

Typeset by TNQ Books and Journals

Cover Credit: Images printed courtesy of Campden BRI

This book is dedicated to Golden Wonder Cheesy Wotsits and all the panellists at Dalgety Plc – without whom this book would not exist.

Contents

Biographies

Lauren Rogers (Main Author)
Lauren Rogers is a freelance sensory science consultant who has over 25 years' experience in sensory science. She worked for Dalgety Plc/DuPont (a food ingredients company) and then moved to work for GlaxoSmithKline working on beverages and health-care products. Lauren has recruited and trained panellists for work on various foods and drinks, as well as several home and personal care products. Lauren is a Fellow of the Institute of Food Science and Technology (IFST) and an active member of the IFST's Sensory Science Group, maintaining their Foundation, Intermediate and Advanced sensory science qualifications. Lauren is also a Registered Sensory Scientist, a member of the Sensometrics Society, the Society of Sensory Professionals and the ASTM E-18 (Sensory) Committee.

Carol Raithatha
Carol is the Director of Carol Raithatha Limited, a UK-based consultancy offering sensory, consumer and food and drink research. Carol has worked in a range of consumer goods sectors including food and drink, personal care and packaging, as well in various contexts such as quality, research and development, new product development and applications and consumer and market insight. Her professional affiliations and accreditations include being a Registered Sensory Scientist, a Fellow of the IFST and committee member of its Sensory Science Group, a Certified member of the Market Research Society and a Member of the Chartered Management Institute and the Institute of Consulting.

Sue Stanley
Sue Stanley was until recently Director of Consumer Science Insight at Unilever Research, Port Sunlight, United Kingdom. She worked for 36 years in Consumer Insight in R&D, including Sensory Science. Her expertise and practical experience encompasses the scientific, statistical and managerial aspects of sensory and consumer science in global product innovation in the Personal Care, Home Care and Foods industries. Research she has managed and advised on has included multinational and single-country projects across Europe, United States, South America, Asia and Africa. Her academic background is in Mathematics (MA University of Cambridge, United Kingdom) and Mathematical Statistics (MSc University of Birmingham, United Kingdom), and for many years she was a Full Member of the UK Market Research Society, a Fellow of the UK Royal Statistical Society and a Chartered Statistician.

Preface

This book is designed for any person who finds themselves recruiting a sensory panel or sensory panellists and/or planning and running sensory panel sessions, in any situation, be it quality control, research or new product development. You might be a sensory panel leader, sensory manager, sensory scientist or quality control manager – basically any person who works with any type of sensory panel. You might be an experienced sensory practitioner looking for additional help and support or someone new to the science. If you are the latter, then welcome to the wonderful world of sensory science!

The book covers many different elements of working with sensory panels including the management, maintenance and motivational aspects as well as elements of ethics, health and safety and human resources. The book only covers human sensory panels: not panels consisting of animals for pet food or pet care products – although that could be a new venture. It is mainly concerned not only with analytical sensory panels (for example, quality control panels taking part in discrimination tests or panels developing momentary or temporal descriptive profiles) but also relevant for sensory consumer scientists. As sensory science covers both food and beverage applications as well as non-food or home and personal care products, examples of panel recruitment, training and maintenance are included for all these product types.

As a sensory panel is made up of individuals who will be different in terms of their sensory abilities, performance and personalities, managing a sensory panel is quite different to using instruments such as gas chromatographs or mass spectrometers. Anyone working with a sensory panel will need to have good skills in communication, organisation and management, to name a few. And the sensory panel needs to be carefully recruited, trained and maintained as it is the main source of your data that will help you reach your business decisions. Therefore taking care in the recruitment, selection of the training level required and management of the panel is critical. This book will give you guidance to help you become the best panel leader you can be and will also help you develop the skills that will enable you to have the best sensory panel you can have.

The information in the book is based on the standard sensory texts (for example, Lawless and Heymann, 2010; Stone et al., 2012; Meilgaard et al., 2016), published sensory standards (for example, from the International Standards Organisation and the ASTM), various publications from journals (such as The Journal of Sensory Studies and Food Quality and Preference) as well as (many) years of experience working as a sensory scientist in the industry and as a consultant. This book is not intended to replace any of the published information sources: instead it is designed as a practical support document to use alongside these invaluable texts. I hope you find the information helpful and constructive: do write and let me know.

Acknowledgments

I am extremely grateful to Carol Raithatha and Sue Stanley for the chapters on panel performance and human resources – I could not have done it without you!

Special thanks to Campden BRI for being kind enough to provide the cover photos.

And thank you to Lawrence Blackburn for the help with creating many of the figures in the book and for supplying me with copious cups of tea.

History of sensory panels

1

We could say that the use of sensory panels in a formal sense began in the 1930s, although of course, there was a large amount of research relating to sensory perception prior to this by the ancient Greeks; Aristotle described five of the senses in 350 BC, through the Middle Ages and the sensory studies conducted with animals, onto Descartes' work on vision in the 1600s, psychophysical research in the 18th and 19th centuries, to the work in the 19th century when touch, pain, hot and cold sensations were documented (Jung, 1984). In the early days of sensory evaluation, many assessments of products were made by company sensory experts using grading methods. Grading generally uses one expert or a small number of experts to make an assessment of the quality of, for example, wines, perfumes and dairy products, some of which are still in use today. An early publication about sensory grading (Crocker and Platt, 1937) stated that expert graders were trained but there was no mention of screening for sensory acuity; however, the experts did check their own assessments against that of colleagues or standard samples.

In 1936, Cover published the first sensory method which she called the 'paired-eating method': similar to the paired comparison discrimination test we use today. Cover recruited a group of people to conduct the paired tests on meat quality. In 1940 she made improvements to her method which included the number, selection and training of judges. As no standards or sensory textbooks were yet available to give her guidance, Cover (1940) states that, 'No method has yet been devised for detecting persons who will make superior judges for using the paired-eating method' (p. 391). Bengtsson and Helm (1946) discuss the choice of people to take part in 'industrial taste testing' (what we might call today analytical sensory tests), stating that there are three groups with differing sensory abilities: a small group of people who have higher sensory acuity, a larger group with average abilities and another small group with lower abilities. They also mention that people who regularly take part in product assessments can develop and improve their abilities to detect and communicate product differences. Bengtsson and Helm suggest the use of the 'triangular test' to select the most sensitive tasters and also mention that their abilities should be checked on a regular basis by monitoring test results. The authors also discuss the number of tasters to 'minimise the effect of chance', stating that 50 to 100 judges assessing unidentified (coded) samples, with forms completed independently of each other, gives excellent results. The Bengtsson and Helm (1946) paper makes an interesting read as it also describes the reasons why staff should not be used for 'mass tests': the terminology used at the time for consumer testing.

Helm went on to write a more detailed paper about the 'Selection of a Taste Panel' in the same year with Trolle (Helm and Trolle, 1946). As one of the first published papers in sensory science and the selection of a taste panel, it is well worth reading.

Sensory Panel Management. https://doi.org/10.1016/B978-0-08-101001-3.00001-X

The reason for the paper's authors' interest in developing better 'taste tests' can be summarised in a couple of sentences taken directly from the paper:

'The traditional manner in which taste tests were conducted was not satisfactory. In most cases we were able to establish only the fact that it was not possible to discern the difference between samples with any certainty' (p. 181).

The traditional manner they refer to is grading, as well as physical and chemical measurements for the various beer experiments they conducted. They set up a committee to develop taste testing so that more reliable results could be gathered, and one of the first investigations they carried out were experiments to determine the tasting abilities of people on the existing taste panel as well as all the staff at the brewery. There is a nice description of how the authors introduced the selection procedure to the staff: '… failure to qualify as an expert taster…' would not be detrimental to their career at the brewery as '… a keen palate is a gift of nature possessed by relatively few persons'. As mentioned earlier, the authors used the triangular test to determine acuity because it was related to the type of test the authors were interested in to determine the differences between beers. They devised a series of tests to find the correct level of 'difference' between the two beers so that the test would be neither too easy nor too hard. The test involved asking which *two* of the three samples were *similar* and also which was preferred, and in the later experiments the test set up was more akin to a 3-alternative forced choice (3-AFC) than a triangle, as the question asked, for example, which samples were strong in bitterness and which were weak in bitterness. The authors recognised that keeping the tasters interested in the experiments was key and so they gave them direct feedback about whether they were 'wrong' or 'right' in their sample choice and also explained what the differences were. They conducted 6878 tests altogether and used the chi-square test to determine which test results indicated statistical significance between the four pairs of products. They used the replicate data for each taster to determine if they could be classified as an 'expert': where the p-value for a pair of samples across all replicates was equal to 0.001. Of the 51 people who completed all the tests, only six were classified as expert for all four pairs of products. Twenty people were selected for the taste panel based on the highest percentages of tests correct, however, as each of the four pairs of products were designed to be different in an important aspect, those tasters who were often correct for a particular pair, say in the pair that were designed to be different in bitterness, were listed for selection for tests where bitterness was the attribute of interest. The authors also investigated the effect of age, occupation, experience in tasting and whether or not the taster was a smoker, as well as improvement in the results over the test, the effect of fatigue and memory.

Interestingly, time-intensity methods began to be developed in the 1930s (Holway and Hurvich, 1937) before descriptive profiling methods, and helped researchers realise that taste intensity was not a static measurement. But measuring the intensity of an attribute over time was fraught with issues before computers arrived in the laboratory. Constructing the curves, comparing the panellists' outputs and reducing the biasing effects of the clock or timekeeper, were all difficulties faced by the sensory scientists at the time. This resulted in some differences in panel recruitment, as panellists needed to be able to use different equipment to aid the collection of the data. Dijksterhuis and Piggott (2001) and Lawless and Heymann (2010) both give good reviews of dynamic flavour profile methods.

Developments in the late 1940s of further discrimination tests lead researchers to consider how best to recruit and train people for their tests, with consideration given to the potential fatigue and health of the 'taster', as well as their memory and sensory acuity. A group from the Carlsberg Brewery refer to their development of the triangle test method in the early 1940s, and it seems that these authors were also concerned about the difference between quality analysis and discrimination tests (Peryam and Swartz, 1950). The authors state that human behaviour *can* be dealt with scientifically, which was often disputed or simply not understood at the time. The authors created three tests: the triangle, duo-trio and dual-standard, for measuring sensory differences because they wanted more objective methods that were discriminative, not judgemental and also that use statistical analysis to give a more simple, direct and actionable answer.

Dove (1947) was also interested in discrimination tests and the choice of the correct panellist for the task as part of the 'Subjective–Objective Approach' suggested by the author. The author uses this terminology to elevate the importance of the 'subjective' assessments, which at the time were being 'discredited' and overlooked by the use of instrumental or 'objective' measures. Dove developed the difference-preference test which is basically the paired comparison with an added preference question using a 10-point scale: 'five equal degrees of acceptability and five equal degrees of non-acceptability are allowed'. The author also lists requirements for the laboratory where the tests are to be conducted (e.g., air conditioned, segregated booths, prescribed lighting), requirements for sample preparation (e.g., controlled quantity and temperature, hidden codes) and requirements for the judges (selection based on vocabulary, experience and ability in detecting small differences, as opposed to screening with basic tastes). Some other authors had begun this task, but this is one of the most complete lists of the time. An interesting aspect to this paper is the description of conducting taste tests with animals instead of humans on products such as lettuce and cabbage, where humans are 'confused' by the taste! It's interesting to consider what might have happened to sensory science had these ideas been extended.

Much of the interest in sensory methods around this time came with economic growth and the huge changes as a result of World War II. The 1940s and 1950s saw a vast amount of work on sensory testing, partly due to focus on nutrition during the war years and also due to the interest in the development of new food products by industry in general. In 1950, in an attempt to collect together all the information and make some recommendations for future food testing, the US Bureau of Human Nutrition and Home Economics held a conference (Dawson and Harris, 1951) which was attended mostly by academics and research associations (Howgate, 2015). The conference proceedings are available to download and really give an insight into their difficulties and dilemmas in the testing of food using sensory methods.

Around this time there was also much discussion about the type of person who was best recruited to be a sensory panellist (Helm and Trolle, 1946; Dawson and Harris, 1951; Ferris, 1956; Platt, 1937; Morse, 1942; Bliss et al., 1943; Dove, 1947).

Many groups advocated the use of trained staff: flavourists, brewers, product developers, due to their knowledge and experience, while others suggested that these people were too close to the product and the reasons for the testing, to be free from bias.

Due to the rapid development of new food products, it became more and more difficult for experts, who at the time were the main source of sensory data, to contribute across all quality and product development projects, and the work of authors such as Dove, as mentioned earlier, helped the industry realise that the expert's view was not necessarily related to the consumers' views. One particularly interesting paper discusses the need for more rigorous consumer testing, stating that previous research appeared inconclusive or transitory and results were not repeatable from study to study (Kiehl and Rhodes, 1956). Kiehl describes the two main research areas working on consumer preference measurements as the 'household' panel and the 'laboratory' panel and makes an important comment on the use of small numbers of people in 'difference-preference' methods, which was pretty much standard at the time, to determine consumer preferences:

> *'The inference of expert preferences to the great mass of consumers required*
> *a heroic assumption about the representativeness of experts' (p1337).*

These changes helped create the need for more detailed, applicable and valid data about the sensory aspects of food, and the Flavor Profile Method was created to help meet these needs (Cairncross and Sjostrom, 1950). This was the first descriptive profile method and is therefore a major landmark in the history of sensory science. The method used a small group of highly trained panellists to create a flavour profile using a consensus scoring method. The panellists do not use a scale as such to mark their judgement of intensity, but rather a number choice: for example, 'I think this is a 2'. Many publications discussed the uses, advantages and disadvantages of the method (Amerine et al., 1965) with the main concerns related to leader bias, the sensitivity of a 0–3 scale and the consensus scoring method.

At this point, we have the real beginnings of the differentiation in sensory science between the use of naïve consumers, trained panellists and experts, ready for the many discussions and heated debates about who actually is the right person to take part in these 'analytical' assessments. The debate is still ongoing and will probably continue, but consideration and discussion about the objectives and action standards for the data gathering and the subsequent use of the information can effectively guide the choice of panel type. Other aspects such as when the data are needed, resources, product type and test schedule will also help cement the decision. For more information on this aspect please see Chapter 13.

The 'contour' method was developed by Hall et al. (1959) for the production of profiles of paired samples, one of which was designated as the control. A small number of panellists then rated the deviation from the control for odour and flavour on a 0–5 scale for a number of samples compared back to the control. However, this method did not get taken up by the industry possibly due to its complexity (Amerine et al., 1965). In 1953, Dove developed a scale to try to standardise intensity measurements for taste. The scale was based on known concentrations of pure chemicals such as sucrose, to allow uniform comparisons across food stuffs. The scale itself has not been used extensively in sensory profiling but perhaps might be regarded as a precursor for the absolute scales used by some later profiling methods.

Other groups developed profiling methods based on the Flavor Profile Method for use on their product category. The first to be published was the Texture Profile Method (Brandt et al., 1963) which was very similar to the Flavor Profile Method but related to the mechanical (i.e., the response of the product to stress, e.g., hardness, chewiness), geometric (i.e., the size, shape and particle composition, e.g., crumbliness, grittiness, flakiness) and mouthfeel (i.e., surface attributes, e.g., oiliness, greasiness, moistness) characteristics. Again the method worked with a consensus scale but later versions of the method use a 15-point scale to measure texture attributes on standard reference scales that cover the entire range for a particular measurement. For example, the standard hardness scale goes from 1.0 (cream cheese) through 7.0 (frankfurters) to 14.5 (hard candy).

Gail Vance Civille developed the Spectrum Descriptive Analysis (SDA) Method in the 1970s based on her experience with the Flavor Profile and Texture Profile methods and hence the Spectrum method has many similarities to these two methods. Around the same time as the Spectrum profiling method was devised, the Quantitative Descriptive Analysis (QDA) profiling method was developed by Stone et al. (1974). This method was very different to the previously published methods. The first main difference is in the use of references. Both Spectrum and QDA use *qualitative* references where required, to ensure that the panellists are in agreement about what element of the product they are measuring, but only Spectrum uses quantitative references. These quantitative references mean that all panellists will agree that a particular reference is, say 5, on the 15-point scale; and therefore comparisons across many types of products and panels can be made. The second major difference is in the panel leader. In the Spectrum method the panel leader is an instructor, whereas in the QDA method the panel leader is a facilitator and does not take part in the product assessments. There are many other differences between the two methods, including panel recruitment criteria and time requirements for conducting the profiles: see Section 7.4.1 for more information.

The next change in sensory panels comes from the use of naïve panellists (e.g., consumers) in sensory tests. Perhaps the most quoted method is Free Choice Profiling (Williams and Langron, 1984). The main advantage of this method in comparison to the other profiling methods is in the sequence of events for the development of the profile. As each panellist works alone to develop their own list of attributes, there is no time requirement for training, moderation, discussion and agreement. Also, the panellists can be scheduled to assess the products at a time that suits them. The main disadvantage is in the analysis of the data and the interpretation of the results.

Many other methods now exist for the development of sensory profiles, perhaps the main method being various in-house hybrids of the published methods adapted for use with particular product types or for particular applications: for example, the Skinfeel Profile method (Schwartz, 1975), the Handfeel SDA (Civille and Dus, 1990), Quantitative Flavor Profile (Stampanoni, 1994) and the Dynamic Flavor Profile (DeRovira, 1996). Other descriptive profile methods exist, such as deviation from reference profile, polarised sensory positioning and pivot profile. For more information on each of the descriptive analysis methods see Section 7.4.1.

Various 'rapid' methods have also been published (of which Free Choice Profiling is often classified), for example, Flash Profiling, Projective Mapping/Napping, Check all that Apply and Free Sorting, to name a few (Valentin et al., 2012; Varela and Ares, 2014; Delarue et al., 2015). Note that although these methods are often referred to as rapid descriptive methods, not all of them are actually describing the products, but they do allow a comparison across products in various ways. Also, although the methods are often classified as rapid in relation to the standard sensory profiling methods, they are not necessarily more rapid than conducting a quantified descriptive profile with a ready-trained and experienced profiling panel (Stone, 2015). A great review of these alternate methods to a full profile is given by Valentin et al. (2012).

Part one

The recruitment of sensory panels

Prerecruitment of sensory panels

2

2.1 Different types of panellists

BS EN ISO 5492:2009+A1:2017 describes three different types of sensory panellist: sensory assessor, selected assessor and expert sensory assessor (see Table 2.1). They also describe a 'taster' as someone who assesses food with his/her mouth but mentions that this term is synonymous with his/her use of the term 'assessor'. The term 'sensory assessor' is further split into two parts: a 'naïve assessor' and an 'initiated assessor'. A naïve assessor is described as 'a person who does not meet any particular criterion', while an initiated assessor is described as someone who 'has already participated in a sensory test'.

The 'sensory assessor' is described as a person who takes part in a sensory test, so this can be, in essence, absolutely anybody. For example, in a study to determine the liking of a range of lagers or washing powders, the consumers would be taking part in a sensory test and might therefore be referred to as a 'sensory assessor'. In this particular case they might be a 'naïve assessor' if they do not meet any 'particular criterion', although generally consumers are selected based on their demographics or shopping habits, for example, or they might be an 'initiated assessor' if they have previously taken part in a sensory test. The standard defines a consumer as a 'person who uses a product' and therefore the definitions for assessor and consumer are not differentiated if both are 'taking part in a sensory test'.

The 'selected assessor' is described as someone who is selected based on their performance in a sensory test. So in this way, someone who has been successfully screened by taking part in sensory tests such as discrimination tests, a description of food odours or the ranking of hair switches, would be referred to as a selected assessor. An expert sensory assessor, on the other hand, is described as someone who is well trained, very experienced and produces reliable and consistent results. This is quite a jump from a selected assessor who may have only taken part in one or two sensory tests.

Incidentally BS EN ISO 5492:2009+A1:2017 also defines an 'expert' as we might read the definition in a dictionary: someone who has knowledge, experience and competence in a particular subject. There is no mention of any sensory assessments, although curiously the definition of 'taster' includes the term 'expert' but possibly should read 'expert assessor' instead.

The ASTM sensory terminology document (ASTM E253 – 17 – see Table 2.1) describes an 'assessor' in a similar way to the BS/ISO terminology document: as any person taking part in a sensory test, and they list the terms assessor, judge, panellist, panel member and respondent as basically meaning the same as an assessor, although they mention that sometimes these words have different meanings depending on the user. They do not list 'subject' with these terms as they describe this word as having a different meaning in home and personal care sensory assessments, where the assessor is the person making the sensory judgements and the 'subject' is the person who is being assessed. For example, the subject may be having their hair washed and assessed or their underarms assessed for deodorant efficacy.

Sensory Panel Management. https://doi.org/10.1016/B978-0-08-101001-3.00002-1

Table 2.1 Definitions of panels and panellists from two perspectives

BS EN ISO 5492:2009+A1:2017	ASTM E253 – 16
Sensory assessor, noun – any person taking part in a sensory test NOTE 1 A naïve assessor is a person who does not meet any particular criterion. NOTE 2 An initiated assessor has already participated in a sensory test.	**Assessor**, *n* – a general term for any individual responding to stimuli in a sensory test. DISCUSSION – The terms *assessor, judge, panellist, panel member* and *respondent* all have the same basic meaning, although sometimes different connotations. Usage of these terms varies with the training and experience of the investigator, habit, tradition, personal preference and other factors.
Selected assessor, noun – assessor chosen for his/her ability to perform a sensory test	**Trained assessor**, *n* – an assessor with an established degree of sensory acuity who has experience with the test procedure and an established ability to make consistent and repeatable sensory assessments. DISCUSSION – A trained assessor functions as a member of a sensory panel.
Expert sensory assessor, noun – selected assessor with a demonstrated sensory sensitivity and with considerable training and experience in sensory testing, who is able to make consistent and repeatable sensory assessments of various products	**Expert assessor**, *n* – an assessor with a high degree of sensory acuity who has experience in the test procedure and established ability to make consistent and repeatable sensory assessments. An expert assessor functions as a member of a sensory panel.
Expert, noun – in the general sense, a person who, through knowledge or experience, has competence to give an opinion in the fields about which he/she is consulted	**Expert**, *n* – a common term for a person with extensive experience in a product category who performs perceptual evaluations to draw conclusions about the effects of variations in raw materials, processing, storage, ageing, etc. Experts often operate alone.
Taster, noun – assessor, selected assessor or expert who evaluates the organoleptic attributes of a food product, mainly with the mouth NOTE The term 'assessor' is usually preferred.	**Judge**, *n* – See **assessor**. **Observer**, *n* – (1) an assessor in a visual sensory test. (2) A person who is watching an individual or group to collect information about behaviour, responses to products, test protocols or processes. **Panellist**, *n* – See **assessor**. **Panel member**, *n* – See **assessor**. **Respondent**, *n* – See **assessor**. **Subject**, *n* – (1) See **assessor**. (2) The individual to whom the stimulus is applied.

Table 2.1 Definitions of panels and panellists from two perspectives—cont'd

BS EN ISO 5492:2009+A1:2017	ASTM E253 – 16
	DISCUSSION – The subject and the assessor may not always be the same individual. In most food and beverage evaluations, the subject and the assessor are the same person. In many personal care evaluations the subject and the assessor may be different people. For example, in a study of shampoo, the subject is the person to whom the shampoo is applied, while the assessor is a different person who evaluates the sensory properties of the shampoo on the subject's hair. In some applications, the subject may not be a person, such as in the evaluation of pet care products.
Sensory panel, noun – group of assessors participating in a sensory test	**Panel**, *n* – a group of assessors chosen to participate in a sensory test. **Sensory panel**, n – a group of assessors used to obtain information concerning the sensory properties of stimuli.
Consumer, noun – person who uses a product	**Consumer**, n – the user or potential user of a product or service, who may participate in research tests to provide opinions of products, concepts or services.

The ASTM uses the terms 'trained assessor' instead of the ISO/BS definition of 'selected assessor'. They describe the trained assessor has having an 'established degree of sensory acuity', experience in the various tests they take part in and, like the BS/ISO definition of 'expert sensory assessor', state that the person is reliable and consistent in their results. The ASTM definition of 'expert assessor' is very similar to their definition of 'trained assessor' except that the word 'established' is replaced with 'high'. However, at the time of writing (2017) the ASTM were considering merging the terms 'trained assessor' and 'expert assessor' under the 'trained assessor' definition.

The ASTM definition of an 'expert' differs from the BS/ISO definition in that they include 'performs perceptual evaluations' in the definition, rather than just the dictionary definition of an expert (for example, a brick layer or orchestral conductor). They also refer to the fact that the 'expert' often works alone and are probably referring to people like sommeliers and graders in this definition.

Your type of sensory panel may also include people with a certain expertise such as chefs, hairdressers or flavourists. These professionals might be classified under experts perhaps, but as they are quite different, in my view, maybe 'specialised panellist/assessor' might be a suitable term to use.

Putting these definitions aside, the most important thing for panel leaders to know is whether their panellists have been screened and how much training and experience the panellists have (or need). There are different levels of training. A 'trained panel' mentioned in the literature may in fact not be a trained panel at all, but simply a group of consumers who have been screened and introduced to the method to be used. This is not necessarily a bad thing (except in that the nomenclature makes it confusing to determine the level of training) as often a panel does not need the same level of training as another panel: it will depend on the objective. Some panellists who have been trained in a particular method may have several hundred hours of experience but still be referred to as a trained panel. It would be useful to have some form of naming that makes it easier to determine the difference in training level. Naïve, informed, semi-trained, trained, highly trained and very experienced might be a good starting point.

Therefore it's a good idea to create your own definitions, especially where the standard definitions do not exactly match your requirements. For your definitions you might like to use some simpler terms to help the wider team understand the type of panellist required for each type of task. For example, if you run discrimination tests with internal staff you might use the terminology such as:

Staff: staff members who have not yet been screened or trained in sensory tests;
Screened assessor: someone who has been screened for sensory assessments, and was successful, but whose data are not currently being used in decision making as they have yet to be proved reliable and consistent;
Discrimination test assessor: someone who has been successfully screened for sensory assessments, and has proved, from at least five[1] validation tests[2], that they are reliable and consistent;
Highly trained discrimination test assessor: someone who has been successfully screened for sensory assessments, and has proved, from the monthly validation tests as well as participation on a weekly basis, that they are reliable and consistent.
Identified highly trained discrimination test assessor: a highly trained discrimination test assessor who has additional skills. For example, they might have skills in detecting a particular taint or off-note or be extremely discriminating. These people are particularly useful to call on for special projects.

The terminology used in this book:

Panellist: general term for someone taking part in analytical sensory tests.
Consumer: general term for someone taking part in consumer sensory tests.
Subject: as ASTM definition.

The training level for panellists can differ method to method and is specified in each of the sections accordingly.

---

[1] This number can be adjusted depending on your requirements.
[2] Validation tests are tests where you know the expected answer. For example, samples that have added ingredients, different processing, documented off-notes by the use of storage times or by 'spiking' (adding) particular chemical compounds. For more information, please see Section 7.7.

2.2   What is a sensory panel?

A sensory panel is simply a group of people who have been collected together to work on a particular sensory study or project. BS EN ISO 5492:2009+A1:2017 describes a sensory panel as a 'group of assessors participating in a sensory test' and also gives definitions for the various types of sensory assessors (see Table 2.1). Panels taking part in discrimination or descriptive tests are generally referred to as **analytical sensory panels,** whereas people taking part in sensory tests to determine elements of product liking or preference tend to be referred to as **consumer sensory panels**. It's not always obvious to someone external to the world of sensory science that sensory science actually includes both analytical and consumer measurements and hence the terminology is required to help highlight the fact.

The people in an analytical panel will generally be screened and trained if they are to regularly take part in sensory discrimination tests or perform many types of descriptive or temporal analyses. A sensory panellist is a skilled profession and the work they perform will have a major impact on the company's success. Therefore it is good practice to develop a screening and training programme that will result in an effective and reliable sensory panel.

The sensory screening tests involve checking that the people are suitable for the work by the use of various tests to check for any impairments and to determine their sensory abilities. For example, the people may be screened to check if they are colour blind or if they are unable to detect odours (anosmic). Figure 2.1 shows some typical screening tests. The screening is also necessary because different people will react differently to different sensory stimuli; however, remember that people selected for a sensory panel based on their sensory abilities can still vary in their output sample to sample, test to test and day to day. More information about screening and training people for analytical sensory panels is given in Chapters 5 and 6.

Asking a regular analytical panel about their likes and dislikes is folly: they will not be representative of the target market; they will be too analytical in their judgements and there probably will not be enough of them to give enough data. A larger

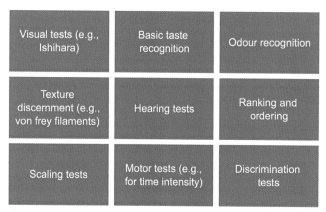

Figure 2.1 An overview of typical sensory screening tests for an analytical sensory panel.

group of people will generally be required to take part in hedonic tests (for example, 100 or more) and they will be selected dependent, for example, on their age, geographical location, buying habits or the types of products they like. If people are to assess products hedonically, i.e., answer questions about how much or why they like a particular product, then the chances are they will not be screened for sensory ability or trained, although there are some tests where screening and training consumers can help.

Screening for a hedonic type test is more related to choosing the right type of person for the test based on their demographics and/or purchasing habits: not their sensory ability. This group of people is generally called a **consumer sensory panel**. There are also smaller groups of consumers that might work together on an ad hoc or regular basis. In a **focus group** they might come together on the one occasion to talk about the project objective, for example, recycling food packages or brands of pasta purchased, and the discussion is led by a highly trained moderator. A more regular panel of consumers might be a **cocreation group** who are working with the company to help design new products or create new ideas. Consumers may also be asked to take part in discriminations tests, descriptive work or some of the more rapid or comparative sensory methods. The consumers in focus groups or cocreation groups are not generally referred to as *sensory* panels but the ideas and suggestions in this book will still be useful for working with these types of panels.

It's a good idea to clearly document the type(s) of panellists that you require for your tests as this will help in the choice of recruitment methods as described in Chapter 5.

2.3   Is a sensory panel required and if so what type of panel is needed?

2.3.1   Analytical sensory panel

Before recruiting an analytical sensory panel, consider how much work there will be for them on a regular basis and how long they will be required for. For example, if you currently run discrimination tests with internal staff and you would like to recruit a panel to replace a proportion of the internal staff in each test, you should be able to calculate the number of tests conducted each week and the time-saving element for the internal staff. Or maybe within the quality control team, which has three members of staff, you run quality checks on a daily basis but would like to extend the number of people in each test and also run discrimination tests at least once a week. If you have created several product profiles with internal staff and the work has been recognised for its importance and relevance to the extent that you need to create one or two profiles every week, you might find that you need to recruit two panels to keep up with the demand.

If there is the need for a regular analytical sensory panel, the next step is to decide what type of panel is required. If there is work for the panel for two or three hours a week or less, it would be worth considering setting up an **internal panel**. An internal analytical sensory panel is made up of people on site that have an interest in helping the company in the sensory assessment of products. They will already be employed by the company and might work in packaging design, finance or administration, for example. They might take part in regular discrimination tests or **quality control** assessments, for example, that are fairly quick to complete, but would not have time for longer studies such as descriptive profiling.

If there is interest in setting up a descriptive profiling panel or a panel working on time intensity measurements, for example, but there may not be regular work for the panel, especially to keep their training at the required level, an option is to use a panel at a sensory agency. The agency might have a suitable panel already employed that has time to complete your company's work or they might recruit a panel specifically for your product type(s). Having this type of **outsourced panel** is a good step towards employing a company panel, as it allows the sensory team to follow the recruitment and training process and to also find out first-hand how much work your company might have for a regular analytical sensory panel.

If there is enough work for a panel every week for around nine or more hours, the recruitment of an **external panel** might be a good option. This does not mean that they will work outside in the cold but simply that they are recruited externally to the company in much the same way as a new employee might be recruited. These panellists might be recruited as a regular permanent member of staff or they might be recruited through an *employment* agency; this mainly depends on your company's employment strategy. Some recruitment companies specialise in the management of sensory panels and will look after the whole process of recruitment for you.

Although this is generally untrue, external panels are often reported to be more expensive than using an internal panel, probably because the associated costs are more visible than those of the 'lost' hours from the packaging design, finance or administration teams.

If, for example, the sensory assessments for the external panel will involve some home and personal care products or products which cause excess sensory fatigue, the assessments may take place at the panellists' homes rather than a company site, and the panel might be referred to as an **external work-from-home panel**. Training programmes for these types of panellists require more care and organisation, as the assessors may find it more difficult to attend sessions during the usual working day. Developing a 'reminder pack' for these panellists to keep at home can help keep the panellists on track. Videos from the training sessions can be especially helpful for work-from-home panels.

The cut-off between analytical and consumer sensory data (and panels) used to be quite clear cut: consumers told us about how much they liked a product (and maybe some diagnostics using just about right scales) and screened and trained panellists told us about differences between samples either by discrimination or descriptive methods. Nowadays, consumers are invited to take part in discrimination tests, do descriptive profiling and are also screened and invited to focus groups where products are assessed. Figure 2.2 shows the main types of sensory panels with a (✓) indicating a possible use and ✓ indicating a typical use. The choice of panellist for the role is key to the correct decision being made. For example, if we recruit consumers for a discrimination test a large proportion will not be able to discriminate even if it is a product they often buy and

Panel focus		Naïve/consumer (Occasionally screened)	Trained Screened	Highly trained Screened
Analytical tests	Quality control		✓	
	Discrimination tests	(✓)	✓	(✓)
	Descriptive profiles	(✓)	✓	✓
	'Rapid' methods	(✓)	✓	(✓)
	Temporal methods	(✓) TDS not TI	✓	✓
Consumer tests	Hedonics (preference and acceptance tests)	✓		
	Attribute diagnostics (just about right scales, intensity ratings, liking of attributes)	✓		
	Understanding emotions	✓		
	Discussion (focus groups, creative groups)	✓		

Figure 2.2 The main types of sensory panels.

use (Stone, 2015). Therefore the tendency is to overrecruit and conduct discrimination tests with large numbers of consumers. However, the issue then is the risk associated with finding a statistically significant difference that has no practical significance. This is not necessarily an issue: it depends on your test objective and experimental design. You just need to be aware of the issues and choose your panel accordingly. A similar issue occurs when using highly trained panellists to conduct rapid methods such as projective mapping. They may not be able to easily view the products holistically as they are so used to dissecting the product by modality and by attribute.

You may need to recruit a panel with a more technical learning, for example, a panel of hairdressers, chefs or fragrance experts. If your objectives are related to technical learnings, for example, in the case of hairdressers, the translation of the technical measurements into formulation changes on the basis of their knowledge and expertise, your approach to recruitment might be quite different to any of the panels listed above. From the starting point, your recruitment may well be internal staff, especially if the panel will be working as fragrance experts or chefs, assessing products for quality control and determining the reasons for samples not meeting sensory specifications. In the hairdresser example, your recruitment may well be external and the advertising process will be different to other internal and external panel recruitment programmes. The reasons behind the use of a '**specialist or technical panel**' versus a conventional panel will need to be well thought out and documented, as the costs associated with this type of panel may well be much higher.

2.3.1.1   Advantages and disadvantages

One of the main advantages of an **internal panel** is that the people will be on site and easy to contact to arrange a panel session: particularly useful when a panel is required at short notice. However, this is also related to the main disadvantage: people are generally focussed on their main job and may view your request to attend a sensory panel as lower priority. They might also be less focussed on the sensory task if they are still thinking about the report they need to finish (that you called them away from). Internal panellists may be harder to motivate to attend panel sessions and tend to be free for shorter periods of time than externally recruited panellists. You might find it hard to find enough employees with the skills you need to build your panel and therefore you might have to settle for less than ideal panellists. They might also be biased about the products: this can be a positive bias in that they find it difficult to describe any product the company makes as bad in any way, or the opposite, where the panellist lists off-notes that do not exist because they have some ulterior motive or are simply fed up with their boss or colleagues. There might also be bias about the project if the panellist knows too much about the objectives. For example, if the panellists know that the project is related to cost savings in a particular ingredient, they may well mention that there is a difference in this aspect, when actually the difference is not perceivable. You will also need to consider issues associated with hierarchy: do not let yourself get lead into the situation where whatever the most senior person in the room says is agreed on by everyone else. However, you would not need to pay your internal panellists yourself and there are no confidentiality issues associated with testing with employees.

A panel made up of **external panellists** is often described in the literature as being more expensive, but when you consider the hourly rate for the finance director to take part in your sensory tests, for example, an externally recruited panel can be cheaper. The panellists tend to be more focussed on the task at hand and are often very motivated to do a good job, particularly in the hands of a good panel leader. You might find that you also have more time available for panel training. The panellists will not be biased by knowledge about the projects or reasons for the sensory tests: but be careful about who else on the site certain panellists might know, as the bias can often creep in when you are least expecting it. There are some disadvantages to an external panel, the main one being the time and effort involved in recruiting and training, which can be even more difficult to justify if you have a high panel turnover. Also, more planning is required to deal with experimental design and panel fatigue, as often panel sessions are planned for two or three consecutive hours of assessments. With an internal panel, the panellists can be sent back to their main role and called back for more sensory tests later. With an external panel you may find yourself finding something for the panel to do while you wait until the next sample can be assessed. An external panel tends to become a quite social group and when one panel member leaves or you have to ask a panel member to leave due to acuity issues, for example, you may find yourself having to deal with the fall out. Table 2.2 summarises the advantages and disadvantages of recruiting internally or externally.

Table 2.2 Advantages and disadvantages of internal and external panels

| Internal panel | | External panel | |
Advantages	Disadvantages	Advantages	Disadvantages
Panellists are easy to contact, and it can be quick to set up a panel	Other jobs may take priority	They can be highly motivated and focussed on the role	Experimental design and fatigue needs to be managed carefully
Confidentiality is kept within the company	Product and project bias	There is more time for panel training	Often seen as more expensive
The panellists are already paid	May be more difficult to motivate to attend panel sessions	They do not know anything about the various reasons for the sensory tests	Can be more difficult to remove poor panellists
Product knowledge can be helpful in identifying references	The pool of people to screen and train may not be large enough to give you the best panellists for the role	The number of people available for the screening is much larger	You will have more paperwork associated with panel payments, holiday planning and other such employment issues
	Issues with hierarchy		Panellists may become 'bored' with the role as it is their only function
	If people do not pass the screening tests, it can be an awkward thing to communicate		
	When people change job roles, you may lose your best panellists		
	It can be difficult to get people to commit to the amount of time needed for descriptive work		
	The person's usual work can be disrupted		

2.3.2 Consumer sensory panels

Consumer sensory panels (BS ISO 11136, 2014) tend to be easier to manage as generally they will only take part in tests infrequently. There is the option to outsource this recruitment through a sensory agency that may have a database of suitable consumers or they might recruit specifically to your needs. Alternatively, the recruitment could be done in-house and this is fairly easy if small numbers are being recruited, for a focus group, for example, but there are several things to consider before you start recruitment. The ASTM, the Market Research Society (MRS) and ESOMAR have some excellent publications on consumer testing, detailing many of the critical things you need to consider before, during and after the tests. For more information about the recruitment of a consumer sensory panel, please see Section 2.6.3 and 5.5.

The first consideration is for health and safety and ethics as well as meeting local regulations about consumer testing. Company procedure relating to the health and safety of visitors on site will need to be checked and whether or not ethics approval is required will depend on the product being assessed. You will need to be careful about the age of the people recruited as there are particular rules about, for example, testing on children. The MRS and the ASTM both have excellent publications about consumer research with children (MRS, 2014; ASTM, 2010, 2011). If you are testing alcohol or novel foods, for example, there will be other considerations to take into account such as informed consent and travel arrangements. The ASTM's *Standard Guide for Sensory Evaluation of Beverages Containing Alcohol* is a very useful document with information about consumer instructions, dealing with alcohol abuse and consumption guidelines. The Advisory Committee on Novel Foods and Processes (ACNFP) have published guidelines on conducting of taste trials involving novel foods or food produced by novel processes (ACNFP, 2017).

The second main consideration is about data storage. Confidentiality is key and all documentation (electronic and paper) must be stored accordingly. There are new guidelines issued by the EU that will affect consumer testing globally. These are called the EU General Data Protection Regulation and you can find more information in Chapter 3 and online on the MRS and ESOMAR websites.

2.3.3 Justification for your sensory panel

You will probably find that you have to justify your panel, whether they are internally or externally recruited, and this will involve gathering various pieces of information. You already have information about the amount of work there is currently for a panel, but you might like to get out your crystal ball and think about future resource requirements. For example, you might think that at present you need just one panel, but what about if the company expands its product portfolio, or there is a requirement for sensory science to be implemented across the globe? Asking colleagues, who are also your customers, for advice can be really helpful, as well as discussing resources with other sensory scientists through a subject interest group or sensory science conference.

You need to have documented why you actually need a sensory panel to help with your justification. This will include the sensory methods you propose to use; the facilities you will need; the team structure as well as the reporting structure; any additional staff that need recruiting and their level of expertise; and the support you will need for the recruitment, training and management of the panel.

For example, you might write:

> *In the past two years, we have been using our internal sensory panellists to take part in discrimination, shelf life and quality control tests. The internal panellists come from the quality control team, finance, administration and human resources. At the beginning of the first year we had requests from our internal clients for around one to three tests per week and this has steadily increased to around eight to ten test requests per week. The internal panellists originally were each spending around 15 to 20 minutes per week taking part in sensory tests and this has recently increased to around 60 minutes each per week. We are finding it increasingly difficult to get the number of panellists required to take part in each test, due to the number of tests and also the recent increase in the number of similarity tests requested, as these tests generally require more participants. We have also had interest in developing sensory profiles/ specifications as part of the new consumer-led quality control programme. We have created sensory profiles/specifications for three of our largest brands and these have been very successful in helping the various factories create a more consistent product. However, the time required from the internal panellists to create these specifications is around three hours per product. We are also concerned about the bias (e.g., stimulus error) that may occur given the fact that the people setting up the samples (QC staff) are also taking part in the assessments. We would therefore like to recruit a sensory panel external to the company. This panel is envisaged working for nine hours each week and the main role will be developing and maintaining the sensory specifications. They will also be used to make up the numbers for the discrimination, shelf life and quality control tests as required. Additional staff [describe] are also required. This has been costed at [insert amount] for the first year.*

If you do not have any sensory facilities, you will need to include this in your budget, but if you are just starting out, you can use a conference room and temporary booths, unless you have products that require complex preparation and the preparation area is not suitably located near to a conference room. You may also have to include additional staff in your justification. For more information see Section 2.7.1 for facilities and Section 2.7.2 for staffing.

2.3.4 Other types of panel

In some cases you might require a panel who actually have a sensory impairment. For example, you might be recruiting a panel to test the efficacy of hearing aids and require people with hearing impairment as well as the standard listening panel. Or maybe you are researching taste and require a panel of anosmics to give you the information you require. If you are recruiting a sensory panel to take part in gas chromatography-olfactometry (GC-O), you might wish to give them a trial on the GC-O to see how they fare.

Whichever type of panel you require, simply select and deselect the modules you require from the various screening tests in Chapter 5 depending on the type of panel you need. However, for some panel types you may not find the screening module you require. For example, if your research is related to studying tastant release during eating, you might need to add in some screening about saliva flow and discuss if the person is happy taking part in the assessments you are proposing. Information about these types of screening tests is not covered in this book, but information is available in the literature (for example, see Salles, 2017 for a good review).

2.4 Developing a job description for a sensory panellist

Before you start on your recruitment process it can be helpful to create a job description for your sensory panellists. You will not need this for consumer sensory panellists as they will only be called on for certain tests. The job description will help the panellists understand what is expected of them and also help you and anyone else who is working with you on the recruitment process, understand what you need. Consider the job role and discuss with your colleagues and human resources to decide what will be included. Figure 5.2 lists the attributes of a good panellist and may be helpful in developing your job description. Figure 2.3 gives an example job description for an externally recruited sensory panellist working on discrimination tests, descriptive profiles and temporal methods. (Albion Mills is a fictitious food factory.)

2.5 Approach to recruitment

2.5.1 Approach to recruitment for an external analytical sensory panel

Many companies who recruit a sensory panel externally follow the route as shown in Figure 2.4, where the employment step directly follows the selection based on the screening results. However, BS EN ISO 8586:2014 recommends a different top-line approach as shown in Figure 2.5, where training and validation comes before employment. The panellists are essentially recruited on a probationary period which ends once the training programme and validation is complete.

There are many advantages to the BS/ISO approach. It allows you to fully assess the panellists, both in terms of their sensory abilities and their personalities, and by having the panellists take part in many different types of activities, discussions and tests, you will be able to build up a good impression of their sensory abilities. You will also get a better understanding of how they might fit into the team, how well they understand and follow instructions and also their abilities at more in-depth skills such as the use of line scales. Panellists may well 'come out of their shells' and show themselves in an unexpected light and this may result in them not being suitable for the panel. Some panellists may also find that actually they are not suited to the role. Maybe they thought they would be assessing products and telling the product

JOB TITLE

Sensory Panellist

MAIN PURPOSE AND SCOPE OF THE JOB

Provide consistent and reliable sensory data to aid research projects for Albion Mills.

POSITION IN ORGANISATION

Reports To: Sensory Panel Leader

DUTIES AND KEY RESPONSIBILITIES

Evaluate the sensory characteristics of Albion Mills' products.

Create descriptions and temporal profiles of products.

Take an active part in sensory training sessions and apply this learning to the role.

Person specification template

This section details the qualities, skills and experience.

Requirements
General good health, good dental health, no allergies, no intolerances
Good sensory abilities
Good descriptive ability
Able to follow instructions
Prepared to try a range of different foods and ingredients
Apply training to job role, take onboard feedback
Commitment to long term role
Good time keeping and flexible
Keen interest in food
Excellent interpersonal and teamworking skills
Computer literate
No skills or experience necessary – all training will be provided

Figure 2.3 Example job description for an external sensory panellist.

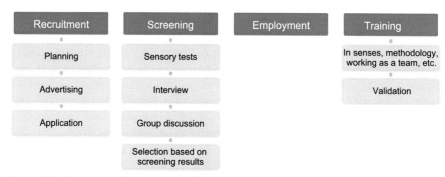

Figure 2.4 One approach to panel recruitment.

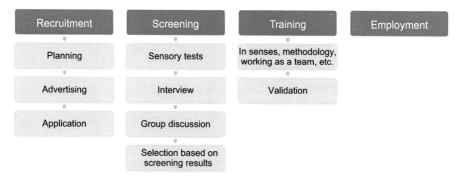

Figure 2.5 ISO 8586:2014 top-level approach to recruitment.

developers what to do, or had another completely different impression of the role, or they just do not seem to fit in with the team. Either way having the probationary period makes it easier for the panellists to bow out gracefully or for you to tell them that they have been unsuccessful. This makes this approach, of employment for a probationary period, seem ideal; however, there are some drawbacks. Firstly, if you are going to lose some panellists through this approach, you will need to recruit more people than you will need. This may not be an issue if you have the facilities and resources to cope with training more panellists. Secondly, people may make friends with the panellist who is not successful, and this can cause bad feelings among the remaining panellists who may think their friend was perfectly capable of doing the role. You might also end up with more paperwork to deal with as you will have recruited more panellists than you need. But I think the benefits outweigh the issues. You need to have the best people you can get on your sensory panel and you do not want to invest a lot of time training them to find that they are, after all, unsuitable or that they leave because it was not quite the role they imagined. In either case you will be in the situation where you have to go through the recruitment steps all over again: not the best outcome.

2.5.2   Approach to recruitment for an internal analytical sensory panel

The approach to the recruitment of an internal analytical sensory panel is more straightforward than the external approach as the potential panellists are already employed by the company. There are, however, some additional things to consider. The first is getting management buy in for the recruitment before you start. This will help because otherwise managers may start to complain about staff leaving their main roles to take part in a sensory assessment. You will also need to consider what to do if people do not pass the screening test. In an old publication when the triangle test was first used in the testing of beer, there is a nice description of how the authors (Helm and Trolle, 1946) introduced the selection procedure to the staff: '… failure to qualify as an expert taster …' would not be detrimental to their career at the brewery as '… a keen palate is a gift of nature possessed by relatively few persons'. An approach like this might be useful when you have to tell internal panellists that they did not quite make the grade. For more information about recruiting an internal panel, see Section 5.2.

2.5.3   Approach to recruitment for a consumer sensory panel

You may have your own database of consumers (also referred to as respondents in some publications) or you might recruit based on each project you conduct. Either way, initial recruitment can be done through many different routes. You might recruit by telephone, advertising (in newspapers for example), through the Internet, by email or post, or by intercept in the local shopping mall or high street. The calibre of the people recruited will have a huge impact on the results of the test. There is no point conducting the test on disinterested and coerced consumers who will simply be box ticking. If you recruit from an existing database be careful that you are not 'training' the consumers on a particular product type. The BS ISO standard suggests leaving three months between tests on the same product type (BS ISO 11136, 2014).

The manner in which the test is conducted depends on the objectives of the study, the products being tested, who you need to recruit, what questions you need to answer, what facilities you might need, the statistical analysis and the action standards for the test. The type and number of people you need will not only depend on the study objective but may also depend on whether the product is already on the market and what segmentation you are planning (the more splits you do, the more people you will need to make sure that each group is large enough to help you make decisions).

The ASTM, the MRS and ESOMAR have some excellent publications on consumer testing, detailing many of the critical things you need to consider before, during and after the tests. For more information about the recruitment of a consumer sensory panel please see Section 5.5.

2.6   Other considerations

2.6.1   Facilities

Apart from the type of panel and panellists, there are many other things that need to be considered prior to setting up a sensory science unit. Perhaps the most important is the facilities required. If the plan is to set up and run a regular internal or quality control panel, the facilities you need may simply be a quiet, clean, well-lit and temperature-controlled room such as a conference room with separate tables to segregate the assessors. If the products require complex preparation or cooking, a food preparation area will also be required and the related hygiene, health and safety rules adhered to.

If the plan is to conduct regular hedonic consumer tests, investment in sensory facilities such as booths and sample preparation areas may be necessary (BS ISO 11136, 2014). Separate tables in a large conference room and collection of data on paper or through tablet computers are another option. If regular focus groups are going to be conducted, a small discussion room with easy chairs or sofas to make the consumers feel at ease may be well worth the investment. Two-way mirrors can also be useful to enable other members of staff or clients watch and listen to the group. Recording equipment can be very helpful with the analysis of the information.

If a regular analytical sensory panel working on descriptive profiling or temporal methods is planned, investment in a fully equipped sensory facility with separate booths, a panel discussion area, a sample preparation room, storage facilities and computers to collect the data will be a must. Controlled temperature and lighting, positive and filtered air handling systems with a quiet and non-distracting working area will also be required. If you are working with a panel for the assessment of sound, you may need a 'listening room' that eliminates external sounds. This might also be the booth for the assessments of sound recordings, although standard design sensory booths can also be used, adapted with headphones. With home and personal care products that may require showering facilities, aroma assessment booths or special sinks, for example, will also need to be planned and budgeted for.

There are many different designs for sensory facilities, even just for food assessments, but the most critical aspect is that the assessments are conducted under 'known and controlled conditions with a minimum of distractions and to reduce the effects that psychological factors and physical conditions can have on human judgement' (BS EN ISO 8589:2010+A1:2014 *Sensory analysis. General guidance for the design of test rooms*). For the sensory booths themselves the most important factor is space: space for the panellists to be comfortable and space for the sample assessments they need to do. The connection between the sample preparation area and the booths can be difficult to design but are incredibly useful to have. Some sites do not have any connection and the samples are delivered via a trolley or by hand from behind the panellist. The choice will depend on the type of samples you have. There are two main ways of allowing samples to be passed to the panellists when the booths and the sample preparation areas do connect: through a bread bin or a sliding door hatch. Figure 2.6 and Figure 2.7 give examples of both types of hatch access between the preparation area and the booths or discussion room.

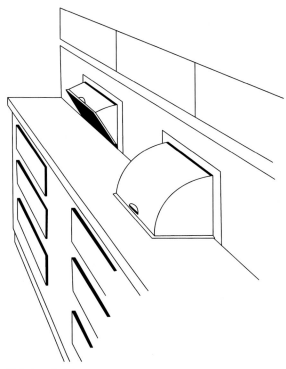

Figure 2.6 Bread bin type hatch access.
Reproduced with kind permission from ASTM Physical requirement guidelines for sensory evaluation laboratories: Special Technical Publication (STP) 913 copyright ASTM International, 100 Barr Harbor Drive, West Conshohocken, PA 19428.

The advantage of the bread bin-type connector is that the panellist cannot see into the preparation area and if you are using coloured lighting in the booths, the bread bin approach prevents any issues with the panellists seeing the samples under white light whilst they are on the preparation area worktop. The disadvantage is that they take up space in the sample preparation area where you might like to lay out samples ready for the next test. But again, it depends on the sample type and whether you would need to do this. The sliding door (sliding horizontally or vertically) seems to be quite popular but it does have issues with the panellist being able to see into the preparation area (unless you stand strategically) and the issue with seeing the samples under white light. Some people have a hatch which opens like a door but this can be difficult to use as it is easy to accidently knock over samples as it opens.

As mentioned earlier, some facilities require specific types of booths for the assessment of their product types. For example, if the facility assesses toilet cleaners, each booth may be fitted with a toilet to enable the assessment. The construction of these types of booths is similar to those for fragrance assessments which are sealed to prevent cross-contamination between the booths. Some booths may be fitted with a small assessment window so that the odour can be assessed without opening the booth fully.

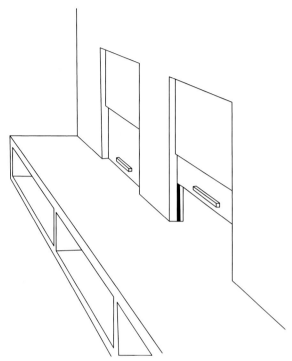

Figure 2.7 Sliding hatch access.
Reproduced with kind permission from ASTM Physical requirement guidelines for sensory evaluation laboratories: Special Technical Publication (STP) 913 copyright ASTM International, 100 Barr Harbor Drive, West Conshohocken, PA 19428.

This is particularly useful for the assessment of masking products such as cat litter and air fresheners. Some booths may be fitted with sinks to allow the panellist to expectorate. If you would like to include sinks in your facilities, be careful to check the set-up of the waste traps as this can sometimes allow malodours to leak back into the booths.

Coloured lighting can sometimes be used to hide differences in appearance which are not pertinent to the test, but be careful that the lighting does not create some other clue for the panellists to pick up on. This is particularly important when running discrimination tests. One approach can be to set up one or two booths with the test you are planning and ask members of staff (who have no issues with colour assessments) if they can complete the test by appearance alone. For example, if they can repeatedly pick the correct odd sample in a triangle test or match the right coded sample to the reference in a duo-trio under white light by appearance alone, you can change the presentation order and ask them to assess the samples again under each of the coloured lights you have access to. Many facilities have white light, red light and blue light. Dimmers can also be useful.

Home and personal care booths are often fully enclosed with an actual 'person-size' door that can open onto a corridor. Samples are given to the panellists through the door of the booth or through a hatch to the side of the door. This allows the easier set up of equipment for sample assessment (for example, trolleys of crockery or laundry to be washed) and larger samples such as cat litter trays to be assessed for aroma or racks of clothes to be ironed. These types of enclosed booths can also be fully evacuated after use which is also very useful for the assessment of odours such as deodorants, pet food and fragrances.

One vital element when considering your connection between the preparation area and the booths is the height of the worktops. If the panellists are sitting on office-type chairs, the worktop height might be similar to that of a computer desk and this can cause issues for people in the preparation area as they have to bend down to the hatches to pass through the samples. One way around this is to have the panellists' worktops at the height of a kitchen worktop and give them tall stools to sit on. This can cause issues for some panellists who find it difficult to climb onto a stool or whose wheelchair does not extend to that height. I recently visited a site where they had neatly got around both issues. The panellists' worktops were at desk height and the hatches were at kitchen worktop height due to a clever design of different floor heights in the two areas.

You may need somewhere for the panellists to wait prior to a test or in between tests. This area does not necessarily need to be as strictly controlled as the discussion room or the booths, but if you think you might use it for sample assessments and discussions in busy times do not decorate it with flowery wallpaper and curtains.

If your panel require somewhere for training or to discuss results, you could use the panellist waiting area or a nearby conference room for example. But if you are going to be creating sensory profiles you will find that you need a designated discussion room. The size of this room is important as it needs to be large enough to hold a round table. A round table is the best option for discussion groups as it ensures that everyone is at the same 'level': no one is at the 'head of the table'. However, they do have drawbacks in that if you are also seated at the table, it's more difficult to make eye contact with the people sitting to your left and right. It can also be difficult to find suitable chairs that fit comfortably around the table.

Some examples of typical sensory science facility layouts are given in Figures 2.8–2.10.

Do not forget to also consider other facilities that a panel may require. Most panels will require somewhere to park their cars, for example, and if you are recruiting different panels who will be working at different times, you might need additional car parking space to account for the overlap in working hours. You may also require someone to act as a receptionist if you are running consumer sensory panels on site and, if your company requires them to be accompanied at all times by a member of staff, there might be a requirement for someone to shepherd the panellists backwards and forward to the site entrance. And of course all panellists will need access to the lavatories.

2.6.2 Staff

If you are recruiting externally for an analytical or consumer sensory panel, the number and type of staff to run the facility will also need to be considered. There might

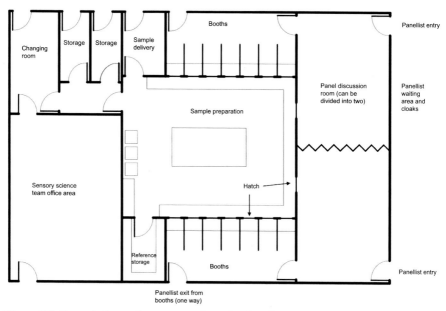

Figure 2.8 Example layout for sensory science facility (1).

be a requirement for administrative staff to manage the panellists' payments, holidays and working hours; chefs or technicians to prepare the samples for assessment; panel leaders or sensory scientists to run the panel sessions and analyse and report the findings; and statisticians to advise on the experimental design or the analysis of the data. There will also be the need for access to an independent ethics panel, especially if products such as tobacco, alcohol or novel foods are to be assessed (see IFST, 2017 and Chapter 3 for more information). The number and type of staff will depend on the size of the organisation, the workload of the facility and the product type. Stone et al. (2012, pp. 37–39) give some useful calculations to determine the number of staff required for a sensory testing facility. BS ISO 13300-1 (BS ISO, 2006) gives some guidance on staff responsibilities in a sensory laboratory. This document is particularly useful if you are just setting up your sensory facility and Table 2.1 on page 7 of the standard gives a very useful outline of the comparison in roles between a sensory manager, panel leader and panel technician.

The chef or technician working in the sensory area has an extremely important role within the team: the preparation of the samples for assessment. If this is performed incorrectly it does not matter how well motivated or trained the sensory panel is: if they assess the wrong sample or a sample stored, prepared or presented incorrectly, the data will be worthless. The chef or technician's other duties may also include maintenance of the facilities, organising and planning panel sessions and preparation for the data collection: all critical for the smooth running of the sensory programme. For sensory teams working in food, having a chef or someone with formal experience in food preparation can be very helpful, especially where complex and numerous product

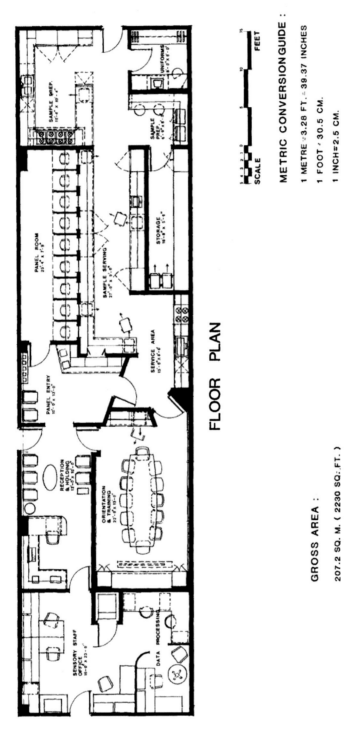

Figure 2.9 Example layout for sensory science facility assessing food (2). Reproduced with kind permission from ASTM Physical requirement guidelines for sensory evaluation laboratories: Special Technical Publication (STP) 913 copyright ASTM International, 100 Barr Harbor Drive, West Conshohocken, PA 19428.

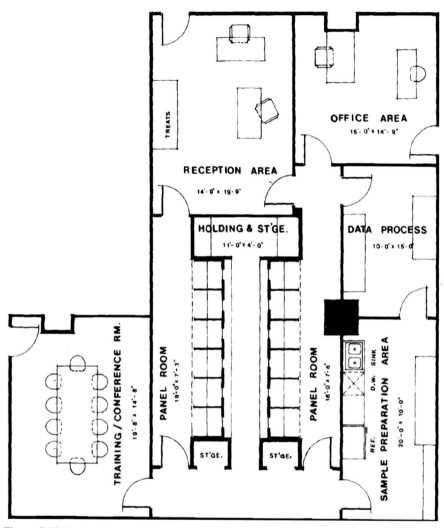

Figure 2.10 Example layout for sensory science facility assessing tobacco products (3). Reproduced with kind permission from ASTM Physical requirement guidelines for sensory evaluation laboratories: Special Technical Publication (STP) 913 copyright ASTM International, 100 Barr Harbor Drive, West Conshohocken, PA 19428.

types are to be assessed: knowing how long it takes to cook eight roast chickens and deliver the correct assigned portion to each panellist is a godsend.

For sensory teams working in flavours or fragrances, having someone on the team who is familiar with these product types and can help prepare samples from the flavour organ can be very helpful, especially in the creation of references. For home and personal care teams, having someone on the team who understands product formulations and how to adjust these to create references for the panel is also very useful.

Panel leaders or sensory scientists will also be responsible for making sure that the appropriate tests are chosen and that the tests are conducted in the correct manner, although they may have advice from a statistician and/or a more senior person with sensory experience. They will have training (see also Section 2.7.3) in sensory science either through work-based learning or through formal education such as the courses run by the Institute of Food Science and Technology (IFST) in the United Kingdom or the University of California, Davis, in the United States. They may also have moved into sensory science via one of the many connecting disciplines such as chemistry, psychology, statistics or marketing. They will require many other skills to run an efficient and successful panel, such as excellent communication, organisation and people management skills, not to mention knowledge of experimental design, measurement techniques, information management, specialist software and statistics.

Even if all the sensory testing is outsourced through a sensory agency, you will still need sensory staff within the company to manage the testing programme. For any of the panels mentioned above, if the recruitment is being conducted in-house, be sure to understand and plan out the whole process before starting. A good place to start would be by reading through the chapter on human resources (Chapter 3) and then move on to Chapter 5 which gives full details of the recruitment process, including interviewing tips, and everything else that needs consideration. In Chapter 5 you will also find some useful plans to help you check that you have covered all the points and have everything in place.

2.6.3 Staff training

The level of knowledge and capabilities for all staff in the sensory facility is critical and therefore recruiting capable staff and developing a training plan for their continued development is a high priority for managers of the team. BS ISO 13300-1 (BS ISO, 2006) and BS ISO 13300-2 (BS ISO, 2006) are useful in developing documentation about responsibilities, job roles and training requirements. For all staff the most critical training will be in food hygiene, ethics and health and safety. There are formal courses and qualifications and it is vital that the staff have training in these areas.

The task of panel leading is often given to junior members of staff, students or sensory scientists starting out on their careers, without a huge amount of thought to the sheer wealth of information and experience that is needed before someone can become an excellent panel leader. In fact BS ISO 13300-2 (2006) states that ideally, experienced panel leaders should be recruited. A good panel leader needs to have not only a good grounding in sensory science and a good knowledge of experimental design and some statistics but also good interpersonal skills, emotional intelligence and a huge dose of organisational ability and thinking! They need to be able to build the right relationships and communicate well with not just the panellists but other members of staff and the client(s). They need to be able to maintain authority and control the panel but in a friendly, motivating and efficient manner. Panel leaders are also leaders: the clue is in the name, and therefore they should be able to command respect, be 'patient, fair, honest and non-judgemental' (BS ISO 13300-2, 2006: p. 2).

Training of a panel leader can be carried out by an experienced panel leader, either from within the company or an external consultant, or they can learn 'on-the-job' through personal experience. Both routes can be enhanced by attending training courses not only about sensory science but also in management, report writing, panel moderating, statistics and experimental design. There are several formal sensory courses available, as well as the related topics, and some are also distance learning or online courses. Access to professional organisations (meetings, conferences and networking), books and journals will also enhance the learning experience.

However, the main part of the training is in learning to run the sensory panel and getting experience in this way. The management of group dynamics is not easy, but it can be learnt through courses and practical experience. It can be really beneficial for the trainee panel leader to see some other panel leaders in action and if this is not possible within your company as you are just starting out, you could try contacting a sensory agency or university with a sensory function to arrange this. Learning to run the panel and getting feedback from an experienced panel leader is a good way to learn quickly and efficiently. For more information about panel leading, see Chapter 4: How to become an excellent panel leader.

As mentioned in Section 2.7.2 the competency and hence training of the technical staff is paramount in the smooth running of the sensory facility. The chef or technician may require training in the specifics of your product type and an understanding of recipes or formulations if they do not already have this knowledge. Sample preparation will be a key part of their role, so they must understand the implications of consistency in approach, experimental design and sensory biases. If the wrong samples are presented to the panellists, or the samples are not prepared in the right way because the technical staff did not understand the reason for the strict protocols, all the hard work in collecting the data will be completely wasted. On a recent visit to a sensory facility, the technician was pouring drinks for a triangle test. He poured sample A with his left hand while simultaneously pouring sample B with his right hand, so that each panellist could be given the samples directly after they were poured because of the possibility of temperature differences. What he was not aware of was that by pouring in this way, the identification of sample A and sample B by visual inspection alone was easily detectable. When questioned about his approach (in the nicest possible way) he told me that he had never participated in a triangle test and was not really quite sure 'how it all worked'. If he had just five minutes of training, his approach would have been improved.

Statistical approaches for experimentation and analysis are also important areas for training, especially if you do not have access to a statistician. All tests are best guided by a statistician, at least the first time each type of test is conducted. There are various training courses in sensory experimental design (run by statisticians) that are invaluable in deciding the best approach for your testing. There are also various books on experimental design and statistical analysis and also other publications in sensory journals that will be useful. See Chapter 14 for some suggestions.

Human resources: legal, regulatory and professional requirements

Sue Stanley
Unilever R&D Port Sunlight (Formerly Director), Wirral, United Kingdom

3.1 Introduction

Once you have decided on the type of panel you need, and how you plan to recruit them, the next step is to consider formal aspects relating to ethics, health and safety.

Your assessors are going to be critical to your success, and it's important to remember that they are people, and not robots! We will discuss the motivational and management aspects of 'assessors as people' later on, but in this chapter we will concentrate on their rights and our responsibilities. This chapter will therefore be about the legal, regulatory and professional frameworks and standards relating to sensory research studies. It is worth noting that these considerations are not optional, and it is wise to cover them at the earliest possible stage. At some point they will appear on the critical path for your work, and it could cause considerable delays later on if they are not in place from the beginning.

Before we get started, one caveat: *there are going to be differences between geographies and institutions.* We will aim to cover the broad framework here, but for your particular situation it will be important to get local knowledge, for example, by talking to people in other companies, institutions or departments to find out what they do. Another source of information could be via a relevant Professional Association in your country (Market Research Society Code of Conduct, 2014), and in certain situations, it may also be prudent to consult with a lawyer who specialises in employment law. There may also be experts in your company who can assist you, such as in human resources (HR) or health and safety.

If you carry out work in other geographies than your own, then you will also need to consider regulations in those countries. For example, if you carry out work in a European Union country, then you will need to consider EU regulations such as those relating to data protection (EU General Data Protection Regulation, 2016).

More information on many of the areas covered in this chapter is also available in Sensory Evaluation: A Practical Handbook (Kemp et al., 2009).

3.2 Employment

3.2.1 Internal staff

3.2.1.1 Existing staff in other departments

In the previous chapter, it has been described how you might recruit staff from other departments in the company for short sensory assessments that require less than two hours of work per week.

Sensory Panel Management. https://doi.org/10.1016/B978-0-08-101001-3.00003-3

Management agreement

Before you start recruitment, it is important to get agreement from either the Senior Management team of your company, or the Senior Management of the specific department/s from which you will be recruiting. This will avoid any misunderstanding later on. A good way to do this is to draft out a short proposal covering the following:

- Purpose of the panel and how it will benefit the company
- Number of panellists required
- Maximum amount of time per day and per week an individual will spend on panelling
- Will this be in work time or break time
- Will agreement from the individual's line manager be sought
- Will there be any small remuneration, e.g., gift voucher or products
- Method of recruitment, e.g., posters, email

When you have drafted the proposal, it will be worth putting it past an HR expert to check it is in line with current company work practices. It will also be useful to have done this beforehand, because when they receive it, Senior Management will be almost certain to ask if HR support the proposal.

Voluntary recruitment

From an ethical point of view it is essential that being a panellist is voluntary, with no coercion either direct or indirect. It has been mentioned earlier that the data from a 'forced' panellist are unlikely to be so useful to you, but in any case, it is not ethical to force an individual to join a sensory panel. This is because human sensory assessment studies may be considered as a type of clinical research involving human subjects, which is governed by the Declaration of Helsinki (World Medical Association Declaration of Helsinki, 1964). Such research is unethical unless the individual has freely volunteered, or else it is a formal part of their job, where they have voluntarily taken on this type of work when they have taken on the work contract. People for whom it is not a formal part of their job therefore need to be free to decline to take part, without giving any reason. They may have private reasons for not doing sensory assessment that they do not wish to communicate to their employer, e.g., health or religion.

It is important to avoid peer pressure or 'hidden coercion', where employees may believe they are expected to volunteer, and will suffer adverse or 'less positive' consequences from their peers or management if they do not. There may need to be specific communication from management on this point, saying that participation is entirely voluntary, and that management and colleagues will not be told who has volunteered and who has not.

3.2.1.2   New staff specialising as sensory panellists

Another option is to recruit new staff, probably part-time, to work solely or mainly as sensory panellists. The decision as to whether to employ them directly by your company, or via an agency (next Section) will be made on business grounds including whether the role will be permanent/ongoing or temporary, and the relative costs, including line management time. You will need to draft out a role specification and consider aspects of their contracts as described in Section 3.3.1.

3.2.2 Agency staff

Agency staff may be recruited by your company specifically for sensory panel work. Their employer is the agency they work for, and you will need to consider aspects such as accounting for the hours they work, the basis on which they will be paid, for example, hourly/daily/weekly, absence due to sickness and holidays. In the briefing to the agency you will also need to cover the types of ethical considerations by which you expect them to work (in the same way that your company works), and how you will interact with the panellists and the agency both day-to-day and over time. For example, you will need to agree how you will deal with behavioural issues, sensory changes (e.g., decline in acuity) and any other issues that may arise with the panellists over time.

It has been found to be very useful to bring the agency on site before they start recruitment, if possible seeing a sensory panel in action or doing some sensory tests themselves. Then they can get a good awareness of the possible issues with health, allergies and so on, and be in a far better position to carry out preliminary screening of job applicants for you, and to help deal with any problems that could arise over time.

3.2.3 External 'consumer volunteers'

Sometimes it may be appropriate to recruit external panellists who are not employees or agency staff but who are 'consumer volunteers'. This might be because the work is preliminary or a one-off, or you are looking for affective responses or you wish to recruit for particular types of home equipment such as WC or washing machine, or specific types of home cooking process or beauty procedure. Such panellists are similar to participants in a market research study or healthy volunteers for a clinical study. They do not have a contract of employment with your company or an agency and instead are rewarded for their participation with a small gratuity and probably their travel expenses.

In terms of the expectations from them, it is useful to have a 'voluntary panellist agreement' (see below), so that it is clear on both sides what to expect. You may wish to confirm with them any professional standards by which you abide, for example, the UK Market Research Society (Market Research Society Code of Conduct, 2014). It will be a positive message to the panellists if you are a member of the relevant professional body in your country, and in any case, the code of conduct for the market research professional society in your geography will be a useful reference point.

In the special case that you wish to include children, not yet legally considered to be adults (under 18s in the United Kingdom), then codes of practice relevant to research involving children will apply. Generally this will involve gaining the permission of the parents or legal guardians, and safeguarding issues should also be taken into account (Market Research Society Code of Conduct, 2014, Market Research Society Guidelines for Research with Children and Young People, 2012). In some countries there may be specific requirements for staff working with children (DBS checks for working with children, 2017). There may also need to be specific parental consent for processing personal data relating to children (EU General Data Protection Regulation, 2016), and it is worth noting that the age at which children are considered to be adults varies widely between regions (ESOMAR Data Protection Checklist, 2016).

3.3   Contracts and panellist agreements

3.3.1   Contracts of employment

If your company will be employing panellists as members of staff, then you will need to work with HR on the relevant type of contract to put in place. This would cover aspects of employment law such as the rights of an individual in different circumstances, e.g., loss of sensory acuity in the future such that they were unable to perform their job.

If you are employing your panellists via the agency, then your contract will be with the agency and needs to cover the work that they will do for you, such as panellist recruitment, screening, payment and line management. It will also need to cover confidentiality (Section 3.4.1) and any code of ethics by which your company abides and which it is expected that your supplier will follow. Your HR expert will be able to advise on this.

3.3.2   Voluntary panellist agreements

It is a good idea to have a specific agreement signed off by both panellist and yourself when they join the panel, so that expectations are clear on both sides. For 'voluntary consumer' panellists this might cover the following, and something similar would also be useful for staff or agency panellists:

* Times of arrival and departure on your site or testing facility and punctuality. Not bringing anyone else, e.g., children or third parties. If testing is to be done at home, prompt completion and return of questionnaires and return of products.
* Treating people with respect and that panel membership can be terminated by the company or the volunteer at any time without providing a reason
* How to cancel appointments
* Confidentiality of test products and other materials such as packs, early advertising materials and concept boards (see below)
* How you will keep confidential their personal information (see below)
* Letting you know about changes in their health status
* How the panellist will be paid. Personal tax responsibility of panellist with regards to payments received. Even small gratuities, gifts or travel expenses could be counted by the tax authorities as earned income, and for regular panellists these can mount up across a tax year.
* Safety, ethics and professional codes of conduct that you follow
* How to report any complaints and how these will be followed up

From 2018 there will be a requirement in the European Union for panellists to give clear and affirmative consent for the processing of their personal data (EU General Data Protection Regulation, 2016), and this can usefully be included in the Panellist Agreement. Similar requirements are good practice everywhere, so that research participants retain control over their personal data.

3.4   Confidentiality and anonymity

3.4.1   Panellists

For external panellists, it is important that they sign an agreement that they will neither talk to third parties about the products or packs/preadvertising they have tested nor will they pass on to anyone else the products they may test at home. This is for reasons of commercial secrecy, so that a business competitor does not get to know what the company is working on, and also for patents and copyright reasons if the work is at an early stage before a new technology has been put forward for patenting or copyright. If a member of the public has not signed a confidentiality agreement, then their testing of a product could prevent a technology being patented in the future because it has already been used in the world at large and is no longer considered 'novel'. A confidentiality agreement is not needed for an internal or agency panellist, if confidentiality has already been covered in their contract of employment.

3.4.2 Company

You must take steps to ensure that personal information about the panellist (e.g., name, address, date of birth, medical, economic, social or cultural identity) is kept securely and is only available to yourself and relevant assistants for the purposes of the work. You should store it for no longer than is needed for the purposes of the project and any future safety issues that might arise (see below). There is more detailed information about client confidentiality available (Market Research Society, 2014) and about digital data storage requirements below under data protection. As mentioned previously, if you carry out work in another country than your own, then you will also need to consider data protection requirements in that country. For example, there are specific regulations in the European Union (EU General Data Protection Regulation, 2016).

If you are going to take voice recordings, photos or video that could identify a particular individual, then it is important to provide an agreement for the panellist and yourself to sign, giving details of how the material will be used in the future (e.g., its purpose and who could view it), that it will not be used publicly or for commercial purposes, that the copyright remains with the company, and how it will be stored. The types of requirement relating to participant anonymity are generally covered in Market Research Codes of Conduct (Market Research Society Code of Conduct, 2014).

3.5 Data protection

For personal data held digitally, i.e., on a computer or other digital device, there is specific legislation in many geographies about how this may be stored and used. Your company IT experts or your professional organisation will be able to advise you (Market Research Society Code of Conduct, 2014). ESOMAR have produced a useful checklist based on OECD principles that will be of particular use to smaller organisations who may not have extensive resources or experience in data protection (ESOMAR Data Protection Checklist, 2016). This aligns most closely to EU requirements and also references requirements in other geographies.

In general terms you should not store personal data for longer than is necessary. As well as considering length of storage in terms of administrative and research requirements, you may also need to consider longer term health and safety, as discussed in Section 3.7.4. This may mean balancing the requirements of 'no longer than necessary' with the potential risk of future health-related issues from any of the products tested. Although the latter is presumably unlikely for sensory testing, it could require keeping information about who tested what products and under what conditions for a longer period of time, in a similar way to clinical testing. This is a risk-related decision that the company would need to make and to document their reasons for keeping the data beyond the immediate research requirements.

Data protection is an area of regulation that is changing rapidly in line with increasing consumer concerns about protection of their personal data. In the European Union, new legislation is coming into force in May 2018, and any company who plans to carry out research in the EU will need to prepare for it before that date (EU General Data Protection, 2016; Regulation of the European Parliament with regards to the processing of personal data and on the free movement of such data, 2016).

3.6 Ethics

It is important to have a legally robust ethical framework for carrying out sensory panel work. This may initially sound like overkill, but as noted earlier, doing sensory assessment may be considered to be a form of clinical research using human subjects, and you need to be aware of people's rights in such types of research, and respect them. Additionally if panellists are external consumer volunteers, then panel work is a public face of the company, and assuring them of the ethical stance of the company is positive in terms of public relations.

It is important to have a robust framework for ethical governance, including an independent ethical review of your studies (Kemp et al., 2009; Sheehan et al., 2014). A larger company may have its own Ethics Panel consisting of relevant external experts to advise on research protocols. A smaller company may use the Ethics Panel of a local University/Medical School (ACNFP: Guidelines on the conduct of taste trials involving novel foods or foods produced by novel processes). If the work is being carried out in an academic environment, then the study will need to be cleared by the Ethics Panel of the University, and if being done on behalf of a company, potentially by the Ethics Panel of the company as well.

The rights of participants in human research include valid informed consent, safety protection, personal data protection and an ethical culture in the company/institution doing the research.

Particular areas requiring close ethical scrutiny would include, for example, studies involving tobacco, alcohol, hygiene products and health and personal care products, changes to normal habits where these might impact on health or wellbeing, psychological measurement, biophysical measurements such as heart-rate or skin conductance and children as participants.

If your studies over time will all be of one type and involve the same protocols and types of product, then it will save work to draft for Ethics Panel review a single generic study proposal that includes all the types of product to be tested, and all the types of assessment and measurement envisaged. Once cleared, you would only then need to go back to your Ethics Panel if anything changes.

3.7   Health and safety

3.7.1   Introduction

The safety of your panellists while carrying out your testing is clearly paramount. As well as ethical aspects, this area is likely to be subject to health and safety legislation.

There can be no short cuts that might potentially put your panellists at risk. This includes the chemical, biological and microbiological aspects of your test products, quality control, freedom from contamination, hygienic methods of preparation and dispensing, storage, labelling and test product disposal, as well as appropriate and continuing screening of panellists for allergies and health conditions.

In addition to *test product* safety, there are also general aspects of health and safety in terms of the panellists and staff moving around the testing room, using computers, manual handling and so on that you will need to take into account, and draw up Risk Assessments for. This will also apply to any procedures that the panellists might be carrying out. If you are unfamiliar with this type of Risk Assessment, more information is available (HSE: Controlling the risks in the workplace, 2017). If your panellists are internal or agency employees, then there may also be specific safety assessments required in your geography for workplace exposure to chemical/biological substances (HSE: Control of substances hazardous to health (COSHH), 2002).

3.7.2   Test product safety

Before any testing can be carried out, you will need to obtain professional expert clearance for the safety of the products according to your test protocol. It cannot be stressed enough that you should seek this clearance as early as possible when planning a test, as it can take considerably longer to obtain than you might expect, and can rapidly become the rate-determining step in how soon the testing can start. If you work in a large company, then your company may have its own professional expert safety consultants. If not, there are sources of external safety clearance, both governmental and private (Food Standards Agency: Importing and testing trade samples, 2016).

If the test products are already marketed in your country, and used in the test according to normal consumer recommendations, then safety clearance is generally straightforward. If imported products are to be used, then special safety considerations might apply (Food Standards Agency: Importing and testing trade samples, 2016).

For novel products being developed in R&D, then you will need to provide in your application for safety clearance the exact formulations and processing details of your test products.

3.7.3   Labelling, adverse effects and informed consent

If the products are to be used at home, then you will need to meet safety labelling requirements, including food or cosmetic and personal care product labelling (Regulation of the European Parliament and of the Council on the Provision of Food Information to Consumers, 2011; Inventory and common nomenclature of ingredients employed in cosmetic products (INCI), 1996), what to do in the event of accidental ingestion of nonfood products, and reporting of and seeking treatment for any adverse effects. For products tested on your site, panellists should also know how to report any adverse effects they might experience after leaving the site, and the seeking of medical advice if they have any symptoms that cause them concern.

Adverse effects would include any physical symptoms experienced by the panellist that they might attribute to the product testing, for example, rashes or other allergic reactions. As well as advising the panellists to seek medical treatment for anything that causes them concern, it is important for you to record such events so that they can be reported back to those providing the safety clearance for the products, to the product developers, to line managers (if the panellist is an employee) and potentially for excluding the panellist from further testing. Your product developers will also find it useful to be informed if there have been any untoward effects on consumers' home equipment, furnishings or clothing. Your company should also consider the insurance they hold, with regards to potential compensation claims from panellists (Kemp et al., 2009).

It has been mentioned earlier that there is a requirement for participants in human studies to give their informed consent. The information provided to your panellists should contain details of any adverse effects that they might potentially experience from testing, as this is an important part of them being sufficiently knowledgeable to give their informed consent. In addition the informed consent should include the processing and storage of their personal data, and you will need to provide information on how this will be handled (EU General Data Protection Regulation, 2016).

3.7.4 Records

Finally, for reasons of health and safety or regulatory requirements, records may need to be kept of all participant names, products they have tested, protocols, dates and adverse effects. This is more likely for products involving tasting or ingestion, inhalation or skin contact. One reason is that new information may become available in the future as to any potential adverse effects relating to ingredients, products or processes, so that panellists who have been exposed could be contacted and advised as appropriate. The information would also be required in the event of potential compensation claims from panellists, or if more detailed follow-up of exposed subjects might be required. The recommendation for novel foods is to keep these records for 30 years (ACNFP: Guidelines on the conduct of taste trials involving novel foods or foods produced by novel processes, 2017). For other novel products, the company may decide to treat the data retention in a similar way to a clinical study, where in the United Kingdom the MHRA guideline for research under good clinical practice specifies at least 5 years for data retention (MHRA Retention of Trial Records, 2015).

As noted in Section 3.5, the requirements for keeping data no longer than necessary need to be balanced with the health and safety requirements, and this is essentially a risk management decision for the company to take.

3.8 Future trends

There are increasing trends for citizens and consumers to require more information about the composition and provenance of the products they consume, for companies to be held accountable for their cultures of safety and ethics, for the privacy and security of personal information, and for legal recourse for compensation for perceived personal injury. Therefore being fully informed about and practising the highest standards of ethics and safety in sensory testing, and keeping informed of changes in legal and professional requirements, is extremely important for the sensory practitioner. This chapter has given what is hopefully a useful introduction to the area to provide a starting point for carrying out professional testing and programmes of research, while avoiding potential pitfalls and risks.

How to become an excellent panel leader

4

4.1 Introduction

This introduction gives a quick overview about the main things you can do to become, or to continue being, an excellent panel leader. Each element below is then described in more detail in the sections that follow. BS ISO 13300-2 (BS ISO, 2006) specifically describes the recruitment and training of panel leaders. The standard defines a panel leader as a 'person whose primary duties are to manage panel activities, and recruit, train and monitor the assessors'. There are two notes to this definition. The first says that the person may design and conduct the tests and also analyse and interpret the data, and the second says that they may be assisted by panel technicians.

Excellent panel leaders are a requirement for the generation of excellent sensory data and therefore should be recruited carefully and trained well. Some panel leaders are involved in all aspects of sensory science from the discussion of the objective through to the analysis of the data and reporting. Others are involved only in the running of the panel sessions, but even then, the majority of the panel leader attributes are still required.

The task of panel leading is often given to junior members of staff, students or sensory scientists starting out on their careers without a huge amount of thought to the sheer wealth of information and experience that is needed before someone can become an excellent panel leader. A good panel leader needs to have not only a good grounding in sensory science and a good knowledge of experimental design and some statistics, but also good interpersonal skills, emotional intelligence and a huge dose of organisational ability and thinking! They need to be able to build the right relationships and communicate well with not just the panellists but other members of staff and the client(s). They need to be able to maintain authority and control the panel but in a friendly, motivating and efficient manner. Panel leaders are also leaders: the clue is in the name and therefore they should be able to command respect, be 'patient, fair, honest and non-judgemental' (BS ISO 13300-2, 2006: p. 2).

Panel leaders also need to consider ethics and professional practices. If you work to a professional body's code of conduct such as the *IFST Guidelines for Ethical and Professional Practices for the Sensory Analysis of Foods* (IFST, 2017) and/or the *Code of Professional Conduct for Members of the Institute of Food Technologists* (IFT, 2017), you cannot go far wrong. Make sure that your first consideration in every test request is the panellists' health and safety. Although most sensory studies are no more or less safe than 'the risks of daily life', you must take into account that you are testing with humans and it is your duty to protect them. Lawless and Heymann (2010, pp. 73–74) give some very useful advice concerning testing with human subjects. An example would be in the sensory testing of pet food. You would need to get ethical approval, informed consent and certification from the manufacturer that the products were safe for human consumption.

It is advisable to have some practice in panel leading. But how do you go about gaining the best experience? Benjamin Franklin is famous for having said, 'Tell me and I forget, teach me and I may remember, involve me and I learn'. So, it follows that by reading about being a good panel leader, watching and working with some good panel leaders, running some sessions and requesting feedback about things you could do even better, might be a good approach to becoming an excellent panel leader. And remember that excellent source of feedback: the panellists. Experience in managing and understanding group interactions is a must. It's not easy to get everyone to contribute or to help the panellists resolve conflicts without causing further issues yourself. I hope this book will give you confidence in many of these areas, as I have found that working on a one-to-one basis with many different panel leaders has helped them achieve their potential as well as the panels'!

As mentioned earlier, there are many aspects to being an excellent panel leader, but let's start with the top two: caring about the panellists and caring about the results. If you focus on these two elements, you will be well on the way to being an excellent panel leader, as the best results will be achieved if the panellists are happy and feel confident in their abilities. Just think about yourself. Do you produce good work when you are demotivated and unhappy in your job? Probably not. You will also need to *demonstrate* that you care about the panellists and care about the results: more about that later.

Next up is setting objectives for your panellists. Good performance starts with setting good objectives: in fact, how will you even know if your panel is performing well if there are no objectives? And more to the point, how will they? If you already have a panel, an interesting exercise might be to ask them to write down what they think you expect from them. At the same time, you need to write down what *you* expect from them. Compare the two lists and see if they match. If they do, then your panel know very well what is expected of them, and if they do not, then you will need to make sure that they do. If you do not already have a panel, it would be a good idea to write down a list of what you will expect from your panel. This can be simple things such as 'arrive on time' or 'attend 85% of all panel sessions' through to more detailed objectives related to attribute generation or performance in quantitative profiling, but either way, they must be things that can be seen and measured (more information on this later in this chapter).

By setting these panel objectives you are also laying the foundations for the next key element which is all about giving feedback. When your panel is new, feedback is critical in helping them learn the job role and what is expected from them; however, negative feedback should be avoided where possible: I have found that it is better to repeat what is expected of them and extend their training. When you have an established panel, feedback about poor behaviour should be done as soon as possible after the event. Do not save up all your gripes until you explode. Never give feedback that comes across as about the person: it should be about what they did, and as anyone can make a mistake, do not berate them for it. And it's best to follow up with a short statement about something they do well; after all you want the panellist to be improving in their role and not unhappy and thinking of leaving – you have invested too much time in their recruitment and training to lose them. For example, if you were discussing poor replication in a certain project (project Y) you might say, 'In project X, your replicates were excellent, I know you are more than capable and I rely on your data, so let's see some more of that'.

It's difficult to decide which is more important: feedback about things the panellist has done wrong or things they have done right; perhaps they are of equal importance. Therefore, remember to give the panellists praise when they have done something well. This can simply be a statement in the middle of a panel session, such as, 'That was a really good description, Kate. I feel like I really understood what you tasted there. Thank you'. Or it might be something more detailed during a feedback session about data. Either way, it needs to be as soon as possible after the good performance and it can work really well if you tell them what was good (as the example) and to keep it up (which I failed to do in the example, but perhaps it was implied).

To summarise the key elements for excellent panel leading:

1. Care about your panellists.
2. Care about your results.
3. Set objectives.
4. Give great feedback.

But maybe there is an overarching element to these and that is to tell your panellists how you like to work and what to expect. If you are open and honest with them, they will be open and honest with you and the respect you both have for each other will help develop a great working relationship.

4.2 The details

4.2.1 Caring about your panellists

One of the best ways to show that you care about your panellists is to listen to them. This means *actually listen* and not be thinking about what you are going to say next or what to cook for dinner. If you demonstrate good listening skills this may well rub off on other members of the panel who really do need to listen if they are to understand what Gerald means by 'herby flavour' or 'greasy feel'. As sensory scientists we should be able to do this easily, after all we know how the ears work, but it's not easy and may take some practice. Good listening is a lot more than just being silent while the other person is speaking: how does silence show you are listening? Good listening is about asking good questions, developing the conversation and certainly not about criticising or becoming defensive. If you are listening just so you can detect the moment the other person creates an opening for you to leap in, then that is not active listening. You need to listen so that you can help the panellist create the good description, definition and explanation. And easy ways to be a good listener include the following: give the person time to say what they want to say; do not interrupt; do not be afraid of silence and do ignore distractions like your phone and emails, as this helps you keep eye contact with the speaker: we listen with our eyes as well as our ears.

It can be very beneficial to start each panel session with an outline of what you hope to get out of the time available and to ask if the panel agree. This way everyone is signed up to the objectives for the session and can raise issues if they think you are too ambitious. By starting each panel session in this way, you will find that the panellists are thinking ahead to what needs to be done in each session and almost planning the work themselves. This can be motivating for many panellists as it gives them autonomy and control over what they do on a day-to-day basis.

It can be very helpful to give a rough outline of each session at the start of each project and when the panellists are expected to do certain parts of the job in hand, although there are some provisos with this approach. On some occasions sharing the bigger picture can be demotivating for the panellists. Consider the project with 30 or more different samples from an experimental design type approach. Telling the panellists the number of samples at the beginning of the project, especially if they are not the most exciting of products, can be demotivating.

By asking for the panellists' input to the plan indicates to them that you care about them and the amount of work they might have each day, and by checking with them about the plan shows that you respect their suggestions for a different approach.

The amount you 'pay' your panellists is also key to the panel feeling like you care, as well as being important for motivation. If they are an external panel, then the hourly rate or salary may be very important to them. In addition to this, your company may offer other benefits such as a holiday allowance, pension scheme, subsidised creche or staff shop. An internal panel can be paid with a treat or with other benefits such as raffle prizes, day trips or social events. These can also be useful for motivating external sensory panellists. One of the best payments for motivation is the 'thank you', particularly if you go out of your way to communicate it. For example, for an internal panel, if you are checking attendance and realise that Martha has attended 7 of the last 10 panels, you could walk to her desk and say, 'Thanks, Martha – I just noticed you attended seven of the last 10 panel sessions. That's really great. Thank you for your time'. You might even notice that she attends 8 out of 10 next month. For more ideas for panel motivation, see Section 12.4.1.

Aspects of pay, respect, motivation, teamwork and enjoyment of the role all impact on panellists' happiness. And, as happy panellists give good data, we are well on the way to achieving both of the main objectives.

4.2.2 Caring about your results

There are many different aspects to achieving good results and these are listed below and discussed in more detail in the following sections:

- Recruiting good panellists
- Checking the panellists' and the panel's performance
- Caring for the panellists (pay, respect, motivation, teamwork and enjoyment of job)
- Having suitable facilities
- Having the correct staff

- Having the right procedures and protocols in place to ensure that the experiment is conducted correctly
- Choosing the right method for the objective and the risk
- Having the right type and the right number of panellists taking part in the test
- Having the right record keeping procedures and data collection devices
- Having the time to complete the experiment in the correct way
- Having a sensory team that are well trained and kept up to date

4.2.2.1   Recruiting good panellists

The recruitment of *good* panellists will obviously be key to having an excellent science function within your business. For full details on the recruitment of sensory panellists, see Chapter 5.

4.2.2.2   Checking the panellists' and the panel's performance

It is important to check the raw data that you collect prior to any statistical analysis. This can help you detect experimental issues such as the assessment of the wrong sample, sample variability or other experimental design issues, as well as check the panellists' performance. If you see patterns in the data that do not stem from simply poor panel performance, you may well have detected an issue with sample delivery to the panellists. See Chapter 11 for detailed information about all aspects of panel performance.

4.2.2.3   Caring for the panellists (pay, respect, motivation, teamwork and enjoyment of job)

You will recruit good panellists if you follow the guidelines in Chapter 5, but they will not be 'good' for long if you do not invest time and effort into their training, motivation and care. For information about motivating your panellists, see Section 12.4. See the earlier Section 4.2.1 for details about caring for the panellists.

4.2.2.4   Having suitable facilities

The sensory facilities are a very important factor in achieving good results and they can also impact the happiness of the panel. There are many requirements for sensory laboratories such as location for easy access, but quiet and odour free, controlled lighting and heating, and control of air circulation. Additional features such as comfortable seating and easily viewable computer screens can mean that panellists are paying their full attention to the test. In the laboratory, one of the most important requirements is space! Space to store samples and space to lay out samples and receptacles prior to the test. Some examples of good facilities are given in Section 2.7.1 and also in many of the standard sensory textbooks (the chapter in Lawless and Heymann, 2010 is very detailed) and sensory standards (BS EN ISO 8589:2010+A1:2014). A particularly good reference is the ASTM Special Technical Publication '*Physical requirement guidelines for sensory evaluation laboratories*' (Eggert and Zook, 1986) which has many examples of sensory laboratories for different product types. If you are just setting up, try and view as many facilities as possible to get a feel for what would work in your situation. Simple additions such as a hatch to pass samples through to the panel discussion room can save time and hassle and make the panel sessions run more smoothly.

4.2.2.5 Having the correct staff

Properly trained, experienced and motivated staff are a must. This is just as important for the recruitment and training of panellists as well as the sensory support staff and scientists. The number and type of staff will depend on the size of the organisation, the workload of the facility and the product type. Stone et al. (2012, pp. 37–39) give some useful calculations to determine the number of staff required for a sensory testing facility. BS ISO 13300-1 (BS ISO, 2006) gives some useful guidance on staff responsibilities. For more information on sensory staff, see Sections 2.7.2 and 2.7.3.

4.2.2.6 Having the right procedures and protocols in place to ensure that the experiment is conducted correctly

Well-documented procedures are very important for the sensory science facility. Sample receipt, storage, preparation and delivery, if not done correctly, can have a major impact on the quality, validity and robustness of the data. The serving temperature, sample containers, quantity of sample, etc., all need to be considered and documented. Even the time of day that the panel sessions are to be conducted needs to be considered. It's not enough to have the procedures if they are not documented and not easily accessible. If staff do not understand the need for a particular step or cannot quite recall exactly what needs to be done, and the access to the paperwork is cumbersome and annoying, they may well decide to ignore the procedure. The information that the test needed to be performed two days after sample receipt then gets missed and the test is conducted five days later and the data are worthless.

Chapter 3 in Lawless and Heymann (2010) is an excellent resource to ensure good practice in procedures and protocols for sample storage, preparation and delivery, which are so critical in the generation of valid data. The chapter also includes information about sample serving, giving the right instructions to panellists, preparation methods, experimental design and palate cleansing. Many of the other sensory textbooks, notably Kemp et al. (2009), also include useful information relating to best practices in the sensory science. ASTM E1871 (2010) also has some very useful advice for serving a wide range of products. For example, the standard includes advice on product carriers, serving temperature and product holding time. There is also a very useful section on serving beverages that includes advice on serving powdered, carbonated and hot products. You may need to use specialised serving containers for some products, such as olive oil tasting glasses (ISO 16657:2006), coloured glasses, or lidded vessels to hide the sample's appearance. Chambers et al. (2016) mention the use of a first cup of coffee to assess aroma and then a second cup to assess the flavour, so that the sample is at same temperature for both assessments.

It can be very helpful for complex products to conduct some preliminary tests on cooking and serving protocols and to also gain input from the client. Sometimes the cooking procedure seems simple enough when serving all the panellists the same sample, for example, in the initial sessions of creating a quantitative sensory profile, but becomes more difficult when working to a balanced design. Another difficulty can be in serving a sample so that it is representative of everyday consumption or use. If you are studying the effect of the change in an ingredient, but by bulk storing the samples, change the essential characteristics of the product, this would be bad practice. However, some assessments just cannot be exactly how a consumer might use the product. Not many consumers create a circular template for the application of sun cream to a section of their skin. You should document these differences in use and include them in your conclusions about the product differences and similarities.

Labelling is a very important aspect for sample serving. Handwritten codes with smelly permanent markers should not be used as they can be confusing, as well as introducing the potential tainting opportunity. Printed labels are easy to create and the use of a template and the 'edit, find and replace' function can make the generation of a new set of labels very quick and easy. Be careful about your choice of codes as presentation of coded samples for ranking labelled 112, 372, 572 and 211, for example, could well be mixed up when completing the test (by the technican as well as the panellist).

Information about biases and context effects, and their implications on experimental design, can also have a huge impact on the quality of the data. It will be worthwhile reading about adaptation, contrast effects, range and frequency effects, 'dumping', anticipation and habituation errors, and the impact these might have on your experimentation. Lawless and Heymann (2010, Chapter 9) have an excellent account of these factors and also some very useful tips for minimising or eliminating the effects.

There are several different options for the order of presentation of the samples for a test. The designs help to reduce sample order effects (where different results are achieved with the same samples when presented in different orders), issues with carry-over from sample to sample, and psychological biases (Kemp et al., 2009). Some tests, such as the triangle test, for example, have a specific number of potential sample combinations (6 for the triangle test) and each presentation design should ideally be presented the same number of times (i.e., the number of panellists taking part should be a multiple of 6: 24, 30, 36…). When constructing a design for a consumer test, it would depend on how many samples it is possible to assess in one sitting and how long each session/sitting can be. For example, if you have six samples to assess but only three can be assessed in one sitting, you might be able to get all the consumers to return and assess the rest of the samples on a different day. For quantitative profiles, the presentation design approaches mentioned earlier are also used. If you are setting up designs for the assessment of samples for ranking experiments, Whelan (2017a) has created some useful designs. The more samples, practical considerations and project limitations you have and the more complicated your design is, the harder it is to give a simple answer about the approach. My advice would be to consult a statistician at the start of each project and to also attend an experimental design course so you can understand the basic principles and the impact of the design on the data analysis. Some useful terms and definitions for experimental design are given in Tables 4.1 and 4.2.

Table 4.1 Terms and definitions used in experimental design

Term	Definition
Block	The name comes from horticultural trials where a field might be divided into areas (blocks) for testing a new herbicide for example. The differences in the blocks themselves (e.g., wind direction) were not the main interest in the experiment, but the experimenters knew, that although the field was quite uniform, each block in the field would differ in some way due to position (e.g., wind direction) and this variation needed to be accounted for in the experiment. A block is simply a subset of data from the complete experiment. For example, the 'sample block' is all the data for a particular sample. If not all samples can be assessed in 1 day, the block might be the samples assessed on Tuesday for example. Then any effect from the blocking of the samples in this way can be checked during the analysis (looking for a 'Tuesday effect'). Blocks are basically any part of the design and could be samples, panellists, days, etc.
Block design	A design for serving samples to panellists where there are many samples. The panellist sees a subset of the samples in the form of a block. The panellist can assess all of the samples in the end (or not) but they will have all been assessed in blocks.
Balanced	Equal numbers of each sample are presented to each panellist.
	Each sample is seen an equal number of times.
	Each sample is seen in combination with every other sample an equal number of times across the session.
Complete	All the parts, full and entire. This is generally viewed as the ideal design.
Incomplete	Not complete. For example, panellists will not assess all the samples in one session.
Randomised	In the presentation design the order of samples is randomly assigned to each panellist. For example, if we had three samples, there are six possible orders of presentation (123, 312, 231, 321, 132 and 213). Each panellist would be randomly given one of these orders. Random does not mean 'haphazard' though. If you had 12 panellists, you might use each of the orders twice. Haphazard would be the first order (123) 3 times, the second order (312) 5 times and the third order (231) 4 times. The haphazard approach completely misses three of the orders and is haphazard! Be careful when using a presentation design that has been generated for a different experiment. If you have 10 panellists and the design is for 100 panellists, the first 10 lines may well be 'haphazard' as the design was created for 100 panellists. The first 10 presentation orders may not be balanced for 10 panellists and might well introduce an unwanted bias.

There is a BS ISO standard relating to balanced incomplete block designs (BS ISO 29842:2011+A1:2015) which gives some useful designs and case studies.

As the experimental design and the plan for the analysis are intricately linked, it is advisable to consult a statistician for any new procedures or just to check that you have things set up correctly. To give a very simple example, if you were conducting

Table 4.2 Experimental design options

Term	Definition	Panellist		Example			
			Product A	**Product B, etc.**			
Completely randomised design	There are two types of completely randomised designs:	P1	✓				
	1. Each person taking part in the test assesses only the one sample and therefore each sample is assessed by different panellists (see example). Generally used in consumer tests for time or product-related reasons.	P2	✓				
		P3	✓				
		P4	✓				
	2. All the samples are given to all the panellists in a random order for each panellist. There is only the one assessment occasion. Therefore, this approach is only suitable for small numbers of samples, where all samples can be assessed in one session without causing panellist fatigue.	P5	✓				
		P6		✓			
		P7		✓			
		P8		✓			
		P9		✓			
		P10, etc.		✓			

Term	Definition	Panellist	Product A	Product B	Product C	Product D	Product E
Randomised complete block design (To keep the same word order as the first term maybe this would be better as 'completely randomised *block design*')	For example: All samples are assessed by all panellists in one session. Each panellist sees the samples in a different randomised order. The panellists return for further sessions to assess all the samples again. In each session the same sample set is seen (see example). In this case the blocks are the sessions. If this was a quantitative descriptive profile there might be three separate sessions to assess five samples three times. These repeated assessments are often called replicates and therefore the sessions and replicates (which are the same) could be called blocks. The five samples would all be assessed in each and every replicate by each and every panellist (in a different randomised order).	**Session 1**					
		P1	✓	✓	✓	✓	✓
		P2	✓	✓	✓	✓	✓
		P3	✓	✓	✓	✓	✓
		P4 etc.	✓	✓	✓	✓	✓
		Session 2					
		P1	✓	✓	✓	✓	✓
		P2	✓	✓	✓	✓	✓
		P3	✓	✓	✓	✓	✓
		P4 etc.	✓	✓	✓	✓	✓

Continued

Table 4.2 Experimental design option—cont'd

Term	Definition	Example Panellist (or block)	Sample number 1	2	3	4	5	6
Balanced incomplete block design (To keep the same word order as the first term, maybe this would be better as 'incomplete balanced *block* design')	The 'incomplete' part refers to one of two situations: 1. The samples are assessed over a number of sessions (not all in one session). Basically there are too many samples to assess in one session so the samples have to be spread out over a number of assessment times. However, each panellist sees all samples.	P1	✓	✓	✓			
	2. Each panellist sees only a subset of samples. For example, imagine we had six samples for a quantitative descriptive profile and we would like to do three replicates. But the panellists are unable to assess all six samples in one sitting, they can only assess three. This means that they only assess three of the six samples.	P2				✓	✓	✓
		P3		✓	✓	✓		
	The 'balanced' part means that each panellist sees the same number of samples and each sample is seen an equal number of times and each sample is seen in combination with every other sample an equal number of times across the session. Note: Because of the missing lines in the example, the sample pairs are not balanced. For example, samples 5 and 6 are seen three times, whereas samples 1 and 2 are only seen together once. When all the rows are shown, the sample pairs are assessed an equal number of times.	P4	✓				✓	✓
		P5	✓		✓	✓		
		P6, etc.		✓			✓	✓

a test where all the panellists assessed only one of two samples, you might analyse it using an *independent* samples t-test (in the absence of any other data). If all panellists had assessed both samples, the analysis would be by a *paired* t-test. The choice of the wrong analysis method ('use a t-test') might well lead to the wrong conclusion and hence the wrong decision.

4.2.2.7 Choosing the right method for the objective and the risk

Choosing the right method is not an easy task. There are many considerations to take into account. The main consideration is to really understand the business and specific study objective and the reasons *behind* the objective. Why does the client need this information? What will be happening as a result of your test report? Is this the first piece of data that is required which will lead onto further studies? What is the business risk associated with the test results? Depending on the risk associated with the test, you may be able to use a different, more rapid approach with less replications, for example (Stone, 2015). Section 4.2.10 gives a number of questions you might like to ask prior to deciding on the test method, as well as information about action standards.

The majority of the standard sensory texts describe the various methods available and also suggest tests for certain scenarios. Meilgaard et al. (2016) include an excellent chapter (Chapter 20) detailing certain practical sensory problems and suggested methodology. The first edition of Lawless and Heymann (1999) has some useful flowcharts to help choose the right sensory method (Chapter 19) which are partially reproduced in Chapter 1 of the second edition (Lawless and Heymann, 2010). Kemp et al. (2009) have some very useful industrial-based case studies that will also help guide your choice of method. Delarue et al. (2015) include some 'critical points in method selection' (1.3.2.3 page 17 onwards) which are very helpful for the choice of method.

One thing is certain: you cannot answer different objectives with just the one test, even if you are only running discrimination tests (Stone et al., 2012).

4.2.2.8 Having the right type and right number of panellists taking part in the test

A recent publication (Ares and Varela, 2017) regarding the use of consumers for analytical sensory testing makes some interesting points. The follow-up commentary papers (for example, Guerrero, L., 2017) are also very interesting. For a summary of some of the points raised see Chapter 13 in this book. Having the right type of panellist take part in your test is not as simple as the choice between trained or naïve. You also need to take into account the person's demographics, attitudes, motivation for the task, product experience and habits, health and sensory acuity, to name a few.

For advice on the number of panellists for each test, please see the relevant method standard (for example, ASTM-E2164 (2016) *Standard Test Method for Directional Difference Test*; BS EN ISO 5495 (2016) *Paired comparison test*; and BS ISO 11136 (2014) *General guidance for conducting hedonic tests with consumers in a controlled area*). Hough (2005) and Hough et al. (2006a,b) give some very useful information about the selection of the right number of panellists for the task.

4.2.2.9   Having the right record keeping procedures and data collection devices

Keeping records for test requests, panel plans, session summaries, data and reports is obviously going to be useful as you will be able to quickly refer back to the test objective and details of what happened in each panel session while writing your report. This will make your report more accurate as well as easier to write. In case of issues, you will be able to check back to see which panellists assessed which samples and when. It is good practice to keep these records for future queries. The right data collection device will also help you keep these types of records. For example, the database of your sensory software may well allow you to query which panellists have assessed which sample and how often. For more information about record keeping, see Sections 4.2.5, 4.2.7, 4.2.8 and 4.2.9.

4.2.2.10   Having the time to complete the experiment in the correct way

Rushing to get data generally means that the data will not be as good as it might have been. In a discrimination test, the samples may have been delivered to the panellists incorrectly. In a consumer test, the wrong type of consumers may be recruited. In a descriptive test the attributes may be poorly defined. All of these mean that the experimental conclusions might not be correct. This can be exacerbated if the experiment is the first of several for the project, as the errors may carry on through to each stage. Sensory teams do tend to be quickly booked up with work because the outputs are so valuable, but do not let that allow you to scrimp on good practice.

4.2.2.11   Having a sensory team that are well trained and kept up to date

A sensory team needs to be well trained and up to date. Training plans for the panellists and the support staff is very important. Plan in the training sessions for the panellists for the year ahead so that the training does not get replaced with requested studies. Ensure that the team have the training they require to do an excellent job. Make sure that there is attendance at conferences and courses through the year and that the information gets disseminated to the whole team. You may well find that three days spent on a course may well save you 10 days in experimentation time! For more information about staff training, please see Section 2.7.3.

4.2.3   Setting good objectives for your panellists

Your panellists will not know what is expected of them unless you tell them. Sounds pretty obvious, but you will be amazed at how many panel leaders do not communicate with the panel about what is expected. If you already run a panel, even if it is a panel made up of internal staff, an interesting exercise might be to ask them to write down what they think you expect from them before, during and maybe even after a panel session. At the same time, write down what you expect from them and then compare the two lists and see what the differences and similarities are. If the panel do not know what is expected of them, then you will need to make sure that they do. Some panel leaders develop a 'panel mission statement' with the panel – 'a reason for being' – that links in to the company's mission statement. This way the panel and the company can recognise and understand the importance of the panel to the business.

But what makes a good objective? Many companies use the acronym 'SMART', but the letters can mean different things in different companies and different locations.

The S stands for specific, so this means the objective cannot be something woolly and vague but simple, sensible and significant. The M stands for measurable and sometimes motivating; both are useful so maybe we should use SMMART. The A is achievable or sometimes agreed; again both are useful: there is no point you setting panellist's goals if that person does not agree and has no intention of completing them. The A is also often 'action orientated' which is another good description of a panellist's objectives, so this might be the best choice. The R is relevant and sometimes realistic or reasonable and here is a good time to mention the number of objectives: a reasonable number would be three or maybe four. T is generally related in some way to time. In can be a good idea to set goals that last three months when you first recruit a panellist, as setting goals that become unattainable due to unforeseen circumstances can be very demoralising.

Another useful way to set objectives that are more behaviour orientated is to use CASE: Context, Action, Standards and Evaluation (Cotton, 2014). This method can be used to encourage further 'good' behaviour and discourage 'bad' behaviour; however, they are generally instated when there is an issue with behaviour.

Context: looking at the behaviour you witnessed and the context in which you witnessed it. For example, you might see a panellist offering to help another panellist when training (good behaviour) or you might see a panellist talk over another panellist in a rude or derisory manner (not good behaviour).

Action: what action should the panellist take? Good behaviour: continue to help panellists as it's very helpful, thank you. Not good behaviour: remember the panellists' rules and the reasons for the rules existing and try not to repeat.

Standards: what are the standards? Refer the panellist to the rules and ways we work to make panel sessions run smoothly.

Evaluation: How will you keep a check on this in future and let the panellist know? By continuing to note good and not-good behaviour and to give timely feedback.

Let's look at setting SMART objectives from the point of view of developing objectives for a panellist. We will start with an internal panellist who works on a quality control panel (Figure 4.1) and finish up with an external panellist on a personal care profiling panel (Figure 4.2). To start with try to think about *why* you want to set an objective. What impact will its achievement have on the results or the team? Why is it important? Consider if it is measurable and relevant to the panel's work.

For more details about each of the panel performance measures and some handy tips on developing your own measures, please see Chapter 11.

Although the title of this section is about *setting* good objectives, there is really no point setting objectives if you then do nothing with them. To make them work you will need to check them (make sure they are still relevant), track them (make

1. Attend 60% of panel sessions in the first quarter.

This QC panellist goal will be very important for the achievement of your goals, as panellist attendance is the only way for you to gather the sensory data you need.

2. Achieve 70% correct results in the daily validation tests in the first quarter.

This goal is as critical as goal number 1: there is no point turning up to a sensory session if they are only 'there in person' without any motivation to do a good job. However, motivation is not all you need from your panellists: you also need to check that the panellist is capable of doing the job. Many companies will insert a hidden validation check test every day that sensory tests are conducted, others are happy to conduct them weekly. See Section 7.7 for a description of how to conduct validation tests for a range of panel types.

The level of correct results could then increase to 75% in the second quarter and 80% in the third quarter. Or you might adjust the difficulty of the validation tests and keep the objective as it is: or a mixture of both.

3. Attend two of the three monthly training/reminder sessions.

Again this objective is an important aspect of assessing the panellist's ability and can be very motivating when run well. See Section 9.1 for a description of how to conduct reminder training sessions for quality control panellists.

Figure 4.1 Quality control panellist goals.

sure the panellists are keeping to them) and update them when things change. This should be done monthly or quarterly to keep the objectives visible to the panellists: if you have written action-orientated objectives the panellists could be checking these each week to ensure they are on track. If the objectives are printed out and then filed away never to see the light of day until the yearly appraisal panic time, you might as well not bother with setting objectives at all. You will also need to explain to the panellists what will happen if they do not meet their objectives. This might be more training, more experience in a particular test or removal from the panel when things do not work out.

4.2.4 Panel monitoring and giving great feedback

Even the word 'feedback' can have negative connotations for people, so you need to approach giving feedback carefully. Consider your objectives for giving the feedback before you start: what does success look like? What do you actually want the panellists to start or stop doing? Obviously you will need to start by monitoring your panel and collecting information and data to base your feedback on. Some feedback needs to be delivered as soon as possible to ensure learning is effective or to nip bad behaviour in the bud, but other information will be collected about the panellists' performance in a particular study or experiment or test. For more information about panel performance monitoring and giving feedback about data, see Chapter 11. For information about giving feedback generally, see Section 12.4.1.

---

1. Attend 90% of panel sessions in the first quarter.

This external profiling panellist goal will be very important for the achievement of your own goals, as panellist attendance is the only way for you to gather data. The level is higher than for an internal panellist as the role is the person's job with the company. They will be expected to attend all sessions unless they are on annual leave or they are unwell. The actual figure used for your panellists can be designed to reflect the number of sessions run per year and the number of annual leave days the panellist is entitled to. The more panel sessions attended, the better the panellist will perform, especially if the role involves the assessment of a wide range of products.

2. Profiling performance:

Replications: have a mean square error (MSE) less than [insert value relevant for your scale and replications].

If you are running a panel that generally uses three replicates and a 100-point scale this might start out at 200 for a training panel and gradually decrease.

Discrimination: has an individual p-value of less than 0.4 (for a training panel and gradually decrease) for each critical attribute in the one-way ANOVA sample effect, if the majority of panellists have discriminated the samples for that attribute.

*You might ignore attributes that were difficult to define for products that were new to the panellists or where everyone had an issue due to experimental or sample variability. (This will now fall into your objectives: helping the panel to create suitable and effective attributes!). A nice simple way to see this is to use the PanelCheck p*MSE plots. More information on PanelCheck is given in Chapter 11. If you would like a 'get started quick' handout for PanelCheck please contact the author.*

Agreement: rank the samples in a similar order to the rest of the panel where the sample effect indicates that there is a statistically significant difference between the samples.

This could be demonstrated visually to the panellist by looking at the profile plots for each attribute and monitored by recording the number of significant attributes that the panellist contributes to the interaction.

Figure 4.2 External profiling panellist (personal care panel).

4.2.5 Make a plan for each session

Working on sensory projects can be quite daunting at first, so it can be useful to break everything down into smaller 'bite-sized' chunks. Take a look at the project objectives and time plan and consider what you might be able to complete within the time constraints. Clients seem to want answers quicker and also in more detail as the pressures of developing new products in several or even global markets increase.

It can be very helpful to make a written plan for each panel session that you run. In fact ISO 13300-2 *Sensory analysis – General guidance for the staff of a sensory evaluation laboratory – Part 2: recruitment and training of panel leaders* states that 'a plan for the preparation and presentation of samples should be developed before the test' (p. 9).

The most critical part is deciding the objective of each session. If you document the objective first, the rest of the plan tends to follow through. When you first start to write session plans, the process can be quite cumbersome, but once you have a few documents under your belt some quick editing gives you exactly what you need in record time. Each session plan should contain:

- The objective(s) for the session
- What needs to be prepared ahead of the session
- Outline of the plan to achieve the objectives
- Attendance at the previous session (and recap for those who were not present)
- Questions to ask to check assessors' understanding
- Time for the assessors to ask any questions
- Outline of next session
- Homework where relevant
- Anything else you usually find useful as a reminder

Carefully consider what you can accomplish in the time you have available. Do not expect to assess all samples and create a lexicon in a one-hour session! Agreement on the definitions and references can often involve quite lengthy discussions which can throw your plan completely. When the panel are new, everything will also take longer. It can be helpful to tackle defined tasks in each session. For example, if you are generating a lexicon, work on the appearance attributes on one day and the flavour attributes the next.

Handy tip: try to imagine the session actually running. What are you likely to need to make it run successfully? Spending 10 minutes hunting down a projector or flipchart paper in the middle of the session will be a huge chunk of your time gone: find it beforehand!

It can also be helpful to have some backup plans for when things go wrong, so think through several scenarios as you write your plan. For example, consider what you might do if you run out of time and do not manage to complete everything you had planned. Will the samples last until tomorrow? If not, it might be a good plan to cut short the attribute discussions and move straight to the sample assessments: the discussion can be done another day.

One of my clients has a good approach to backup plans. They have several panel training sessions planned in advance with everything set up ready to go: from the labelling of the sample vessels to the software sessions. This way, if the current project samples do not arrive on time or there are issues with the microbiology analysis, for example, they have a sensory session ready to go and the panellists are often not even aware that there is a change of plan.

An example session plan is given in Section 4.2.5.1 for a group of new assessors joining an existing work-from-home hair care panel. The session described is the third session of their product training, having had two previous training sessions introducing them to sensory work. The session described is related to the learning of the lexicon for the assessment of shampoos and will be followed by three sessions about conditioner use, three sessions about the assessment of dry hair and for those assessors joining the styling products panel, a further six sessions regarding hair sprays, gels and leave on conditioners. For more information about panel training please see Part II.

4.2.5.1 Example session plan for training a group of new assessors to join an existing work-from-home panel

Training session 3 outline
Session objectives
To make sure that all assessors understand the shampooing lexicon and each individual attribute protocol. To identify problem attributes and work through these so that all shampoo attributes are finalised. The plan for the next session is to run through all attributes to check the complete profile so we need to have the majority of attributes understood and working well. To make sure that the panellists understand the volume of shampoo they require and that this volume cannot be changed between products.

In advance of activity
1. Print out copies of the shampoo lexicon for each panellist.
2. Prepare demonstration area with mannequin and sample 1 or 5.
3. Set up the booths with two mannequin heads, a comb, two towels, the activity sheet and a rating sheet for each assessor (line scales on paper for each shampooing attribute).
4. In each booth place a set of samples in small pots and 2 syringes for each pair:
   a. Sample S1 = XXX Shampoo
   b. Sample S5 = YYY Shampoo
   c. Sample C2 = ZZZ Conditioner just in case we get to this stage
5. Print out the activity sheet for the panellists.
6. Print out the conditioning lexicon for homework.

Panel leader's instructions for the session
A. Feedback from homework: Shampooing Attributes (30 minutes):
   After the last session the panel were asked to assess the shampooing attributes using two quite different products (Shampoo 2 and Shampoo 3). The two products should have given different results and enabled them to practice the shampooing lexicon at home.
   In the feedback session ask:
   a. Did anyone have any problems using the protocol and rating any of the attributes? Discuss any problem attributes and demonstrate with a mannequin head how to make the assessment.
   b. How were the products different? Check that the panellists are in agreement about key attributes such as speed to foam and amount of foam.
   c. How much shampoo did you need to use? Record this for each panellist for future reference.
B. Demonstrating the shampooing lexicon (20 minutes)
   a. Demonstrate lathering attributes to the whole panel using a wet mannequin head sat on a towel and a willing assessor. It is not possible to demonstrate rinsing to all panellists: this will need to be done in the booths to pairs of assessors if the homework feedback and questions indicate that this is necessary.
   b. Answer any further questions about the protocol, process or attributes.
   c. Talk through the activity sheet and check everyone knows what to do by asking one or two assessors to explain the process.

C. Practising the protocol and rating shampooing attributes (60 minutes)
    a. Ask the group to work in pairs and to go through the shampooing lexicon with sample 1 and then sample 5.
    b. One assessor should do all of sample 5 whilst the other assessor rates on the paper form based on the other assessor's ratings.
    c. The assessor capturing the rating should also make notes and check that they are using the protocol in the same way as the assessor making the assessment.
    d. Then the assessors should swap roles and complete the next sample.
    e. Make sure that half the group start with sample 5.
    f. Once back in the discussion room ask the group:
        i. Which attributes were the most difficult or easy to carry out and assess?
        ii. What were the key differences between the products?
        iii. Discuss the rankings of the two samples on the scales.

Let them know that the next session will involve the use of the conditioner – hence the homework.

D. Presentation of homework (10 minutes).

Homework: To take home a copy of the conditioning lexicon
    a. A copy of the conditioning lexicon

Ask the assessors to read through the conditioning lexicon at home and then try it out using their standard products. It would be helpful if they could try to work out how much conditioner would be a suitable volume for their hair. Ask them to consider each attribute and protocol and come back with their questions.

Activity sheet for the panellists
1. Working in pairs in one of the wet booths, follow the protocol exactly as you have been shown. Thoroughly wet the mannequin head including the underneath areas.
2. One of you should shampoo the whole mannequin head with either sample 1 or sample 5.
3. During the shampooing concentrate on any difficult attributes identified in the homework and previous discussion.
4. During your turn at shampooing, your partner will write down any comments you have in your note book. Even if you are not applying and rinsing the shampoo, make sure you watch and feel the hair so that you can appreciate the difference between the two products.
5. Still in booths, complete the rating sheet for the first product. Complete your own rating sheet so you both have your own copy.
6. Swap places and assess the other sample, rating it on the same sheet – remember to label your rating marks!
7. Return to the discussion room with your wet mannequin heads wrapped in a towel.

4.2.6   Running panel sessions

Running panel sessions will probably be a key part of your role as a panel leader or sensory scientist. The sessions might be training sessions or reminder sessions, or if your panel works on the development of product profiles, they might be something you get involved in every time the panel is working. Section 4.2.5 includes lots of useful information about what to do prior to a panel session and an example of a session plan. Chapter 12 includes ideas for motivating panellists and dealing with tricky situations.

Before you start each session make sure you have everything you need and run through the plan in your head, imagining each stage. This is a good way to realise that you had forgotten the flipchart pens or the panel time sheets. Also check that the room is at a comfortable temperature and in the right layout for your session.

As mentioned earlier, it can be very beneficial to start each panel session with an outline of what you hope to get out of the time available and to ask if the panel agree. This way everyone is signed up to the objectives for the session and can raise issues if they think you are too ambitious or you forgot that today's session had been cut by 30 minutes for whatever reason. So, for example, you might start by saying, 'We have six samples in total to assess this morning. I'd like to get descriptions written for each sample and start making a list of attributes for appearance and aroma. What do people think? Is that a sensible plan?' You might get full agreement when you are working with a new panel, but later you might find a panellist is brave enough to say, 'What about getting the protocol finished, Lauren? We had real issues with that yesterday. It'd be good to get that finalised before we assess any more samples', which, of course, you had completely forgotten about. By starting each panel session in this way, you will find that the panellists are thinking ahead to what needs to be done in each session and almost planning the work themselves.

As the session progresses, remind the panellists where you are heading, what has already been achieved and what is coming up next. This helps everyone (including you!) see the bigger picture and keep the session objectives in mind. Assign a panellist as a timekeeper so that you can concentrate on your tasks, but remember to appoint someone different each time. Assigning a panellist as a note-taker can also be useful as you can type the attributes, comments or findings directly into your laptop, while the panellist writes the information for everyone to see on the board or flipchart.

When you are checking understanding of an attribute or a test procedure, you can ask the group for input, but it can also work really well if you ask a panellist by name. That way you can get contributions from everyone in every session.

Make a note of any 'bad' behaviour and address it at the end of the session. For example, if Camille has not been contributing as well as usual, despite you asking her a couple of questions, ask if everything is OK at the end of the session. Do not start the discussion by saying, 'Do you have a minute?' or 'May I have a word?' as these discussion starters can be enough to send anyone into a spiral of panic. Just say something like, 'Just wanted to check everything was OK with you?'

It can be very helpful to give a rough outline of each session at the start of each project and when the panellists are expected to do certain parts of the job in hand, although there are some provisos with this approach – see next paragraph. For example, if you are about to start a quantitative profile on seven products and you have nine sessions, you might begin by saying, 'There are seven samples and I hope to be able to assess them in triplicate. I think with this product type you might be able to assess all seven samples in a day, but let me know if not. Two sessions will be needed for the practice profile and feedback session and so that leaves four sessions to develop the lexicon. It's a similar product to the project before last so we might be able to use some of our learnings from that project. What do you think? Does it sound sensible or not?'

When you are running the panel session one of your main objectives is to create an atmosphere whereby the panellists feel at ease, they feel that you respect their opinions and that most things are agreed through discussion. You need to get the balance right so that you are running the session but not preventing the panellists from contributing nor allowing them to overcontribute! You are acting as a facilitator for the panel: planning, guiding, documenting and managing the session.

You will need to stay alert during the session, listen actively and follow the discussions. By asking relevant questions and checking your understanding, you set a good example to the panellists who will hopefully show the same level of engagement. You will need to be enthusiastic about the task at hand and help to energise the panel to create great work.

Consider different techniques for achieving the various tasks that the panel has to achieve. For example, if the panel have recently worked on the development of a profile, and the generation of a new attribute list for a similar product has made them sigh with boredom, consider collecting the attributes on post-its and get the panellists to group the similar attributes in small groups by modality. You can then go through all the suggestions with the whole group to finish things off.

If you are in the middle of a difficult discussion about an attribute and things just seem to be going round in circles, either take a five-minute break or ask the panellists to assess a sample that might help. If you ask the panellists which sample might help solve the issue, this can be very motivating for them.

A handy tip when there is a dispute about the presence or otherwise of an attribute is to present two unknown samples and ask the panellists if they can identify which sample is which. For example, you might present samples X and Y. Sample X is actually sample 5 from the set of samples you are working on and is the one that some panellists feel has, for example, an off-note. Sample Y is another sample from the set. If the panellists are able to tell you that sample X is sample 5, that can indicate that sample 5 does indeed have the disputed note.

4.2.7   Recording information from each panel session

If you make a note of what happened in each panel session as you go along (i.e., during the panel session) this can be really helpful to refer back to while documenting the attribute list, summing up the findings or writing your report. When planning in refresher training for the panel, the records of previous work can be assessed to determine the need for additional training. This document will also prove useful to your colleagues if you are unable to attend the next panel session. Some typical headers are given in Table 4.3. In the Description of the analysis section you could include experimental design, test location, type of test, training, documents arising from session with names and locations,

Table 4.3 Typical headers for a panel session record sheet

Panel name	Date	Panel leader
Project name	Samples tested	Panellist attendance
Description of the analysis	Comments	Next session plan
References assessed	Issues	

any problems encountered, suggestions for future analyses, etc. In the references section you could list the references assessed and whether they were useful or not.

4.2.8 Working with your 'clients'

You may have several different clients depending on the type of company you work for. If you work in industry, your clients might be various project managers or maybe the research and development manager. Your client might be the factory manager as well or even several different factory managers. If you work in academia, you might find that your clients are students as well as your manager or the faculty professor. In a sensory agency you might have clients in several different industries as well as their managers who may visit and see your panel in action. Whoever your client happens to be it's a good idea to develop a good working relationship with them: more like partnerships if you can. The key to this is good communication. If you are able to clearly understand what is expected and they clearly understand what your output is going to be, that is a very good start (see also Section 4.2.10).

One way to do this is to document every sensory request and then meet to talk about the client's requirements. Create a sensory request form that your client can fill in to give you the relevant information you need. Some ideas to include in your sensory request form are shown in Table 4.4. You might also like to create a sensory

Table 4.4 Header suggestions for your sensory request form

Project leader and contact details	Product type (e.g., sports drink, hand cream, shampoo)	Date of request
Priority level (e.g., project priority rating or business risk rating)	Background to the project (e.g., what the project aims are and where this sensory test request fits into the bigger picture)	Test objective (e.g., why this test is being requested)
Action standards (e.g., what is the standard and what will happen if it is met or not met)	Sample details (e.g., list of samples, names and ingredients)	Project timings (e.g., when can the work start and date report required by)
Details of previous assessments (e.g., bench assessments, instrumental analyses, etc.)	Health and safety information (e.g., novel ingredients, cleared for assessment (microbiological tests, etc.))	Sample delivery and preparation details (e.g., sample arrival date, responsible person, cook time and method, quantity to be applied, etc.)
Panel performance checks (e.g., does the client require these to be documented in the report)	Sample constraints (e.g., consumption restrictions)	Sample storage information and disposal after test
Report format and distribution (e.g., slides, written report and simple email summary)	Toplines requirement (e.g., does the urgency of the project require topline reporting in X hours)	

test plan form as a means of replying to the client, as that can help make the choice of method and approach clear for both parties, but communicating in person is critical for success: even if it's a five-minute phone call.

When you are chatting to your client, one question you can ask that can really help get to the bottom of things is, 'What does success look like?' or in other words ask them: 'You have your sensory report in front of you, what is it telling you?' If the client begins by saying, 'The sensory report tells me that the new product is preferred over the current product', and you had been thinking of conducting a discrimination test, it will be very clear that your plan would not have given the client the information he/she needs. Further ideas for questions to ask prior to a sensory project are given in Section 4.2.10. More information about action standards can also be found in that section.

If you are working with a client in a different location, do not rely on the method name to tell them what you are planning: give them some detail on the method and how the panellists will assess the samples. Not all methods are called the same thing by everyone: even in the same country! Sharing the panellist questionnaire/test sheet can be very beneficial, as this is when you realise that when the client conducts the tetrad test in his/her region to look for sample differences, he/she uses the instructions 'select the two most similar samples', whereas your test asks the panellists to group the samples into two groups based on similarity. This way you can explain to the client that research has shown that the 'select the two most similar samples' option reduces the power of the test (Rousseau and Ennis, 2013).

After meeting with your client, it can be useful to summarise any conclusions and actions in a short email so that everyone is clear about what is changed or what has been planned and when. This way, if there are any issues, you will get to hear about them before you start working on collecting the data, rather than after you have completed all the hard work.

4.2.9   Writing reports

There are many different ways to construct a report and it will depend on the objective, the product type and the method. You might have different types of report for different clients or methods, such as a short email or a longer detailed presentation. However, there are certain headings (see Table 4.5) that are worth mentioning that could be useful for you to include. They are not necessarily in the order that you will need them in and you probably will not need to include them all. For the report title it is worth thinking about this carefully. Calling the report 'Shelf life study for Project India' might mean everything to you at the moment, but in three years will you remember what project India was related to? And which time point in the shelf life the report refers to? If you are lucky enough to have sensory software that creates reports for you, remember that you still need to check and validate the raw data before pressing the 'analyse' button(s).

It can be helpful to have a database or dashboard of sensory reports for all clients within your company to view. This way if the client wants to fully understand a particular factory line, product ingredient, or consumer base, he/she can easily search, find and read the relevant information for himself/herself. This will help in terms of the workload of the team as you will not be asked to do studies that 'reinvent the wheel'.

Table 4.5 Report heading suggestions

Report title (future proofed)	Author(s)	Date
Objective(s) and background to the study (why you did it)	Action standards	Summary of the results
Next steps	Product type and sample list (including date codes and full details)	Sensory method (not just the name but the actual detail), attribute list/lexicon and sample assessment protocols
Panel details/demographics (but generally not the names of individual panellists)	Date(s) and location(s) of test(s)	Test constraints
Sample storage and preparation	Sample serving method (e.g., type of plate, type of application)	Sample issues (e.g., particular sample that was very variable)
Data checks and panel performance	Results and discussion	Tables, diagrams, photos, videos
Conclusions and recommendations	Appendices – you might have to include the raw data or analyses output in some cases	

Sample	Sample1	Sample 2	Sample 3	Sample 4
Sample 1		26	87	46
Sample 2			79	27
Sample 3				40
Sample 4				

Figure 4.3 Sample matrix comparison.

One useful addition for reports, where you are trying to compare several samples across several measures but you need the fine detail, is to create a sample matrix comparison sheet. An example is shown in Figure 4.3. Count the number of times the two samples are similar (or different) across the range of attributes or questions, and then express this as a percentage of the total number (or just include the counts). In the example 12 samples were compared in a sensory profile. Only four of the samples are shown in the matrix. We can see that sample 1 and sample 2 were found to be statistically significantly different (at $p < 0.05$) for 26% of the attributes assessed, so these two samples were quite similar. Whereas sample 1 and sample 3 were different for 87% of the attributes, so these two samples were quite different. These matrices can be drawn up by modality or question type. They can give a more detailed overview than multivariate statistical methods.

4.2.10 Asking the right questions and creating the right action standards

Before deciding how to approach a study, it's a good idea to think about the questions you might ask to make sure that you choose the best method and approach to answer the question. Some questions are given in Table 4.6 which might help. A great question to ask is simply, 'Why?' Getting to the bottom of *why* your client needs the information can be very enlightening – for you and your client! Another way to think about the different approaches you might take to tackle the question your client has, is to consider what will happen once your client/customer has got your data. These are called action standards because they detail the *action* to be taken when a certain standard is met or unmet.

Table 4.6 Question suggestions to ask prior to a sensory study

1. What is the client's objective?
2. What does success look like?
3. What are the action standards?
4. What is the client doing with the data?
5. What are the next steps?
6. What sensory methods might we consider to meet the objective and action standards?
7. What is the budget?
8. Resource availability in terms of staff. Panellist availability/panellist holiday issues?
9. Where are we going to conduct the test? Do we need to book facilities? Does the test need to be conducted in several locations?
10. Do we need specific facilities? For example, shower booths or powerful air extraction.
11. The design for the analysis of the data
12. What is the time frame?
13. What is the deadline?
14. How many sessions do we have?
15. What type of report is required?
16. How many samples are there?
17. What type of product is it?
18. Is the product being supplied by client? Are they supplying training samples or just product set?
19. Are any samples needing to be imported?
20. Are the samples served hot or cold/how is the sample prepared?
21. What will the samples be served in/how will samples be applied/used?
22. How are the samples to be stored?
23. Are the samples all different flavours/fragrances?
24. Are the samples served on their own or with a carrier?
25. How much sample is available for training?
26. How much sample is available for assessment/rating?
27. How long can a sample be served before it goes stale/changes/cools down?
28. Are the products meant to be served whole?

Table 4.6 Question suggestions to ask prior to a sensory study—cont'd

29. What do we do about broken or damaged samples (for example, squashed bags of tortilla chips and crisps/perfumes where bottle is left open)?
30. How much time do we need to leave between each sample assessment – fatigue/adaptation/carryover?
31. What type of palate cleansers might be needed?
32. How are the panellists going to clean fingers/skin between samples?
33. Are the samples spicy? Are they likely to cause fatigue?
34. What is the serving size?
35. How big are the product packs? How many will need to be opened?
36. What is the product variability?
37. Are the samples novel? Ask for ingredients list.
38. Are the samples alcoholic?
39. When are the samples going to arrive?
40. Is there a maximum number of samples we can assess in a session?
41. What other examples of this product are already on the market?
42. Is it quality control/cheaper ingredients… taints and references might be required?
43. Should we use lights to mask appearance differences?
44. Do we need photos of the products?
45. What do we do with the left over samples?
46. What panel performance measures am I working to?
47. What do we need to consider in terms of health and safety/ethics?
48. Do we need to take into account sensitivities/aversions?
49. Have we done any previous studies on this product?
50. Has anyone worked with this type of product/project before in a previous life?
51. Have we done any previous work for this client – do they have standard practices?
52. Have we used any training methods in the past related to this product we can transfer information from?
53. Does the client have a specified (profiling) method they have to use?
54. Does the client have a specified scale they have to use?
55. What questions do we need to include in the questionnaire?
56. Are there any known product references from client (for quantitative profiling)?
57. Are there any specific attributes/modalities the client wants to focus on?
58. Does the client have a protocol or lexicon to work to? Or do we have one already within the team?
59. How many replications is the client expecting?
60. Does the client want the scale to represent their samples or wider market?
61. Do we need to consider the product context?
62. How many panellists/consumers will be required for the test?
63. What type of panellist/consumers do we need?
64. Do we need to conduct sensory acuity screening prior to the test?
65. Are there any existing papers/standards/ASTM guidelines to refer to?
66. Contact with client – what do they want?
67. Are there going to be any clients visits/observations?
68. What disposable materials might we need? Do we have enough?

Action standards can be incredibly helpful in designing the test because they allow you to think about what will be done with the data or results. An action standard (AS) not only defines the aim of the overall experiment but also states the action or next steps to be taken dependent on the results. If you can also include the business risk level, this can be even more helpful. For example, the study may be a focus group looking at the potential or the scope of a new idea and the information will be used to feed into and develop company communications. Compare this to a preference mapping study with quantitative consumer research and decisions to be made about the direction for a major brand. The risks would be quite different and the approaches can therefore be different too.

Let's look at an example of an AS where a new supplier was being reviewed for a textural ingredient. The AS might have read:

AS1: 'If the sensory test confirms that there is a difference in texture between the new supplier and our existing supplier for product X, we will not proceed with the new supplier'.

The objective is clear: the client wishes to know not only if there is a difference in texture, but also, if there is a difference, they will be rejecting the new supplier. In this case a simple discrimination test could have been selected to determine if there was a texture difference. The next AS might have resulted in a totally different sensory approach:

AS2: 'If the sensory test confirms that there is a textural difference between the new supplier and our existing supplier for product X, we will need to understand what the textural differences are and if there are any other changes to product X as a result of the potential supplier change. Our existing supplier will stop producing at the end of December so it is critical we find a new supplier'.

The sensory scientist may well have chosen to ask the client how likely they thought a difference might be between the new supplier's ingredient and the existing supplier. If it was likely that there would be a large difference the scientist may have decided to conduct a full profile to understand what the differences were. These differences would be critical in understanding the effect of this change on consumers' reactions to product X. If the change threatened key drivers of liking, further discussions with the new supplier may be required to achieve a match.

Let's look at another action standard:

AS3: 'If the sensory test confirms that there is a textural difference between the new supplier and our existing supplier for product X, we will need to understand what the textural differences are and if there are any other changes to product X as a result of the supplier change. We also need to understand how this difference is related to the natural variance in our product'.

In this case the sensory scientist may well decide to carry out some batch-to-batch variability tests and determine where the batch with the new ingredient fits among the sample set. The difference from control method (Whelan, 2017b) might be a useful starting point to gather data for this test as several batches can be included in one test.

The next example (AS4) is an example of a poor action standard:

AS4: 'Is there a textural difference between the new supplier and our existing supplier for product X?'

The stage after understanding if there was a difference or not is not documented and therefore the choice of test is a difficult one. It is more likely that the wrong test would be chosen and then, when the results are reported, the client would be questioning what the next steps should be and the sensory scientist will probably not have the information at hand to guide the client. I can recall a particular project I was working on many, many years ago which was a great example of this. The client asked, 'Can you triangle test this?' and I complied. When reporting that there was a difference between the standard product and the new process, the client replied, 'Yes – I knew the samples would be different. I wanted to know in what way they were different'. After profiling the samples, I reported how the samples were different and the client said, 'Yes – I knew the new process would create that difference, what I needed to know is how consumers will perceive that difference'.

These examples indicate how knowing the objective and knowing what the next steps will be as an outcome of the sensory study are vital in the choice of sensory test.

4.2.11   Palate cleansers for food and cleansers for other products

It's important to choose the right cleanser for the palate or other working sense as part of the assessment protocol. The cleanser can help to reduce adaptation and clean the area of the previous product. Lucak and Delwiche (2009) found that the only palate cleanser that worked for seven food categories (sweet, bitter, fatty, astringent, hot/ spicy, cooling and non-lingering) was table water crackers. There is also a saying in sensory science: 'the only real palate cleanser is time'. Some cleansers you could evaluate to see if they might help speed things up are given below in Table 4.7. Be aware that sometimes external factors can impact your work. Check that the soaps or hand treatments in the facilities local to your panel rooms are unfragranced and if you are working with panels that assess hand creams or hand sanitisers or similar, that these soaps and hand treatments are not too harsh. You may need to buy specific products for your panellists to use. You may also need to train the panel on the correct use of palate cleansers. They need to be used according to a protocol to prevent some panellists, for example, consuming three packs of water biscuits every session.

4.2.12   Other options (for being an excellent panel leader)

Reviewing: Once a project is finished spend some time reviewing what worked well and what did not. Ask all the different people involved. List the pros and cons for the method you chose. Consider how the results might have turned out if you had approached the project in a different way. Write a short one page of notes that you can look over the next time a similar objective turns up: you can then read this and refresh your memory and this will help you make even better decisions next time.

Reading: Keep up to date with the literature. 'The Journal of Sensory Studies' and 'Food Quality and Preference' are the 'go to' journals for sensory scientists, but there are many others that have some very interesting publications. The standard sensory textbooks will also be very useful. And do read the sensory standards related to your area, as these contain a wealth of information and further references that are helpful. Reading sensory standards outside of your area and contributing to the development of these standards, for example through joining the ASTM, can help your scientific, and career, development a huge amount. A list of useful references is given in Chapter 14.

Table 4.7 Palate cleansers and cleansers

Palate cleanser	Details/works with?
Resting and time	Everything. Resting the eyes can be important especially if staring at a production line or similar
Plain crackers or water biscuits	These are very popular along with water for many foods and beverages
Water	Sipping water between samples can help cleanse the palate. If the product being assessed is dissolved in water, as in the case of basic taste assessments, it's important to use the same water for palate cleansing. Sparkling water can also be useful for greasy food
Steaming water	Can be useful to rehydrate nostrils after gas chromatography-olfactometry for example
Diluted lime juice/warm water	Lime juice at a concentration of around 10% in water can be very useful for clearing grease from the mouth and warm water can also be effective. Warm water can also be useful between assessments of very cold products like sorbets and ice creams
Fudge	Fudge can be helpful for removing minty tastes and effects after the assessment of menthol, toothpaste and mouthwash
Lemon juice in water	A solution of lemon juice in water can be useful for rinsing fingers when assessing greasy food but be careful it does not introduce lemon notes to the aroma or flavour profile
Parsley	Parsley is an excellent garlic palate cleanser
Bread and apple	Can be very useful for products that are oily, especially fish oil
Cheese, yoghurt/milk, cucumber/apple, plain bread, crackers	These products can all be useful for spicy foods but beware of the introduction of other tastes and effects. The best palate cleanser for spicy food is time
Back of hand	It's said that smelling the back of your hand or the inside of your lower arm is a good cleanser for the assessment of fragrances or odours
Leave the room	If the extraction system in the assessment area is not quite doing its job, or the assessments are being made in a local hall as part of a central location test, then leaving the room before assessing the next fragrance can be helpful
Creams	Creams or lotions applied to the skin can be notoriously difficult to cleanse. If possible, select different regions on the skin area for assessment rather than using a cleanser. A gentle fragrance-free soap can work, but recovery time from the soap and water will be needed before the next assessment
Hair products	If these are being assessed on the head by the assessor themselves, the only cleanser is time. It could be that only one or two assessments can be made in a week. If the assessments of the products are being made on switches, assessors can clean their hands with gentle fragrance-free soap and the next assessment can be performed once the assessor is ready. If it is only the dry hair that is being assessed, several assessments can be made by the same assessor without any cleansing, unless the products leave any residue. In that case the assessor can clean their hands with gentle fragrance-free soap and the next assessment can be performed once the assessor is ready. If other hair products such as gels or hair sprays are being used, the hair may need to be cleaned with a base shampoo to remove all residues
Aroma assessments	Breathing through a warm, clean, cotton terry cloth to filter nasal passages (see Chambers et al., 2016)

Networking: You can also find information about updates in sensory science by joining a professional organisation and reading their published articles and attending their events. The Institute of Food Science and Technology in the United Kingdom has the Sensory Science Group which oversees the institute's sensory science training courses and accreditation at an advanced level. The Society of Sensory Professionals is another nonprofit organisation and is devoted to developing and promoting the field of sensory science. The Sensometric Society's aims are to increase the awareness that the field of sensory and consumer science needs its own special methodology and statistical methods. Another useful group to join is the ASTM Committee E18 on Sensory Evaluation. Attending sensory conferences such as Pangborn, SenseAsia, Eurosense and Sensometrics, which are held all over the world, is invaluable for keeping up to date and making contacts.

For more information on all these areas see Chapter 14.

4.3   Training to be a panel leader

There are four main ways that people become panel leaders:

1. Previous job role
2. Training by an experienced panel leader
3. On-the-job training
4. Training from being a sensory panellist

Having a panel leader join the company from a previous job role running panels can be very beneficial. They will already know the basics and will have had plenty of experience in group dynamics so there will be less training required. However, training may be required in the product type and the specific methods that the new company uses, as it can be quite disastrous, for example, if a panel leader implements quantitative references for a panel that has never used them before.

A new graduate or employee from another section can also be trained to be a panel leader. They might be trained by an existing experienced panel leader or someone who has been hired on a consultancy basis specifically to train them (ISO, 2006). They might be trained by the experienced person but will also need guidance on the right books and journals to read, or even good networking groups to join. The training may involve some time working as a panellist or might be on a one-to-one basis with guidance and feedback. Sensory courses like those run by the Institute of Food Science and Technology, the Institute of Food Technologists or the University of California, Davis can be a good starting point if the new panel leader requires training in sensory science. The panel leader gradually learns all aspects of the role from product preparation through to report writing (as required) under the guidance of the experienced panel leader.

A panel leader may also be 'thrown in at the deep end' and learns from actually doing the job. The panellists themselves do a lot of the training of the (frantically learning to swim) new panel leader. The learning is often done on a trial and error basis, and although it can be a very good way to learn (once you have messed up on sample delivery once, it is unlikely to happen again), the errors associated with this training method make it less than ideal. However, if the training also incorporates formal experiences such as watching other panel leaders, attending courses and reading relevant publications, an excellent panel leader may still emerge.

The final route to becoming a panel leader is from being a sensory panellist. The person will be very experienced in the company's product(s), will understand the group dynamics, and will have witnessed the company's methodology on many occasions. They will probably need training in sensory science, statistics and test controls but these are easier to learn than some of the aspects they will be well versed in. In my experience panel leaders who were once panellists tend to make excellent panel leaders.

One thing to keep in mind with any of the training that is run internally is that sometimes the learning can be diluted to such an extent that it becomes worthless or even wrong. Imagine the situation where an experienced panel leader makes an excellent job of training a new panel leader, but where all the learning was verbal: nothing was written down and there were no standard operating procedures (SOPs) or documented instructions. This new panel leader then becomes an experienced panel leader (let's call them experienced panel leader 2) and a few years down the line gets the opportunity to train a new panel leader. The new panel leader asks questions like, 'Why do we use three-digit codes?' or 'Why do we check if the odd sample can be determined by appearance alone before running a triangle test?' and experienced panel leader 2 replies 'I'm not sure. I doubt it's important. Let's say we stop doing that from now on and save ourselves some time'. And it all goes downhill from then on.

Do not ignore the importance of documenting the learning via SOPs or instructions. These can be helpful, time-saving and might well mean that excellent rather than poor data are collected.

Panel leaders can also benefit from training in the use of the sensory software, office applications (spreadsheets, documents and presentations) and, perhaps the two most important elements of all – health and safety! In food companies a food hygiene certificate may be a legal requirement and in home and personal care panels there will be other legal requirements for working with human panels (see Chapter 3 for more information).

As panel leaders may well need to train existing and new panellists, they might benefit from attending a 'train the trainer' course. These courses are specifically designed to help people train colleagues in a workplace situation. The course teaches elements such as designing a training programme, producing teaching aids and materials, communication and presentation skills, managing questions and dealing with difficult trainees. It should also help the trainer become more confident in their abilities and give them skills in evaluating their own teaching ability so that they can continuously improve.

Recruitment of a sensory panel

5.1 Introduction

The sensory panel is an essential analytical tool for the research and development, quality and consumer insights teams in many industries and therefore will require in-depth planning, the right resources and the right approach for success. There is an ISO standard that covers the recruitment, training and monitoring of panellists (BS EN ISO 8586:2014) and also an ASTM publication: STP758, which although it is quite old, includes some very useful information (ASTM, 1981). There are four main checks of a person's suitability to become a sensory panellist: their health (e.g., diet, allergies); personal characteristics (e.g., motivation for the role and personality); ability to follow instructions; and their sensory abilities. You will also need to check the applicant's age and comply with local regulations regarding testing with children (for example, see Market Research Society, 2017).

The sequence of events for the recruitment of a sensory panel is shown in Figure 5.1 and therefore, you will also need to read various other chapters in this book to understand the complete process. For example, aspects of training are including in Chapters 6, 7 and 8, but you might also wish to read Chapter 4 as that includes hints and tips to make your training work hard for you and to help you be the best panel leader.

This chapter has been split into several different modules so that specific sections can be chosen depending on the requirements for the type of panel needed. For example, there is a section/module on 'screening for basic tastes', which will be essential for a panel assessing foodstuffs, but if the plan involves recruiting a home and personal care panel, will probably not be needed. If you are recruiting a panel for the assessment of just one or two modalities, consider adapting the different tests included in this chapter for your needs. The chapter includes full details on how to conduct the recruitment and suggested criteria for the employment of the different types of panel (for example, quality control, internal/external, descriptive, consumer) for home and personal care, as well as food products. It includes advertising for panellists, application forms, pre-screening, screening, interviewing and probationary periods. It also includes handy tips, how to organise everything, different approaches to the tests and how to finish up with the right panel and the right panellists.

For some panel types you may not find the screening module you require. For example, if your research is related to studying tastant release during eating, you might need to add in some screening about saliva flow and discuss if the person is happy taking part in the assessments you are proposing. Information about these types

Figure 5.1 Steps in the recruitment of a sensory panel.

Sensory Panel Management. https://doi.org/10.1016/B978-0-08-101001-3.00005-7

of screening tests is not covered in this book, but information is available in the literature (for example, see Salles, 2017 for a good review).

There is currently interest in moving away from simply conducting basic taste tests and recruiting those panellists that pass. Many people feel that the assessment of panellists using these tests alone does not necessarily create the best panel for the job. It's important for panellists to be able to detect, describe and sometimes measure various aspects of the product's sensory characteristics, and these skills are not easily checked just by conducting basic taste tests. The BSI standard that covers the recruitment and training (as well as monitoring) of panellists (BS EN ISO 8586:2014) recommends that the selection of sensory assessors should come after some initial training. This is a very useful idea for all panellists that require training of some sort, as often panellists are recruited and it is only during the training that you realise that their ability to replicate their output is not quite as good as you hoped, or that they do not really fit in with the rest of the team. From the panellist's point of view, they might also realise that the job is not quite what they thought and actually they do not like working in the booths, or working as a team, or taking part in discrimination tests, etc. Therefore, following the standard's recommendation is a good plan. However, this can be quite difficult if you are recruiting an external panel, as you may need to recruit more panellists than required to cover this fallout and the human resources (HR) processes might not work easily in this situation. One way to get around this is to extend the screening to cover the initial training sessions, paying the panellists for attending the training sessions in the same way that you might pay consumers for taking part in a test for example, with shopping vouchers. Another option is to recruit more panellists than you need and employ them for a probationary period, say six months. That way, the panellist can decide to leave up to the six months time point without any additional hassle and you can also ask them to leave if they are not working out. The main disadvantage of this is that you will need to have the facilities (booths and room) for the additional bodies. Another disadvantage is that some panellists may form a firm friendship with the panellist(s) who is leaving and also opt to leave. But this can happen anyway and is something you will need to learn to deal with.

If you are recruiting sensory panellists to work with instrumentation, say with functional magnetic resonance imaging (fMRI) or gas chromatography-olfactometry (GC-O), it might be useful to include an actual assessment with the instrumentation within your recruitment procedure if ethics and legal restrictions permit this. This will give you the chance to check if the panellist is happy to take part in these types of tests. The same applies to products or assessments that are likely to be less pleasant, such as the assessment of pet foods or malodours. If it is possible to include the products in the screening tests, you are likely to recruit panellists who are not averse to these types of assessments. You could also consider including an overall liking question alongside the test questions if you are able to arrange for your product type to be assessed in the screening. This might give you an indicator of how likely the applicant is put off by the product type and therefore lose motivation and interest in the role.

Before you start the actual recruitment of the panellists, note that you will be collecting personal information and therefore you will need to consider how you are going to keep the information secure and also plan what will you do with the information once you have made your final choices. Your HR team should be able to help you deal with local regulations, and there is more information about the regulations you may need to deal with in Chapter 3.

5.1.1 What do you need in a sensory panellist?

There are several elements that are required and several that are 'nice to have'. Let's start with the required elements first. The main requirement is that the applicants are interested in the role. Maybe you thought I would say their sensory ability was going to come first, but people can have very good abilities in recognising differences between products, describing and defining attributes or replicating their results, but if they are disinterested, bored or cajoled into taking part, they will not make good panellists. They simply will not be focussed and motivated to do a good job.

The next aspects to check are that they are available when you need them; they are healthy without any allergies or illnesses that might impact on the new role; they are happy to eat the foods you are going to present or use the products that need assessment; and if there any other points that might restrict them joining the panel or taking part in the test. If you are recruiting a new sensory panel, plan your sensory panel training session dates prior to the recruitment, as that way you will be able to check that the panellists are all available for the training sessions.

If an applicant has any allergy it is best to remove them from your recruitment list. Even if they are not allergic to anything you plan to assess, your company may start to make products that contain that ingredient in the future and the company may well be held liable if there was any adverse reaction. All of these elements can be checked by the use of an application form and examples are given in Section 5.4 below. Checking that the panellist is able to communicate what they perceive to you (and other panellists where necessary) is best checked in the screening phase, however, you can begin the assessment of this in the application form by asking questions such as 'Describe the flavour of a cola' or 'How would you describe the texture of a hand-wash soap?' depending on your product type and the work the panellists might be asked to do.

Panellists will also need to have a good memory, especially if they are to join a descriptive panel, to enable them to recall previous products they have assessed and generate the descriptors they will need for the assessment of various products. This is quite difficult to check in the application form or the screening process, but the screening test for the description of aromas (see Section 5.4.5.6) can give you some information about the panellist's memory abilities. The screw cap jars containing the various aromas for this test give no clues as to what product the aroma originates from, so this is a situation that relies on the applicant's sensory ability to detect the aroma and also their memory, in the absence of these clues, to dredge up descriptions for the product.

And of course, the final element is the panellist's sensory ability. If you are recruiting people for a consumer test for acceptability or preference you probably will not need to screen for sensory acuity, although this can be very helpful in several test types. Often people think that analytical sensory panellists need to be 'supertasters' or have a very keen sense of smell, but this is not true of all panels. Generally, you need to recruit panellists that have a 'normal' sensory acuity, i.e., recruit people with the same sensitivity to sensory stimuli as that of 50% of a given population of individuals, so that their results will be similar to that of the average consumer. If this sounds counterintuitive, consider a discrimination panel that is sensitive to the slightest of changes to an ingredient, be the change for cost saving or supplier issues. The panel would never allow any change to the product to 'pass' even when the majority of consumers would never even notice the difference. For a descriptive panel working on projects to understand which attributes and how the changes in their intensity affect the consumer's liking of the product, if the panel create and measure tens of attributes that consumers might not even realise exist in the product, their output would not be very representative of the average consumer's liking of that product.

If you took a group of 100 consumers and screened them for their sensory abilities, you would find a large number of differences between them – especially in terms of their sensitivity to bitter compounds. These differences in sensory ability can be due to many things including physiology, gender, age, experience and genetics. One genetic difference that is particularly interesting is the 'taste blindness' to some bitter compounds such as phenylthiourea (known as phenylthiocarbamide (PTC)) and 6-n-propylthiouracil (known as PROP). In 1932, Arthur Fox first discovered a taste 'anomaly' with PTC in his famous 'dust flying' experiment. He was weighing out PTC when a colleague, Dr. C.R. Noller, mentioned that the dust from the weighing was causing a bitter taste in his mouth. Fox had not noticed any such taste and proceeded to taste some crystals and still felt they were tasteless. Fox and Noller went on to investigate a large number of people and found that there was a group of people who, like Fox, did not describe PTC as bitter. The group had no other identifying characteristics (gender, race, age, nationality) and further work (Blakeslee, 1932) showed that the differences were related to genetics. People were classified as being tasters and non-tasters of PTC and more investigations with other bitter compounds were also conducted. However, testing your panellists' sensitivity to PROP (PTC was found to be detectable by odour and so studies tend to focus on PROP nowadays) does not mean that you will be able to select the best panellists: they may not be sensitive to compounds that you need them to be sensitive too, just PROP! Another factor to take into account when recruiting panellists is that what people do not sense may well enhance what they do sense (Lawless and Heymann, 2010).

There are certain rules for panellists to follow if they are to take part in a sensory test or join a sensory panel. Figure 5.2 summarises these.

5.2   Introduction to the recruitment of an internal panel

There are several different types of internal panels, but they are generally involved in shorter sensory tests such as quality control, discrimination tests or perhaps informal ('bench' or small group assessments) or qualitative sensory assessments. The people who take part in the panel sessions are generally already employed by the company and may work in different areas such as finance, research, or health and safety. Performing sensory tests will not be their only focus so it's important to recruit people who are interested in the work: someone who is forced to take part will not give the type of data needed. The first step in this type of recruitment process is to stress the importance of the panel's work and their role in the process of ensuring product quality. Only invite people for screening who are interested in taking part, are generally in

Before attending a panel:

- Does not wear any perfumes of any kind;
- Does not eat strong curries before a panel session;
- Does not drink strong coffee, chew gum or smoke before a panel.

In a session:

- Is punctual;
- Is a good listener;
- Able to take part in the discussions but does NOT run the discussions;
- Allows **everyone** time to express their opinion; does not talk at the same time as someone else;
- Expresses **their** opinion;
- Respects everybody;
- Listens to the panel leader and follows instructions.

Figure 5.2 Attributes of a good panellist.

the office and are not located too far away from where the tests are to be conducted. There is one exception to this rule. It's generally a good idea to recruit some senior managers to take part in the tests as this sets a good example to the rest of the staff. Be aware of any issues with hierarchy as often a senior manager will be allowed to speak first and everyone else will agree with her. Note that, as they might not be able to take part as regularly as might be required to keep their training and experience up to the same level as the rest of the panel, you may sometimes have to disregard their results. If these issues, hierarchy and lack of training, cause you too many problems, another way to get management support is to justify the sensory panel by 'selling the sizzle and not the sausage'. This basically means that instead of stating what the panel might do and the results they might produce, you will need to document the benefits of taking part in the sensory tests. For example, people working in administration may find taking part in the tests motivating and enjoyable and they might also make new friends within the business. But in terms of the 'sizzle', you might describe this to management as lower turnover in staff and hence lower recruitment costs.

The potential panellists will need to be screened for sensory acuity (ability) and trained to take part in the type of tests planned. The following modules may be required: internal panel recruitment, internal advertising, application process, sight acuity tests, basic taste acuity tests, olfactory screening, discrimination tests and ranking tests. For example, if part of the internal panel's work is in the assessment of the correct colour of a product, the assessment of each panellist's ability to see colours is imperative. However, if colour is not an important aspect of the sensory tests you conduct, for example, in the assessment of spice level in powdered curry mixes, there may not be the need to conduct the sight acuity tests as part of the recruitment process. If the colour or appearance of the products is an important measure, several of the sight acuity tests listed in that module may be required and also further discrimination tests to fully check each potential panellist might be advisable.

5.3   Introduction to the recruitment of an external panel

There are several types of external panels such as

- panels that might do descriptive type tests (e.g., qualitative descriptions, quantitative descriptive profiling, temporal methods, shelf life tests) on one type of product only;
- panels that work on a mix of quality, discrimination and descriptive work;
- panels that work on a large range of products for an ingredients company or sensory agency;
- panels that work with subjects to make their assessments such as deodorant or hair-care panels (the subjects are the people who attend panel sessions and have their hair or armpits assessed, for example);
- panels that work at home assessing home and personal care products or high fatiguing products such as tobacco, gum or spices;
- panels that require a particular type of trained panellist such as a hair stylist or flavour chemist for the type of output and integration needed within the company.

As for the internal panel, the potential panellists will need to be screened for sensory acuity and trained to take part in the type of tests planned. The following modules may be required: external panel recruitment, external advertising, application process, sight acuity tests, basic taste acuity tests, olfactory screening, discrimination tests and ranking tests. You might also want to test for memory and ability to concentrate. Further modules to test for descriptive ability, ranking and scaling tests as well as a one-to-one interview will be required for a panel whose work involves any descriptive output. However, one-to-one interviews need not be planned for all applicants, just those that pass the initial screening tests. For panellists that will be working on descriptive profiling methods that require work as a team, such as Quantitative Descriptive Analysis (QDA), Spectrum or some hybrid or adapted method, it might be beneficial to conduct a mini panel session, to see how the panellists might work together and with the panel leader.

The most important part of external panel recruitment is to assess each potential panellist's motivation and personality. This is critical because if someone is recruited who causes issues within the panel, the data can be adversely affected. Therefore, if someone passes all the sensory parts of the test, even doing very well, but does not pass the interview or mini panel session, do not invite them to join the panel.

5.4   Recruitment modules for analytical panels

5.4.1   Recruitment Module 1: Internal panel advertising

Advertising for internal staff to join the panel can be a fairly easy task if a company-wide email can be sent, however, these are often not read and other methods may be more effective. Posters in the reception, canteen or coffee areas can attract attention, but if there is a panel already in place, word of mouth can be very helpful. Existing panellists can be asked to hand out a flyer to colleagues, or managers can be asked to nominate interested people from their teams to take part in the screening tests. The advert should describe the screening process and it's particularly important to make clear the commitment required, especially if there are some longer training sessions to attend, so that staff know what they will be signing up for. Some potential adverts are shown in Figure 5.3. (Albion Mills is a fictitious food factory.)

5.4.2 Recruitment Module 2: External panel recruitment planning

External panel recruitment is more complex than internal panel recruitment as all the steps involved in recruiting any other member of staff will have to be performed, as well as those related to sensory acuity. The first step is to involve the HR team as their support will be needed in all stages of the recruitment process. One of the major hurdles that may need to be jumped is in the set-up of the screening process itself. It is worth having the process fully documented before you begin, so that the HR team understands what you will be doing at each step. Figures 5.4 and 5.5 give an overview of the process and more detail is given in the following sections and chapters.

Things to be considered in the recruitment and training of an external panel will include employment process steps; what will be included on the advert; where the advert will be placed; recruitment other than by advert; visitors coming on site process; review of application forms; communication process with potential applicants; where the screening will take place; room booking; tests to be conducted during the screening; equipment required for the screening process; staff required for the screening (which may include existing panellists); mark scheme for the screening; plan for the one-to-one interviews; feedback to the applicants; training programme; staff required to implement training; existing panel sessions while training is being conducted; facilities for training; equipment for training; process for leavers; probation period: moving from trainee to panellist or exit; integration with existing panellists; ongoing monitoring and performance. More information on each of these aspects is provided in this book. Letters can be drafted ahead of the process to send to successful applicants to invite them for the screening. Letters also need to be sent to unsuccessful applicants.

Development of the training programme is not an easy task. It may take several days to determine what needs to be included and to develop the relevant documentation for each step. The training section includes several examples of training programmes for different panel types and is laid out in a modular fashion so that the programme can be built more easily. Several of the documents are available on the author's website (www.laurenlrogers.com) to save reinvention of the wheel.

Integrating new panellists within an existing panel can cause some issues: Section 6.6 gives some solutions and ideas to help.

5.4.3 Recruitment Module 3: External advertising

Adverts for new panellists can be placed in the local press, placed on notice boards in local supermarkets and shops (with permission) or handed out at local interest groups (e.g., mother and toddler groups or sports clubs). You could also advertise at the local job centre, online job sites or via a radio article about your panel. Recruitment via

Are you interested in helping our company continue to make first class products?

Can you attend panel sessions at 10am most days for 10 minutes?

Please email makeadifference@companyname.com to receive an application form.

You may subsequently be invited to attend a one-hour sensory screening test and if you are successful, training will take place on two consecutive afternoons (dates tbc) from 1pm until 4pm.

As you know, the quality of our products is critically important to our consumers.

Would you like to join our existing sensory panel that is key to ensuring our products' success?

Please email makeadifference@companyname.com to receive an application form.

You may subsequently be invited to attend a two-hour sensory screening test and if you are successful, training will take place on four afternoons (dates tbc) from 1pm until 3pm.

Making sense of quality

Be part of our exciting new sensory panel helping to maintain excellent quality products!

Email newsensorypanel@companyname.com for an application form and an invitation to the sensory screening test. If you're successful you will be invited to join the sensory panel which will meet most mornings at 10.30 for 30 minutes.

What to find out more? Come along to chat to us every lunchtime this week in the canteen. You can try out some of the sensory methods we will be using and there will be freebies for everyone!

Dear colleague

You will be aware from the management communication earlier this week, that there are opportunities to join our sensory panel at the Albion site. Joining the sensory panel is one way to complete your 'commitment to quality' objective that makes up our personal development plans. The sensory team will be available in the Albion canteen every lunchtime next week to answer your questions. Come along and pick up an information leaflet to learn more. We will also be holding a 'guess the aroma' competition and everyone who takes part will get a small prize.

Looking forward to seeing you there,

The sensory team

Figure 5.3 Example adverts for an internal panel.

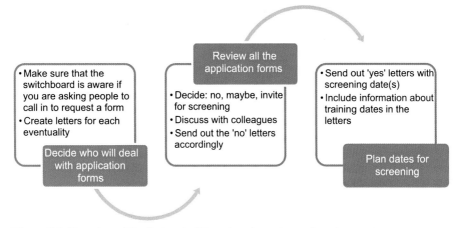

Figure 5.4 Overview of the first part of the external panel recruitment process.

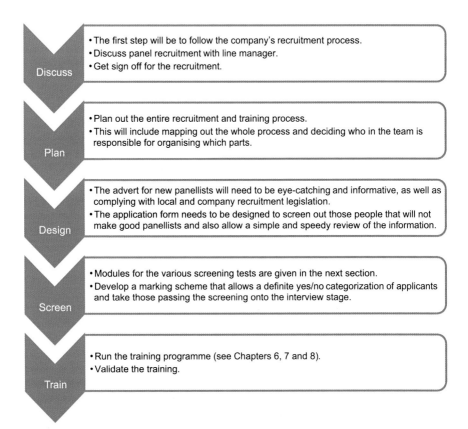

Figure 5.5 Overview of external panel recruitment process.

existing panellists can be both helpful and unhelpful: helpful in that similarly interested people can be recruited, and unhelpful when the friend does not pass the sensory acuity test and the existing panellist feels bad and leaves. If the friend passes the test and joins the panel, this can also cause issues: think carefully before taking this route for recruitment. If you are recruiting through an employment agency, be very clear about the type of people you are looking for. If they have never recruited sensory panellists before, you may have to explain in detail exactly what you need them to do.

There are several different ways to deal with the number of applicants interested in becoming a sensory panellist. If previous experience has resulted in too many applicants to deal with, it might be worth considering asking people to write in for an application form. If time is of the essence, asking people to phone in for an application form can speed things up, as long as the receptionist can cope with the number of calls. Applications online can make the whole process streamlined and resource friendly. You may even be able to use your existing sensory software or one of the free online survey tools (but remember to check data protection regulations in all cases).

Some example adverts for external analytical sensory panels are shown below in Figure 5.6.

5.4.4   Recruitment Module 4 (RM4): Application process

The application form can be designed to suit the panel type. Some companies may well have a standard application form and you will need to adapt this to suit the recruitment of a sensory panellist. Various questions are shown below; some questions may be more pertinent to different panels. Pick and choose questions as required from the five sections below. Section 1 asks for personal details and Section 2 asks the important questions about health and allergies. Sections 3 and 4 ask about food, and home and personal care habits, respectively, while Section 5 asks about interest and availability. After each section header there is some additional information that will help in selecting the right questions and also dealing with the answers when the forms are being reviewed. It's often useful to pilot your application form with colleagues to make sure that the relevant questions have been selected and that the application form is easy to complete. Order the questions so that the most important ones are asked first, as this will save time reading answers that are less relevant. For example, asking the questions about allergies on the first page enables a quick screen of the forms if you receive 100 or more applicants.

Figure 5.7 gives an outline of a typical paper application form. An online form would be similar, and it can be very useful if the questionnaire can be created using the sensory software the panel will be using, as this gives you an additional check of panellist suitability.

5.4.4.1   RM4 Section 1: The applicant

Choose from the questions below to develop the application form and add any questions that might also be required by your particular situation or company. If the answer to Q8 is yes, it would be worthwhile asking some more questions during the interview to determine if there is any conflict of interest. Check that the applicant is local as it

---

Tickle your taste buds with this sensational opportunity
Sensory panellists: part-time
Location: [insert location here]
[Insert company information here (e.g. Company X is a world leader in X and produce famous brands Y and Z…)]
You too could have a taste of our success. Join our friendly sensory science team at our Albion site, and you'll be fully trained to assess products from a range of our premium brands including X, Y and Z. The information you provide will help ensure that our [such-and-such] products remain amongst the leading brands in the UK and across the world.
The positions will suit friendly people who enjoy food/cooking with a good general education to GCSE level (or equivalent) and who enjoy working as part of a team. Taste a more fulfilling future!
If you are interested and can dedicate three mornings OR afternoons every week (9 hours per week), please telephone the hotline number 01234 567890 or write to The HR Officer, Albion Factory, Arthur Street, England for an application form.
Closing date: 1st April 2020
[Insert Data Protection Act statement here.]

EXCITING OPPORTUNITY TO JOIN OUR FOOD TESTING PANEL

Location, Location

£X.X0 an hour

Sensory Panellist

Position: Permanent, part time

Hours of work: Monday, Tuesday, Wednesday and Thursday fixed hours from 9.30 am to 12:30pm

Albion Mill is looking for new panellists to join our trained sensory panel in [insert location] to assess various dairy and meat products including cheeses, soups and ready meals. Experience is not necessary as full training will be given. You will join a sensory panel that evaluates a range of products for our company to help ensure high quality daily production as well as help with new product development. This role would suit those who can work well in a team and have excellent communication skills. You must be in good health, with no allergies or intolerances. You must also be an excellent time keeper, conscientious, reliable and willing to commit long term.

To find out more information or to apply please email your CV to makeadifference@companyname.com and we will contact you.

Job Type: Part-time

Salary: £X.X0 /hour

Job Type: Permanent

[Insert provisos here e.g. eligible to work in country X, data protection, interviews will be held w/c 3/3/20 etc.]

Figure 5.6 Example adverts for an external panel.

Work from home!

Site location, home

Sensory panellists

Salary:

Job type: Permanent, part-time (4 hours per week)

This is not one of those cheesy 'work from home' adverts that do not deliver. This is a genuine advert for a major blue-chip company who require people to work at home assessing home and personal care products such as floor and worktop cleaners. If you are successful in your application you will be invited to take part in an evening screening session, which, if you pass, will lead on to training to be a sensory panellist. The training sessions will be held three evenings a week for three weeks. Further training sessions will be held once a month, every month, generally on a Tuesday evening. You must be in good health, with no allergies or intolerances. You must also be conscientious, reliable and willing to commit long term.

[Insert provisos here e.g. eligible to work in country X, data protection, interviews will be held w/c 7/10/20 etc.]

Part time product evaluator

$X.XX per hour, 9 hours per week

Prettiville, US

This part time role would suit someone with an interest in skin care. Working as part of a friendly team you will assess existing and new skin care products. Full training will be given but you must be able to commit to a permanent part-time role. The hours of work are every Tuesday, Wednesday and Thursday from 1:30 to 4:30.

If you are interested [and insert legal requirements here] please apply with a full CV via the website link below, where you will also find more information about this interesting and varied role.

Figure 5.6 *(Continued)* Example adverts for an external panel.

[Insert Company Header]

PRE-SCREENING QUESTIONNAIRE FOR RECRUITMENT OF SENSORY PANELLISTS

Please complete this application in your own handwriting.
All the information recorded in this questionnaire will be treated in the strictest confidence.

[Insert questions required from Sections 5.4.4.1 through 5.4.4.5.]

I hereby certify that the information given above is correct to the best of my knowledge.

Signed: _____   Date:_____

Please return this questionnaire to: Sensory scientist
                                          Albion Mills
                                          Arthur Street
                                          England

Figure 5.7 Example application form.

is unlikely that a panellist's pay will cover long-distance travel costs. Look through the applicant's CV to understand previous roles and likelihood of stability in the role.

Q1. Full name
Q2. Address
Q3. Telephone number
Q4. Email address
Q5. National insurance number
Q6. Other HR requirements, e.g., gender, age, date of birth, driving status (important if work site is remote).
Q7. Please attach your CV or describe your current and previous employment below.
Q8. Do any members of your immediate family or close friends work in any of the following occupations?
Advertising, Journalism, Marketing, Market Research, Radio, TV, Food Research Company.
If yes, please state names of company, type of job and relation to self.

5.4.4.2   RM4 Section 2: Health

Check with HR before including any health-related questions as there is different legislation relating to this in different countries' employment laws. Choose from the questions below to develop the application form and add any questions or points that might also be required. For example, the question regarding oral/dental surgery may only be applicable to the recruitment of a panel assessing toothpastes, toothbrushes or mouthwash. Smoking does not necessarily exclude the applicant (especially for a panel assessing smoking products!), but a discussion in the interview about refraining from smoking prior to a panel session will be necessary. If the applicant has dentures, this again may not prevent employment if they pass all the sensory acuity tests (dentures can lower sensitivity to taste if the person has a denture that covers the palate), but if the panel assesses products such as toffees, nuts or toothbrushes, for example, these might cause issues for the wearer. If the applicant wears light sensitive or tinted glasses and the panel work includes the assessment of product colour, it would be worth asking the applicant if they have a plain pair of glasses they might wear during panel sessions. And, as mentioned earlier, if the description of colour is vital to the work the panel take part in, colour blindness will result in applicant exclusion. It's generally advisable to exclude anyone with any allergies from the panel and, if you are recruiting a food or drink panel, also anyone who takes long-term medication that affects the sense of taste. This can be checked online or through the company doctor if available.

Q1. Please answer the following questions with a yes or a no as appropriate.	
Do you smoke?	yes/no
Do you wear full or partial dentures?	yes/no
Do you wear light sensitive/tinted glasses?	yes/no
Are you colour blind?	yes/no
Have you recently had oral/dental surgery?	yes/no

Q2. Do you suffer from any of the following?	
Allergies (describe _____)	yes/no
Asthma	yes/no
Celiac disease	yes/no
Diabetes	yes/no
Digestive complaints (e.g., ulcers)	yes/no
Frequent mouth infections	yes/no
Frequent nasal infections	yes/no
Hay fever	yes/no
Heart complaints	yes/no
High blood pressure	yes/no
Sensitive teeth	yes/no
Sinus problems	yes/no
Skin problems	yes/no
Sore throats	yes/no
Unusually cold or warm hands	yes/no
Q3. Are you taking any medication as part of a long-term course of treatment? yes/no	
If yes, please describe:	
Q4. Do you have a known history of hearing damage?	
Q5. Do you have reduced sensitivity in your fingers?	

5.4.4.3   RM4 Section 3: Food habits

This section may not be relevant for home and personal care panels. Choose from the questions below to develop the application form and add any points that might also be required. If the panel assess meat products and the answer to Question 1 is yes, and vegetarian or vegan as the explanation, the applicant will need to be excluded. Vegetarians do not eat any foods that 'consist of, or have been produced with the aid of products consisting of or created from, any part of the body of a living or dead animal' (source: Vegetarian Society). As this includes gelatine, for example, it may exclude applicants from panels that work on some confectionary products, jellies or chocolate puddings. There are different types of vegetarians and you might like to ask which type in the interview section if your products are not primarily meat:

- Lacto-ovo-vegetarian. Eats both dairy products and eggs. This is the most common type of vegetarian diet.
- Lacto-vegetarian. Eats dairy products but not eggs.
- Ovo-vegetarian. Eats eggs but not dairy products.
- Vegan. Does not eat dairy products, eggs or any other animal product (e.g., honey).

Obviously, if the applicant answers Q3 (What foods or drinks would you never consume?) with any of the foods the panel will be required to assess, the applicant will be unsuccessful. When I was working for GlaxoSmithKline we were recruiting a new panel to assess a range of soft drinks. One of the applicants answered Q3 with 'soft drinks'. I'm still trying to work that one out.

Questions 4, 5, 6 and 7 will give an idea of the applicant's interest in food and eating. If they never eat out and never try new products, this might be an indication of food neophobia, and this can be investigated further during the interview.

Q1. Are you currently on a restricted diet?	
If yes, please explain:	
Q2. What is (are) your favourite food(s)?	
Q3. What foods or drinks would you never consume (eat or drink)?	
Q4. If you eat out, how often do you eat out in a month?	
Q5. If you eat out, what type of establishment have you visited on your last three or four meals out?	
(Please tick the types of restaurant you have eaten out at in the last few months)	
Chinese restaurant	[ ]
Indian restaurant	[ ]
Italian restaurant	[ ]
French restaurant	[ ]
Fast food restaurant	[ ]
Hotel restaurant	[ ]
Pub restaurant	[ ]
None	[ ]
Other (please specify)	
Q6. Do any of the following statements apply to you?	
I am always looking for good quality products.	yes/no
I eat fresh fruit most days.	yes/no
I like to try new products when I see them in the shops.	yes/no
I think it's important to eat plenty of fibre.	yes/no
I like watching TV advertising.	yes/no
I buy and eat (insert relevant product type) most weeks.	yes/no
I prefer to stick with products that I know and rarely try new products unless recommended by friends.	yes/no
Q7. Which of the following items (if any) would you say that you normally make or cook from basic or fresh ingredients (as opposed to buying as a ready-to-eat/cook item)?	
Bread	yes/no
Casseroles	yes/no
Lasagne	yes/no
Pastry for pies	yes/no
Pizza	yes/no
Soup	yes/no

5.4.4.4   RM4 Section 4: Home and personal care habits

This section may not be relevant for food and beverage panels. Choose from the questions below to develop the application form and add any points that might also be required. Obviously, if the applicant answers Q2 with any of the product types or brands the panel will be required to assess, the applicant will be unsuccessful. Question 3 will give an idea of the applicant's interest in new home and personal care products. If they never try new products, this can be investigated further during the interview. Questions 4 to 7 are questions relating to recruiting a panel to assess toothpaste, toothbrushes or mouthwash. Questions 8 and 9 might be useful for the recruitment of a listening panel, and Q10 for a panel assessing hand creams.

Q1. What do you currently use to clean your [insert relevant word, e.g., floor, bathroom, face, hands]?	
Q2. What types cleaning products do you tend to avoid, if any?	
Q3. Do any of the following statements apply to you?	
I am always looking for good-quality products.	yes/no
I like to try new products when I see them in the shops.	yes/no
I like watching TV advertising.	yes/no
I use [insert relevant product type] most weeks.	yes/no
I prefer to stick with products that I know and rarely try new products unless recommended by friends.	yes/no
Q4. Do you regularly attend the dentists?	yes/no
Q5. Do you have sensitive teeth?	yes/no
Q6. Overall, how would you rate the health of your teeth? (Excellent, good, fair, poor.)	
Q7. In the last 7 days, how many times did you use mouthwash or other dental rinse product? (Insert number.)	
Q8. Do you listen to music?	yes/no
Q9. Do you play a musical instrument or sing?	yes/no
Q10. In the last 7 days, how often did you use hand cream/lotion? (Insert number)	

5.4.4.5   RM4 Section 5: Interest and availability

This section asks if the applicant has ever been a member of a product testing/ sensory panel before. If the applicant answers yes, this can be both beneficial and detrimental depending on the type of panel they previously belonged to and the type of tests they used to be involved in: make a note to ask further questions in the interview. Questions 3 to 8 can help highlight applicants with good (or bad!) descriptive and language skills. Questions 9 and 10 check the applicant's availability and additional questions can be added here, for example, if recruiting for a work from home panel and you would like to check on availability for regular training sessions. Questions 11 and 12 may help in identifying the applicants that are the most interested in the role, and Question 13 may be useful if the site is remote or the hours are awkward to rely on public transport. If you are recruiting a panel to assess pet foods you might like to include a question relating to pet ownership, as panellists who own a pet may be more motivated to take part in aroma and flavour assessments of pet foods.

The scaling exercise will give early warning for applicants who either are 'unable' to read instructions or cannot scale. However, do not put too much emphasis on the results of this question, as training can improve accuracy.

Q1. Have you ever been a member of a product testing panel?	yes/no

Q2. If yes, please give details below:

Q3. How would you describe the aroma/smell of [insert relevant product, e.g., cheddar cheese, your favourite shampoo/soap, a bakery...]

Q4. Describe the flavour of [insert suitable product, e.g., cola, chocolate, coffee...]

Q5. Describe the texture of [insert suitable product, e.g., crackers, chewing gum, suntan lotion...]

Q6. Describe the sound of [insert suitable product, e.g., opening a bottle of carbonated (fizzy/sparkling) drink, snapping off a piece of chocolate, cleaning liquid from a spray bottle...]

Q7. Describe the appearance of [insert suitable product, e.g., a glass of sparkling water, shampoo foam, a lychee...]

Q8. List some of your favourite brands of [insert suitable product] and describe which aspects of the [insert relevant sensory modality, e.g., aroma, appearance, flavour, texture] you like best.

Q9. Please indicate **all** days you are available for a [insert length of panel session] hour panel session (tick all that apply).

	Monday	Tuesday	Wednesday	Thursday	Friday
Morning					
Afternoon					

Q10. Are you able to work during school holidays?	yes/no

Q11. Why are you interested in this role?

Q12. Is there any other information you consider relevant to your application?

Q13. How would you anticipate travelling to Albion Mills?

Q14. Please complete the scaling exercise on page [insert page number] (see Figure 5.8).

Scaling exercise

Rate the area SHADED BLACK on the line below each shape. You do not need to calculate the area, just look at the shaded area and use the line scale to indicate approximately how much of the shape is shaded in. Question 1 has been completed for you to show you what you need to do.

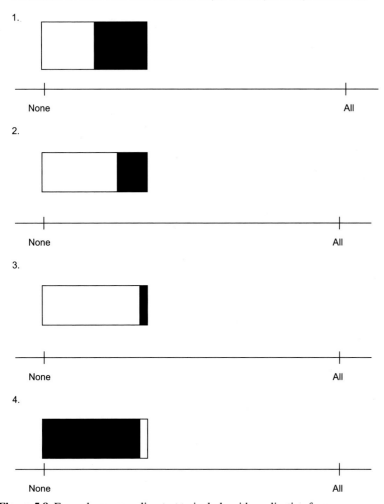

Figure 5.8 Example paper scaling test to include with application form.

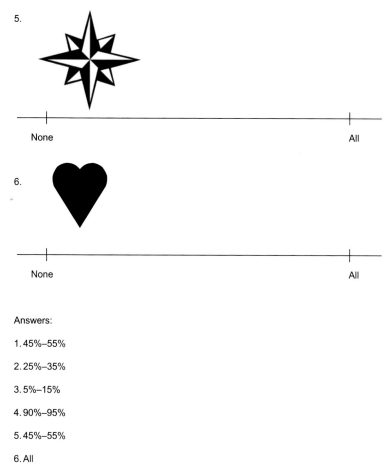

5.

None   All

6.

None   All

Answers:

1. 45%–55%

2. 25%–35%

3. 5%–15%

4. 90%–95%

5. 45%–55%

6. All

Figure 5.8 *(Continued)* Example paper scaling test to include with application form.

5.4.5 Recruitment Module 5 (RM5): Sensory screening

5.4.5.1 Introduction: Why do we need to screen people for our sensory panels?

Screening is an important step in the recruitment of a panel working in sensory assessments because we are starting with consumers who all have very different sensory abilities. One of the reasons for these differences amongst people is down to the different levels of thresholds; some people have very low thresholds for certain product attributes (e.g., they might be very sensitive to bitter) but high thresholds for other attributes (e.g., they may not be able to detect the odour of vanilla until it is at a very high concentration). In fact, around 30% of consumers cannot discriminate changes in products they consume (Stone et al., 2012) and it's a good idea to be able to understand our panellists' particular sensory abilities.

Generally, in the recruitment of a food and drink panel, the aim is to recruit people who have normal sensitivity to the type of product attributes they will be used to assess, however, in some cases there might be a requirement to recruit people with higher sensitivity, for example, if the panel's main role is in the detection and prevention of shipping tainted products. If you recruit a panellist with very low thresholds (see later for more information on different types of threshold) you might find that they detect differences between the samples that the rest of the panel cannot. They will probably soon become frustrated and unmotivated. On the other hand, if you recruit a panellist with very high thresholds, they will find it difficult to describe and detect the attributes they are required to. Therefore, the preferred approach is to recruit panellists with normal sensitivity. For more information about individual differences in perception, Chapter 15 in *Foundations of Sensation and Perception is very informative* (Mather, 2016). There are also several publications about the differences in perception with age (for example, see Mojet et al., 2005).

Everyone has different sensitivities to different elements of products and we also differ during assessments, between assessments and over time. And, to complicate matters, we all have different levels of different types of thresholds. There are four main typesof threshold:

Detection threshold: this is the lowest concentration that will result in a perceivable sensation. For example, let's say we had two glasses of water and one of them had a very low concentration of sucrose (just a few grains). You might taste the water from both glasses and notice that the water with the few grains of sucrose added tasted odd. You would not be able to say that it tasted sweet, just that it was peculiar. This threshold is measured in sensory science because, for example, flavour chemists need to determine the impact made by a certain compound. And when these values are quoted it can be a bit confusing to understand, as the *lowest* quoted threshold actually means that compound has *more* impact. To add a little more complication to the mix, we also know that this detection threshold is not at all like an on/off switch. If your detection threshold was, say, 2 g of sucrose in a litre of water and I gave you this solution 20 times (obviously in a nicely designed and timed experiment with the right number of 'blanks') about half the time you would not detect that I had added the sucrose to the water. This is because of the background level of nerve activity and is also related to the statistical probability based on the normal distribution. (If your statistics could do with a brush up, there are several great books out there on the topic but one of the best, because of its sensory applications, is O'Mahony, 1986). The threshold is therefore affected not only by the person's variability but also any variability from the stimulus (the sucrose solution) itself.

Recognition threshold: this is the lowest concentration that will result in a perceivable recognisable sensation. Let's go back to our two glasses of water. This time we have more sucrose in the second glass and as you taste it, you recognise it and say, 'How odd – I think this water tastes sweet!'

Difference threshold: this is the lowest concentration that will result in a perceivable recognisable difference in intensity between stimuli (when we use the word stimuli this basically means a thing, e.g., flavour compound, odourant, that creates a reaction in the human doing the assessing of the stimuli). Going back to our two glasses of water, now imagine that they both contain sucrose at the same level: you would probably say they were the same intensity of sweetness if asked. If one glass had 20 g of sucrose per litre of water and the other 21 g per litre, you might not be able to detect a difference, but a colleague might be able to. This 1 g per litre difference might be your colleague's difference threshold for sucrose, while you might need a difference of 2 g to be able to tell the samples apart. This difference threshold is important for detecting weaker and stronger levels of the stimulus.

The final threshold is called the **terminal threshold** and is the concentration for some sensory systems at which no further increase in concentration results in any change in intensity levels. The terminal threshold is less important in food and beverage sensory science as it's rare that products would contain ingredients to the level of saturation. However, this threshold might be important for some panel types working on strong odours, for example.

For more information about thresholds see Bartoshuk (1978), ASTM E679 (2011), ASTM E544 (2010) and BS ISO 3972 (2011).

There are three main types of sensory screening tests: looking for impairments (for example, ageusia or anosmia); determining the person's sensory acuity (for example, their ability to detect small differences between samples) and checking their descriptive and communication skills (BS EN ISO 8586:2014). Checking for impairments is important, as you do not want to recruit someone who is unable to perceive important aspects of your product (e.g., bitterness in beer or a certain aroma in washing powders) or is unable to perceive differences in samples that the rest of the panel are easily able to detect. Conversely, you do not want to recruit someone who is extremely sensitive to certain product aspects nor supersensitive to very small differences between products (for more information on this see Section 5.1.1).

In fairly recent publications, where the authors are recruiting panellists for descriptive analysis, there are several different approaches to screening. Some authors conducted no screening (for example, McMahon et al., 2017), others have used discrimination tests to choose panellists for their sensory profiling tests (for example, Imamura, 2016; da Silva et al., 2013). Others have used the textbook/BS ISO screening tests or adapted BS ISO screening tests for their product type (Vidal et al., 2016; Chen and Chung, 2016; Hwang and Hong, 2015; Pereira et al., 2015) and others have used experts for their assessments (Coulon-Leroy et al., 2017; Jose-Coutinho et al., 2015). Many papers omit to mention the screening methods they used for the recruitment of the panel.

Table 5.1 gives an overview of the sensory screening tests that might be useful for different sensory panels taken from various sensory standards and textbooks. If there is a tick in the cell, this means that generally thatscreening test, for example, taste acuity tests, is seen as a useful step in finding the best panellists for that particular role. If the tick is in brackets, this indicates that the screening test might prove useful in some situations and not others. For example, if a quality control panel checks the product appearance, it might be useful to check potential panellists' sight, but if they only ever check the texture of the products coming off the line, then these sight acuity tests may not be necessary. Sometimes there is a tick in brackets because that aspect may be important for your specific needs. If there is no tick in a cell, this indicates that

Table 5.1 Suggested screening tests for the recruitment of a sensory panel

Examples of Panel type	Sight acuity tests	Hearing acuity tests	Taste acuity tests	Olfactory acuity tests	Texture acuity tests	Discrimination tests	Ranking tests	Descriptive ability	Mini panel session	Interview
Analytical panels for food										
Internal discrimination testing panel	(✓)	(✓)	(✓)	(✓)	(✓)	✓	(✓)			
Internal quality control panel	(✓)	(✓)	(✓)	(✓)	(✓)	✓	✓	(✓)		
External panel used for food texture profiles	(✓)	✓	(✓)		✓	✓	✓	✓	✓	✓
External panel used for discrimination and profiling (and time intensity studies)	✓	(✓)	✓	✓	✓	✓	✓	✓	✓	✓
Analytical panels for home and personal care										
Internal discrimination testing panel	(✓)			(✓)		✓	(✓)			
Internal quality control panel	(✓)			(✓)	(✓)	✓	✓	(✓)	✓	
External panel used for discrimination and profiling (and time intensity studies)	✓	✓[a]		(✓)	(✓)	✓	✓	✓	✓	✓
Consumer panels										
Analytical sensory focus group	(✓)		(✓)	(✓)		✓		✓		
Hedonic and preference testing										

[a]Particularly important for the recruitment of listening panels.

generally this type of screening is not required for that type of panel. But be careful and consider your particular type of panel.

You might find that some different types of screening might be required for your specific panel. For example, it's less common for the sensory panel recruitment standards compiled by ISO and sensory textbooks to include texture acuity tests in the panel screening, therefore, if you are recruiting for panellists to assess products where the assessment of texture is critical, for example, meat products or fabric conditioners, it might be a good idea to give priority to texture tests over the other screening tests. This could be easily incorporated into the discrimination tests or descriptive ability sections of the screening. Some additional ideas for testing texture acuity are given in Section 5.4.5.7. The ASTM publication '*Guidelines for the Selection and Training of Sensory Panel Members*' has a useful section for the screening of panellists for texture (ASTM, 1981). The ISO standard for the Texture Profile has a very rapid panel screening tool which involves asking the panellist to rank four sets of samples for various texture attributes (e.g., hardness, gumminess) in the correct order. For more information on texture sreening see Section 5.4.5.7.

If you are recruiting for a discrimination panel, design the tests to use the products that the panel will be working on: you might find that some people are very discriminating on one product but not at all discriminating for others. Start by explaining the tests to the applicants, as this way you will be testing the person's sensory ability and not just their ability to do the test. You might like to treat the first result for each test type differently, as the panellist will be learning more about what to do and not concentrating so much on the product differences. Use several of the discrimination tests that are in current use as this gives you a head start on the training programme. You will need to repeat each test to determine if the applicant can actually detect the difference or if they are just very lucky. The ASTM Selection Guidelines (ASTM, 1981) suggests repeating four sample sets six times: conducting 24 trials at least with each applicant.

5.4.5.2 Developing the screening tests

Screening tests can be split into four different categories:

1. Tests to check for impairment;
2. Tests to ascertain sensory acuity;
3. Tests to understand the applicant's descriptive and language abilities;
4. Interviews to ascertain suitable levels of interest, motivation and personality.

Impairment tests include checking the applicant's senses to ensure they are not colour-, odour- or taste blind and that their ability to assess textures or sounds by the senses of hearing and touch is sensitive enough for the role. Acuity tests can also give information about impairment but are really designed to assess the applicant's ability to discriminate between samples using sensory discrimination or descriptive tests. For example, if the panel will be assessing various soups, a ranking test on salt-water solutions might be helpful, or a 3-alternative forced choice (3-AFC) test where the applicant has to choose the saltiest sample from a set of three salty solutions. However, conducting similar tests using different salt levels in soup itself might give you a better idea of the person's ability to take part in tests on your products.

Asking an applicant to describe a product can be a simple test to check for language and descriptive ability. For example, if the panel's role is in the assessment of shampoos, foams can be created for the panellists to assess by the use of dishwashing sponges and various foaming products (e.g., shampoos, laundry washing liquids and washing-up liquids) and ranked in order of amount of foam, speed to foam or bubble size. If you are recruiting a panel to assess less desirable samples such as pet food or malodours, it can be useful to include these types of products in your screening tests if local regulations and ethics allow. This way you get an idea of the applicants' response to these types of products, and the applicant also gets an idea of the type of work they might be asked to do. Including an overall liking question for these types of panels can be helpful in gauging suitability for the role.

It may take some time to develop the screening plan as there are several things to consider. Firstly, the screening tests need to be devised and checked, which if there is an existing panel, can be more easily accomplished. If there is no panel currently, consider testing the screening programme on colleagues to determine if the tests are easy to understand, that the levels you have chosen are suitable and how long the screening might take.

Pick and choose from the sections below for suitable screening tests. Within each section, ideas for screening tests for various product types are also given as well as answers that will indicate a potentially good panellist.

5.4.5.3 Running the screening tests

More details on each of the screening tests are given in each of the sections below, along with examples of questions asked and how to assess the answers. It can be useful to run the screening with batches of applicants, say 8 to 10 at a time depending on the amount of work required to lay out the tests. Set up the screening so that there is time between each section to allow for any holdups or issues. A typical plan is shown in Figure 5.9. Go through the process and each of the tests prior to the panellists taking part to make sure that panellists understand what they need to do (ASTM, 1981). You want to give them every chance of passing.

5.4.5.4 RM5 Section 1: Sight acuity tests

A fair proportion of people have issues with colour vision. There is variation amongst humans generally: differences between males and females and differences due to age. About 8% of males and 0.4% of females are colour blind. There are several types of colour blindness and the most common is deuteranopia, known as red/green colour blindness, where, for example, blue and purple will be confused, as the red element of the purple cannot be perceived. The standard test for checking for normal colour vision is Ishihara's tests for colour blindness (Ishihara, 1994). This is a small hardback book containing colour plates of discs made up of lots of different coloured dots. The location of some of the coloured dots make up shapes or numbers, and the tracing of the shapes or the identification of the numbers allows the tester to determine if the subject is colour blind or not. This test can be carried out by printing out sheets of Ishihara plates or projecting discs within a slide presentation and asking applicants

What?	Where?	How long?	Who?	What?
Applicants arrive: Batch 1	Reception	5–10 minutes	Receptionist and existing panellist A	Panellist A and receptionist to feedback any comments
Batch 1: Applicants are taken for screening introduction	Panel waiting room/ conference room 1	5 minutes	Panellist A	Panellist A explains what life is like as a sensory panellist
		5 minutes	Sensory scientist	Explains the recruitment procedure (e.g., screening, mark results, invite successful applicants for interview, training dates)
Batch 1: Screening introduction		10 minutes	Technician, sensory scientist and panellists A and B	Slides and demonstration for various tests. Show the coded cups and vials. Give handy tips. Hand out/show screening questionnaire and check if any questions.
Batch 1: Screening	Booths/ assessment area/ conference room 2	1 hour	Technician, sensory scientist and panellists A and B	Conduct screening tests
Applicants arrive: Batch 2	Reception	5–10 minutes	Receptionist and existing panellist C	Panellist C and receptionist to feedback any comments
Batch 2: Applicants are taken for screening introduction	Panel waiting room/ conference room 1	5 minutes	Panellist C	Panellist C explains what life is like as a sensory panellist
Batch 1: Finishing screening	Reception	Until all finished	Panellist B takes batches of finished applicants to reception	Panellist B explains the next steps (marking forms, interview, training programme) and applicants leave
Batch 2: Screening introduction		10 minutes	Technician, sensory scientist and panellists A and C	Slides and demonstration for various tests. Hand out/show screening questionnaire and check if any questions
Batch 2: Screening	Booths/ assessment area/ conference room (2)	1 hour	Technician, sensory scientist and panellists A and C	Conduct screening tests *Continue…*

Figure 5.9 Typical outline of screening programme.

to write down the numbers they see. There are even applications (apps) that can be downloaded onto tablet computers and mobile phones, but unfortunately, in all these cases, it's difficult to guarantee that the colours will print accurately or will be shown as intended. It's best to conduct this test within the interview section using the actual Ishihara book of plates if it's an important area for the sensory assessments.

Other useful vision tests are the Farnsworth-Munsell D-15 dichotomous and 100 hue colour vision tests which checks a subject's ability to discriminate colour hues. Various colour plates are presented to the applicant who is required to sort them into the correct order, e.g., from one colour to the next colour as they gradually change in hue. Searching online for 'Farnsworth-Munsell' will bring up several versions which can be carried out using a computer screen, but again the colours may not be accurate. BS EN ISO 8586:2014 includes some useful 'recipes' for creating coloured solutions rather than the Munsell colour plates. If colour assessments are an important area for the panel's work, it will be worth investing in a Farnsworth-Munsell kit, which can be used during the interview section to assess the applicant's colour acuity. You might also like to include some appearance and colour descriptions or matching tests in your screening. Munsell (2017) stock several different kits to run these types of tests and also have some interesting and useful posters regarding colour measurements.

5.4.5.5   RM5 Section 2: Basic taste acuity tests

There are five basic tastes: salty, sour, sweet, bitter and umami and a few others are waiting to join the list such as fatty and metallic. To be classified as a taste there must be a component in the food or drink that is soluble in saliva to activate the receptors on the taste cells. The tests described below to check for taste blindness, or ageusia, are useful if the panel's work will involve assessments of any of the basic tastes. For example, if the panel work on beer, their ability to discriminate and measure the levels of bitterness in the beer might be a critical part of the assessment procedure. There are probably more than ten thousand bitter-tasting molecules (Laffitte et al., 2017), and choice will depend on the products to be assessed, therefore, take care with the choice of compound used to check for taste blindness. For example, it will not be helpful to use caffeine to check if a beer panel can assess bitterness; iso-alpha acids (e.g., humulone, cohumulone, adhumulone) will probably be required. Another example would be in the assessment of pet foods, as you will probably need to change the compounds for acidity and bitterness if the panellists are going to be involved in tasting the products (Pickering, 2009).

If a panel works on products that contain various artificial sweeteners, it might be worth considering a check on the applicant's ability to detect and discriminate between various types, such as aspartame, saccharin and sucralose, as well as natural sugars such as sucrose, lactose and maltose. If you have a specific interest in the measurement of aftertaste of certain attributes, you may wish to adjust some of the tests to take this into consideration. In the product description part of the screening you could specify that the applicant also write a description of the aftertaste over a particular time period. Or if you are interested in a specific aftertaste like the aftertaste of various sweeteners, you could present two or three solutions of sweeteners and ask the applicant to firstly describe the aftertaste (one sample at a time with a break in between) and then ask them to compare the aftertastes of the three samples. You could do this test instead of the descriptive test, for example. Alternatively, you might like to try a simple temporal dominance of sensations test by asking the applicant to try two solutions of sweeteners and write down what they perceive every ten seconds or so over a certain time period. This can be set up using sensory software or by use of a timer.

The message here is to target your screening tests to your requirements.

It is important to choose the correct bottled water for the tests. Hoehl et al. (2010) found that the type of water used had an impact on the threshold levels determined in their tests: tap water resulted in higher recognition thresholds than spring water, and deionised water gave the lowest thresholds of all. However, deionised water is not always found to be pleasant to consume and production can be variable. Therefore, it might be better to use good quality bottled water, but one low in minerals. A good option is to review several locally available water sources and determine which one has the least 'mineral taste' and use this water for all of your tests, including as a palate cleanser.

Food grade basic taste compounds should be used and these are available to purchase from flavour houses, although salt and sugar, for example, can be purchased from a local supermarket. The preparation of caffeine solutions requires special care as caffeine is toxic; however, care should be taken when making up all the solutions. It's wise to check with local hygiene and health and safety regulations about the purchase, storage and use of the basic taste compounds. An instruction sheet giving details to prepare solutions of the various basic taste compounds: sucrose (sweet); sodium chloride (salt); citric acid (sour) and caffeine (bitter) is available from the author on request.

It is possible to purchase kits for the preparation of the basic tastes. Search 'basic taste kit' on the internet to access lists of basic tastes in capsule, tasting sticks and pre-weighed form, which can help minimise errors with solution preparation and ensure the correct levels are used. Campden BRI (2017) produces some very helpful kits for sensory screening: https://www.campdenbri.co.uk/training/sensory-training-aids.php.

There are several different tests that can be used to assess applicant's impairments and acuity for basic tastes. The first step is to check recognition of the tastes. Present each applicant with each of the basic tastes selected for the screening in labelled cups. For example, if the plan is to screen for sweet, salty, sour, bitter and umami, present five cups labelled sweet, salty, sour, bitter and umami containing around 50 mL of each solution at the recognition concentration (see Table 5.2). These concentrations are for recognition of the basic tastes. The level for aspartame is from Mojet et al. (2005). The other levels are similar to the levels in the ISO standards for basic taste recognition. Give the applicants time to assess the solutions and ask questions if required. It can be easier to conduct this part of the test in the conference room or panel discussion room prior to starting the tests themselves in the booth room (or second conference room). Then, for the test itself, present the same solutions again but in three-digit coded cups and ask the applicant to identify each basic taste. Obviously, if just the five solutions are presented, the applicant will be able to guess the last taste that they might not be capable of identifying, and therefore repeats and blanks are required for this test. If the applicant does not identify each of the solutions correctly, you might like to consider

Table 5.2 Suggested concentrations for basic taste recognition tests

Solution	Concentration (g per 1000 mL/g water)
Sweet (sucrose)	10
Sweet (aspartame)	0.3
Sour (citric acid)	0.4
Salty (sodium chloride)	2.0
Bitter (caffeine)	0.3
Umami (MSG)	0.6

TEST # – BASIC TASTES

- In front of you are solutions that represent the basic taste sensations.

- One or more of these may be a blank (water) or some solutions may be repeats.

- Please taste each solution in turn and identify the dominant taste in each sample.

- Please describe each sample below as SWEET, SALTY, SOUR, BITTER, UMAMI or BLANK/WATER against the appropriate code number.

- Ensure that you cleanse your palate thoroughly between each sample.

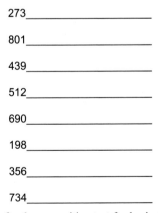

273_____

801_____

439_____

512_____

690_____

198_____

356_____

734_____

Figure 5.10 Questionnaire for the recognition test for basic tastes.

conducting the familiarisation step again and then repeating the test. This is easier to do with an internal panel, but with an external panel this can be difficult to accommodate. One solution is to proceed through to the interview stage and those people who you would like to join the panel, but got some answers wrong in this test, can repeat the familiarisation and recognition test.

A typical applicant questionnaire is given in Figure 5.10. It's a good idea to make sure that each of the three-digit codes start and finish with a different digit for this test, especially if it's the first test the applicants will be performing, as this helps prevent errors when completing the form. As bitter and sour are often confused, it can be well worthwhile presenting the sour and bitter solutions twice. Remind the applicants to take their time in the test and to cleanse their palate with the water provided between solutions: adaptation in the mouth to one of the solutions can suppress the taste or even create the illusion of another taste. For the mark scheme see Figure 5.21: Mark scheme for a typical food and drink panel recruitment screening test.

Basic tastes can also be used to check the panellist's ranking ability. Sorting samples in order of their concentration is a must for panellists working on descriptive tests, quality control and discrimination tests. It can also be a useful screening test for panellists working with temporal methods. Generally, the ranking tests are performed once, however, the ASTM publication STP758 '*Guidelines for the Selection and Training of Sensory Panel Members*' suggests conducting these tests three times (ASTM, 1981). The suggested concentrations for this test are given in Table 5.3 and are based on STP758, Meilgaard et al. (2016) and experience. The panellists' questionnaire for this test is given in Figure 5.11.

Other solutions that might be worthwhile testing at this stage, depending on the panel type, are astringent, metallic and fatty. For astringent solutions, potassium aluminium sulphate (alum) can be used. For metallic a good option is iron(II) sulphate heptahydrate ($FeSO_4 \cdot 7H_2O$: ferrous sulphate), and to assess fattiness, oleic acid has been used (Stewart and Keast, 2012).

5.4.5.6 RM5 Section 3: olfactory screening

Before starting to prepare for odour screening, check local health and safety regulations to ensure that applicants are not being exposed to higher levels of substances than those allowed by law. You will also need to check local and company regulations associated with the purchase, storage and use of odour chemicals.

Table 5.3 Suggested concentrations for basic taste ranking tests

Concentration sucrose (sweet)	Concentration sodium chloride (salty)	Concentration citric acid (sour)	Concentration caffeine (bitter)
0 g sucrose (water)	0 g salt (water)	0 g citric acid (water)	0 g caffeine (water)
10 g sugar per 1 L water	1.0 g salt per 1 L water	0.35 g citric acid per 1 L water	0.2 g caffeine per 1 L water
20 g sugar per 1 L water	2.0 g salt per 1 L water	0.7 g citric acid per 1 L water	0.5 g caffeine per 1 L water
50 g sugar per 1 L water	4.0 g salt per 1 L water	1.0 g citric acid per 1 L water	1.4 g caffeine per 1 L water
100 g sugar per 1 L water	7.0 g salt per 1 L water	1.5 g citric acid per 1 L water	2.6 g caffeine per 1 L water

TEST # - INTENSITY RANKING

In front of you there are five solutions of different strengths. Please taste each one and then put them in order of increasing strength of taste. Remember to cleanse your palate thoroughly between each sample.

Least taste/   _____
weakest

                     _____

                     _____

                     _____

Most taste/   _____
strongest

How would you describe the taste of these solutions?

Figure 5.11 Panellist questionnaire for ranking of basic taste solutions.

If the detection and/or description of odour are an important aspect for the panel, olfactory screening will be required. Olfactory screening is generally a must for food and beverage panels and critical for some home and personal care panels where the assessment of the efficacy of deodorants or odour masking is part of the panel's work. But it may be totally irrelevant for the assessment of some other home and personal care panels such as those that assess fabrics or tissues. Olfactory screening can also be used for the assessment of memory, and also to assess the ability to recognise and describe certain aspects of products from a product that is made up of many descriptors.

Olfactory screening can check for both impairment and acuity and is important in the recruitment of a sensory panel, as a large number of people have difficulty detecting aromas due to a complete or partial loss of olfactory function. This complete or partial loss of olfactory function is known as anosmia and can be caused by surgery, disease, viruses, brain injury or it can be congenital (present from birth). Some people are unable to detect any odours at all, while others may have a specific anosmia: the inability to detect a specific odour although they have otherwise normal olfactory function. Research in the 1970s indicated that, of the 60 or so odourants assessed, all exhibited specific anosmias; however, different people exhibited specific anosmias to different odourants. This suggests that perhaps most people might have a specific anosmia if we consider the number of odourants available (Croy et al., 2015). For example, around 50% of people cannot smell androstenone, a key contributor to boar taint in pork, while 35% describe the odourant as stale urine and 15% describe it as floral! Recent research regarding olfactory dysfunction in older adults found that a fifth of participants had olfactory deficits and were unaware of the issue (Adams et al., 2017). Therefore, it's important to check for specific anosmias as many of the applicants may have issues.

It is also important to use odourants or fragrances that are relevant to the products the panel will be assessing: there is no point presenting applicants with various fruity aromas to assess if they are going to be assessing meat pies (unless of course, you are making traditional Cornish pasties!).

There are a number of ways of obtaining and preparing odourants to assess applicants' impairment and acuity dependent on the type of work they may do in the future and the facilities available at the panel site. If a flavour organ is accessible, the preparation will be very simple, as specific odourants can be selected and prepared for the screening session. The majority of sensory scientists do not have this luxury, but there are ways around this. One of the simplest ways is to purchase food grade liquid flavours from a supermarket (they can often be found in the baking aisle). If a fragrance or flavour house provides you with fragrances or flavours for your products, you may be able to request these and/or the individual ingredients for use with the panel. There are also proprietary kits for the assessment of odour impairments in the form of 'pens', vials and capsules. Search the internet for 'olfactory performance' for the pens and 'aroma kit' or 'aroma capsules' for the vials and capsules.

One way to prepare the odourants for assessment is to create stock solutions of the odourants in ethanol and then dilute to a specified concentration (ISO 8586:12; Jellinek, 1985). When the concentration of the odours is important for the assessment of products, this preparation route, although the most complex, is recommended. You might also have access to a dynamic dilution olfactometer which can deliver a certain odour at a known volume and speed (ASTM E679 – 04) which can make life a lot easier for the determination of odour thresholds or difference thresholds. The ASTM standard includes some interesting case studies and will be very helpful if your aim is to determine odour thresholds.

In many cases, knowing the concentration of the odour is not critical and the odours can be simply prepared by the use of a fragrance blotter or perfume test strip, dipped in the odourant, dried and placed in a small screw-cap jar. Another method is to place a cotton wool ball in a small screw-cap jar and use a disposable plastic pipette to drip three or more drops of the odourant onto the cotton wool ball. Place another cotton wool ball on top of the liquid so that any liquid colour cannot bias the applicant, but do leave some space in the jar for the odour to build up in. These jars containing small amount of liquid flavours (on blotters or cotton wool) may last several days if the lids are tightened each evening. If the product type to be assessed is quite odourous, for example, pet food, another option for setting up jars of odours is to place small parts of the meat or gravy of various pet foods, wrapped in tissue or cotton wool, into the jars. In this way, you will be able to determine if the applicants can describe the difference between lamb and beef flavours, for example, and also assess their reaction to the assessment of these odour types.

In all of these cases, check that the odour is at the right level to be detected and described by asking several colleagues to pilot this part of the screening test. Their results can also be used to develop a marking scheme for interpreting the results.

A large part of these odour tests is related to memory, as it can be difficult to find the correct words to describe an aroma when presented in a vial or a flavour blotter that is so remote from the product itself. To help with this, a list of potential descriptors with some additional terms can also be given to the applicant with the odourants to prompt correct identification. However, if the panellists are being recruited for a descriptive panel, a test of the memory may be part of the assessment and therefore just the odourants need be presented.

The number of odours presented will depend on how critical odours are in your sensory panel work. If you are assessing odours as part of a food panel that is also assessing appearance, texture and tastes, six or so odours may be enough. However, if you produce a wide range of different products you may wish to include more odour checks to cover various odour types. If your panel's main focus is on odour, you will need to include several odour tests and maybe an odour descriptive test or pair of odours to compare as well.

A typical questionnaire using small bottles or jars containing either the blotters or the cotton wool balls for the odour impairment test is shown in Figure 5.12. To test memory as well as odour ability it is better not to include lists of odour descriptors that the panellist can select from. Including a list of descriptors can also cause issues with overlapping identifications, so if you do wish to take this option be careful about your choice of the 'wrong' descriptors (Fjaeldstad et al., 2017).

Examples of some of the odours and fragrances that might be used are given in Table 5.4 with an example of the type of panel that might assess the odour. BS ISO 5496:2006 also has some useful suggestions and relevant descriptors in Table A.2 which can be found in the standard.

When compiling the screening questionnaire, alternate odour tests with other tests to allow the panellist time to recover between assessments.

5.4.5.7 RM5 Section 4: texture screening

The BS ISO standard for the recruitment of panellists (BS EN ISO 8586:2014) does not include any tests for the screening of texture, however, as the standard suggests a two layered approach to recruitment which involves the screening and training being completed prior to the employment of the panellists, much of the texture 'screening' is performed in the training section of the standard.

You might wish to screen for texture specifically, especially if you are working on products where texture is a critical element or if you are working in new product development where you are trying to match a particular texture, for example, in the creation of meat analogues or less greasy feeling topical creams (Beeren, 2016). The BS ISO standard for the recruitment of panellists (BS EN ISO 8586:2014) suggests that discrimination tests or a descriptive test can be used to determine panellists' suitability for texture assessments: this is included in the training section of the standard. The discrimination tests can be prepared by slight adjustments to the texture of various products (for examples, see Table 5.5). For the descriptive tests, you could simply present two or three of your key products and ask the panellists to describe the texture only. By asking them to describe the texture only, rather than also including aroma and flavour, you will get a more focussed description. BS EN ISO 8586:2014 also includes some ideas for products for texture descriptions such as oranges, granulated sugar and squid. You could also use this descriptive test as the starter for the mini panel session (see Section 5.4.5.12). The standard also includes a description for the preparation of a range of gelatine samples for ranking in order of touching firmness.

Odour assessments

- Please assess and describe the odour of each of the coded samples following the instructions below.
- Describe each odour in the relevant box below.
- Remove the cap from the first bottle and gently sniff in the space above the bottle. If you can detect and describe the odour, replace the cap and write down your description in the box next to the code of the sample you assessed.
- If you cannot detect any odour, bring the bottle a little closer to your nose and again, sniff gently. If you can detect and describe the odour, replace the cap and write down your description in the box next to the code of the sample you assessed.
- If you cannot detect any odour, bring the bottle under your nose and sniff gently. If you can detect and describe the odour, replace the cap and write down your description. If you cannot detect any odour, replace the cap and move onto the next bottle.
- DO NOT sniff too hard if you cannot detect an odour, as this may affect your ability to detect the odours in the later bottles.
- Remember to replace the cap on each bottle before moving to the next.
- Take a short break (around 30 seconds) before assessing the next sample.

Code	Description
972	
134	
580	

Figure 5.12 Typical questionnaire for an olfactory test.

Another useful BS ISO standard for texture is the Texture Profile standard ISO 11036 (ISO, 1994) which has some very useful information about the method itself, as well as definitions and techniques for the measurement of the various attributes classified as mechanical, geometrical and other (moisture/fat). The standard also includes scales of reference products and advice on how to adopt and adapt for use. For example, it comments that the hardness scale cannot be used as published for the assessment of products which are all soft, and suggests that the lower part of the scale is 'expanded and other portions deleted' (p. 6).

Table 5.4 Examples of odours and fragrances for olfactory tests

Odour descriptor	Origin/stimulus	Panel assessing:	Typical descriptions
Almond	Flavour house: benzaldehyde	Food or beverage, beer, home and personal care products	Almonds, marzipan, Christmas cake, fruit cake, vanilla
Caramel	Liquid food flavouring*: caramel	Food or beverage	Caramel, nutty, earthy, maple syrup, sweet, confectionary
Chicken	Flavour house: Chicken fat flavour	Pet food, food or beverage	Chicken, fatty, lard
Clove	Aromatherapy oil: eugenol or clove oil	Home and personal care, fragrance	Dentists, cloves, medical, spicy
Earthy	Flavour house/aroma-therapy oil: Patchouli	Fragrance	Earthy, musky, spicy
Floral	Flavour house: Linalool	Food or beverage, fragrance	Floral, sweet, lemon, orange
Floral	Flavour house: Bourgeonal or 3-(4-tert-Butylphenyl) propanal	Home and personal care, e.g., fabric care	Floral, green, fresh
Lavender	Flavour house/aroma-therapy oil: lavender	Fragrance	Lavender, flowers
Lemon	Liquid food flavouring: lemon	Food or beverage, fragrance	Citrus, lemon, confectionary (lemon sherbets)
Malodour	Flavour house: dimethyl disulphide	Beer, malodour (e.g., masking odours), deodorant	Rotten vegetables, sewage, garlic
Menthol	Liquid food flavouring: menthol	Food or beverage, fra-grance, tobacco, home and personal care	Woody, minty, cooling
Peppermint	Flavour house/aroma-therapy oil: peppermint	Food or beverage, fragrance, home and personal care	Peppermint, minty
Plastic	Flavour house: Styrene	Food or beverage, fragrance panel	Plastic, polystyrene, chemical
Smoky	Flavour house: guaiacol	Food or beverage, beer panel	Smoky, fishy, medicinal, woody, smoked cheese
Sweaty	Flavour house: 3-methylbutanoic acid	Deodorant	Sweaty, cheese, fatty
Vanilla	Liquid food flavouring: vanilla extract	Food or beverage, fragrance panel	Vanilla, sweet, custard, confectionary

*Available from supermarkets.

The ASTM special publication '*Guidelines for Sensory Panel Selection and Training*' (ASTM, 1981) has some useful suggestions for screening texture and assessing the panellists' discrimination ability. They suggest presenting a range of products such as cream cheese, cubed cheese, olives, carrots and hardboiled sweets and asking the panellists to rank these in order of hardness. They also show ranking suggestions for viscosity (e.g., water through to condensed milk). Another useful test is the matching texture test where several products are presented and the panellists have to match the correct geometric texture descriptors. For example, canned chicken meat matched to the term 'fibrous' and boiled haddock to 'flaky'. The products you choose will depend on the product range that you plan to assess. All of the test ideas from the ASTM special publication (ASTM, 1981) could also be applied to home and personal care products. For example, a range of foams could be created for ranking or descriptive assessment, or two or three swatches of material could be treated with different levels of fabric conditioner in pairs and the panellist asked to match the pairs of swatches. You could also include a rating scale after samples have been ranked.

You may wish to assess your panellists' chewing patterns using products such as chewing gum (Prinz, 1999; Schimmel et al., 2007) or instrumental measures such as electromyography (Brown, 1994; Brown et al., 1994). The chewing gum method uses a two-colour gum and determines the extent of mixing at various time points. Electromyography can also be used to understand chewing behaviour by recording the electrical activity occurring in the facial muscles.

To determine a panellist's sensitivity in the mouth, oral stereognosis assessments can be employed (Boyar and Kilcast, 1986). This method involves the assessment and identification of two-dimensional or three-dimensional shapes in the mouth. For example, sugar letters and shapes, such as those used in the preparation of celebration cakes, can be placed on the tongue and the panellist has to correctly identify the letter or shape (Beeren, 2016). Some authors have used von Frey filaments to assess in mouth sensitivity (Nachtsheim and Schlich, 2013).

The stereognosis method can also be employed to test the sensitivity of the hands (Boyar and Kilcast, 1986) with the letters or shapes constructed from paper or card. Other methods to assess hand or finger sensitivity include assessments of Braille, JVP domes (Remblay et al., 2000) and grating orientation tasks (for example, see Van Boven and Johnson, 1994). In the grating orientation tasks the panellists have to run their index finger over surfaces with various parallel grooves etched in them and describe the orientation and spacing.

For detailed reviews on food texture perception with some useful history, explanations, physiology, definitions, and diagrams explaining the various stages of texture in the mouth, see Koç et al. (2013), Boyar and Kilcast (1986) and Lillford (2017).

When screening panellists for home and personal care products, there are several other things to consider (Greenaway, 2016). For example, if you are recruiting for a panel to assess shaving products you will need to assess the panellists' frequency of shaving or hair removal and the amount of hair the potential panellist has. For skin care panellists you will need to take into account the current skin texture and skin type. All of these aspects may well change the panellists' perception of texture. If you are assessing skin in the screening tests you will need to consider the cleansing stage: panellists may only be able to assess a small number of products before needing to cleanse or rest until the skin returns to normal.

5.4.5.8 RM5 Section 5: hearing tests

Legarth and Zacharov (2009) describe the three-stage process for the recruitment of panellists for listening tests: questionnaire to determine interest and suitability, auditory/visual acuity tests and four screening tests. These panellists are recruited for the assessment of audio from, for example, headphones, speakers and concert halls. These tests might also prove useful for the assessment of panellists for texture assessments. The questionnaire includes asking questions about excessive exposure to loud noise and whether they have any hearing issues or tinnitus, for example.

The first of the auditory tests involves the applicant wearing headphones and listening for very quiet sounds, and then clicking the mouse or pressing a button every time they hear the noise, and releasing the device when the noise subsides. The sounds can be directed to each ear if required. The starting sound is usually clearly audible to ensure the applicant has understood the task.

The triangle test can also be used to determine if the applicant can tell the difference between various loudness of sounds, pitches and distortions, for example, depending on the work the panellists will be taking part in. Triangle tests with music or speech created under various compression rates can also be used. Other discrimination test can also be employed and can test the differences in sound recordings from fresh and stale food products, for example. See the next section for more details about suitable discrimination tests.

5.4.5.9 RM5 Section 6: Discrimination tests for screening

Discrimination tests are incredibly important for screening applicants for any type of sensory panel. In fact, some authors (Stone, 2015; Minoza-Gatchalian et al., 1990) suggest that all other screening tests (e.g., basic tastes, olfactory and scaling tests) do not give reliable results for the choice of panellists (unless of course, the panel's role was only in the assessment of basic taste solutions and odours!). Discrimination tests can give critical information for the sensory scientist about the ability of the applicant to discern differences between relevant products for the panel (Zook and Wessman, 1977). Panellists can be recruited if they achieve a success rate of over 50%–65% in the discrimination tests, especially if they were more successful in the later, more difficult, sets of products. If the differences in the products that you select for the tests are similar to those that will be experienced by the panel, this can give you 'real-life' information of how the applicant might perform if recruited. These products will also make good training products for the profiling validation trials once the panel have been recruited, as you will have data on which differences were detectable. The tests and test results will also prove invaluable for future screening experiments. Selecting those panellists who are all able to discern the same differences in your products gives you good information to base the results of future tests on. And finally, after completing the discrimination tests, the panellists will have had a good chance to experience a range of differences in your product(s) and a range of different discrimination tests before they even start the training programme.

Before conducting any discrimination tests, it's a good idea to familiarise the potential panellists with the tests themselves, otherwise you may find you are testing the person's ability to take part in the test and not their sensory ability, when really you wish to check both. One way to do this is to use an example that the panellists can easily see and demonstrate the test whilst they are able to read through the instructions on a printed handout or tablet screen. The Farnsworth-Munsell Dichotomous D-15 Test colour chips can be a useful way of demonstrating a particular sensory discrimination test. You can also demonstrate many of the discrimination tests with simple examples such as pens with different colour lids or squares of paper cut from different paint charts (e.g., like those found in DIY stores).

There are many different discrimination tests available, but the tests most useful for screening potential panellists are the directional difference test, duo-trio and ranking. Other tests might be useful for the screening of panellists dependent on the product and type of work the panel might be involved with. For example, if the panel will be working on the development of malodour masking, the ABX task might be a sensible option (Greenaway, 2017), and if your product is quite fatiguing it might be sensible to avoid the duo-trio and use only tests that require the assessment of two samples (e.g., A-not-A and same-different tests). If there is a particular test that is used widely within your organisation, it would be wise to use this for the screening; at least for some of the tests. Some industries use the triangle test (see Figure 5.13, please note the diagram is *not* a description of how to run the test – this is simply a quick way to demonstrate the test in a diagram) as their default discrimination procedure, but this test can take the panellist some time to understand and develop skills in, as the test involves more sample-to-sample comparisons to be made than the 2-alternative forced choice test (2-AFC) or duo-trio. For example, the subject would need to assess the sample coded 207 (see Figure 5.13) and then the sample coded 643. They would compare 207 to 643 and consider if they were the same or different. They would then assess the sample coded 451 and decide how this compared to both 207 and 643; making a total of three comparisons, while the 2-AFC and the duo-trio only require two comparisons.

In the directional difference test (also known as the directional paired comparison and the 2-AFC) the subject is presented with a pair of samples and asked 'which sample is more bitter?' or 'which sample is more smooth?' (Yang and Ng, 2017). An example is shown in Figure 5.14. The attribute of interest (i.e., bitter or smooth as in the previous examples) is modified in panel screening so that the scientist knows which of the samples

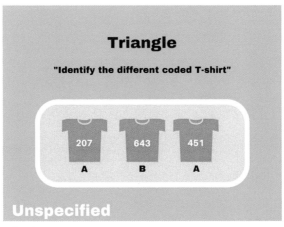

Figure 5.13 The triangle test.

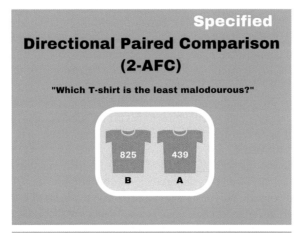

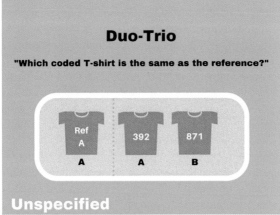

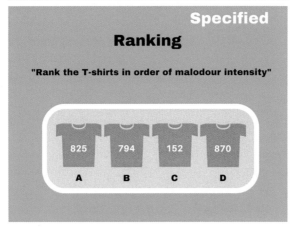

Figure 5.14 Three discrimination tests useful for screening.

is more bitter or more smooth. Products can be adjusted, for example, by the addition of ingredients, changes to the cooking or preparation process, dilution, spiking with known standards or different storage conditions: more information about these product edits is given later. In the duo-trio (see Figure 5.14), subjects are again presented with two products, but this time there are three samples: two of the samples are the same product, one of which is labelled as a 'reference' (Purcell, 2017). After assessment of the reference, the subject's task is to decide which of the two coded samples is the same as the reference. In ranking, the subject is presented with a series of samples and asked to order them in the intensity of a specific attribute (Whelan, 2017a). An example is shown in Figure 5.14.

It is better to start with products that are quite different and gradually increase the difficulty of the tests. This allows the panellists to learn the test procedures, familiarise themselves with the product type(s) and the mechanism by which they will assess the products and do the test, before the differences between the products becomes too difficult to detect and hence dents their motivation. It might also be useful to present each set of products twice with two A products and one B in one of the tests and then one A product and two B products in the other test: particularly if you are using the triangle test, as some combinations with some products can be easier to complete correctly than others (Zook and Wessman, 1977). Table 5.5 gives some suggestions for various categories of products for use in discrimination tests for screening applicants.

Products chosen for these discrimination tests must be identical in all ways other than the attribute(s) or modality of interest. For example, there is no point asking subjects to take part in a duo-trio test with standard and reduced salt crisps (potato chips) if the reduced salt product looks completely different to the standard product. Sometimes appearance differences can be disguised by the use of different coloured lighting, assessment of liquids from lidded cups or dark containers, or assessing the feel of fabrics or tissues contained behind a screen. Basic taste solutions can also be used to introduce the concept of the 2-AFC test to the applicants. Choose concentrations from Table 5.3 that are more than one step apart, for example, for sweet select 0 and 20 g of sucrose and for salty 1 and 4 g of sodium chloride. These tests should be quite easy for the applicants and will give them confidence in taking part in the rest of the tests.

Figures 5.15–5.17 give example questionnaires for use with the discrimination tests suggested in Table 5.5.

Choosing between duo-trio and 2-AFC: the duo-trio should be used if you do not know in advance which attributes may change as a result of the product edits or you are unsure what the differences between the two products might be, whereas the 2-AFC requires an attribute to be specified and is useful if you are aware of the change in the product. For example, if you were purchasing an own label or shop brand and comparing it to a branded product that looked identical, there might be differences in aroma, flavour or texture and therefore a duo-trio would be a good option. If you had diluted a juice product, a 2-AFC with the instruction 'which is the most intense in flavour?' would be one choice, however, the duo-trio could also be used.

Using basic taste solutions for screening is not especially comparable to the testing of real products, but it can be a useful way to introduce discrimination tests and give the applicants the chance to understand what they are being asked to do rather than think too much about the products they are being given to assess. If you used the basic

Table 5.5 Example products for discrimination tests

Products compared	Discrimination test	Difficulty	Notes
1. Farnsworth-Munsell Dichotomous D-15 Test	2-AFC or ranking	Easy to difficult	Using the colours from a Farnsworth-Munsell kit can be a good way of introducing discrimination tests, as you will be able to actually demonstrate the process prior to the applicants conducting any tests themselves.
2. Various levels of basic taste solutions	2-AFC or ranking	Easy to difficult	The difficulty depends on the levels of basic taste compounds used. See Table 5.3 for suggested concentrations and Figure 5.11 for the panellist questionnaire.
3. Edited versions of company products	2-AFC or duo-trio	Create to cover a range of difficulties	If your company creates various juices, for example, you can create products that are different very easily by the use of dilution, addition of different ingredients (e.g., sucrose, citric acid), addition of other flavoured juices, or the use of flavour capsules for 'spiking' various off notes. Using juices in this way can be a useful first step for any food panel as an introduction to discrimination tests.
4. Products created specifically for the tests	2-AFC, duo-trio and ranking	Easy to difficult	When your company has the facilities to produce various versions of your product, for example, if the pilot plant can make a series of crackers that result in slight differences in flavour or texture, or products that can be cooked for slightly longer or at a different temperature.
5. Products across shelf life	Duo-trio or ranking	Some easy, some more difficult	Products can be selected from different shelf life points or aged artificially.
6. Standard versus reduced fat, salt, sugar, etc., variant	2-AFC or duo-trio. Ranking if you can source several versions.	Some easy, some more difficult	For example, fat reduced custard or yoghurts, sugar reduced cereals or soft drinks, salt reduced soups or cheeses.

Table 5.5 Example products for discrimination tests—cont'd

Products compared	Discrimination test	Difficulty	Notes
7. Standard versus increased fragrance	2-AFC	Some easy, some more difficult	For example, laundry liquids or fabric conditioners with differing levels of added fragrance.
8. Own/store brand versus Brand A of relevant product	Duo-trio – use the balanced reference technique (Purcell, 2017)	Generally easy	For example, baked beans, soups, fabric conditioners, pet foods. These tests are generally quite easy but some might be harder than others.
9. Competing brands	Duo-trio – use the constant reference technique	Dependent on products	For example, Pepsi cola versus Coca cola or Persil versus Ariel washing powders (aromas or fabrics after use).
10. Different batches of company products	2-AFC or duo-trio	Difficult	You may have batches from different lines, time points in production, factories or days that are very similar that could be used to create a difficult test for the applicants.

<u>Directional paired comparison test</u>

Assessor Number:

You are provided with two hand creams, each labelled with a three-digit code.

Please assess each product in the order provided, from left to right.

Please indicate which of the samples below has the most cooling effect on the skin surface by circling the corresponding sample code below.

Protocol:
- Wipe the hand cream over the back of your right hand using 3 fingers and leave it for 15 seconds.
- Assess the overall cooling feel of the product on the skin surface.
- Once completed, repeat the same for the second sample on your left hand.

Sample 203 / 831 has the most cooling effect on the skin surface

Comment: _____

Figure 5.15 Questionnaire for the directional paired comparison/2-AFC test (Yang and Ng, 2017).

Duo-Trio Test		
Assessor Name:	Assessor No.:	Date:

You have been provided with an identified reference sample and two coded samples.

Taste the reference sample and then the two coded samples from left to right.

Circle the sample you identify to be the same as the reference.

Explain why the other sample is NOT a match and indicate the intensity of difference (very slight, slight, moderate, obvious, or very obvious).

If you cannot determine which the matched sample is, please make a guess.

You must rinse your mouth between each sample.

| Reference | : | 853 | 394 |

Comments:

Figure 5.16 Questionnaire for the duo-trio test (Purcell, 2017).

taste solutions earlier in the screening, the applicants will also be familiar with the solutions and the look of the cups and labels, which can help them feel at ease in conducting a test that may well appear very alien to them. Figure 5.18 gives an example questionnaire for ranking one of the basic tastes, and Table 5.3 gives some suggested concentrations for sour, sweet, salty and bitter. If the ranking involved bitter or sour, asking the final question about how the taste might be described can be helpful in helping decide about specific taste impairments as these two tastes are often mixed up, especially in people with a higher threshold for bitter.

5.4.5.10   RM5 Section 7: Scaling tests

If you have given your potential panellists a paper scaling test as part of your application form (see Figure 5.8) this might be enough to begin with, however, you might wish to include a scaling test in your screening assessment. One easy way to do this is to use one of the sets of solutions for one of the basic taste ranking tests and repeat the test just after the ranking so that the panellists can recall the range of the samples, but ask the panellists to scale the solutions (see Figure 7.9 for an example). Another useful way to assess scaling ability is to give the panellist one or two 'known' standards and ask them to scale some samples relative to these. Again the basic taste solutions can be useful here. An example is given in Figure 5.19. If you already have some intensity references available, you could use any of these for this test.

5.4.5.11   RM5 Section 8: assessing descriptive ability

If your panel is going to be describing products, this test will be crucial to assess their abilities. Descriptive abilities will apply to internal and external panellists working in

Ranking Test

Assessor ID:_____   Date:_____

Instructions:

a) You are provided with four samples, each labelled with a three-digit code.
b) Evaluate the samples in the order presented from left to right, cleansing your palate between samples before evaluating the next sample.

648	561	140	937

c) Rank the samples in order of increasing <u>bitterness</u>. You may re-assess any of the samples again as often as you wish until you have made your mind up.
d) Then please write down the codes of the samples in the order from least to most bitter in the table below.
e) If two samples appear the same, make a best guess as to their rank order and note down in the comments section that it was a forced choice.

	Rank order			
	1	**2**	**3**	**4**
	Least Bitter			Most Bitter
Sample Code				

Comments:
_____
_____
_____

Figure 5.17 Example questionnaire for the ranking test.
Adapted from Whelan, V.J., 2017. Ranking test. In: Rogers, L. (Ed.), Discrimination Testing in Sensory Science. A Practical Handbook.

quality control, profiling tests, temporal methods and even discriminations tests where descriptions of the differences might be useful. Choose an easy product to describe, as the panellists will find this a difficult test: it is not something people are very used to doing. Also choose something that will be easy to prepare for each individual panellist as you will find people will work through the previous tests at very different rates.

Basic taste ranking test

- In front of you there are five solutions of different strengths.

- Please rinse your mouth with water between each sample.

- Please taste each sample and order them in **increasing** strength of taste.

- Handy tip: as you taste each solution, place the cup in front of you to the left if you think it's quite weak and to the right if you think it's quite strong.

- As you assess each sample, place the cup where you think it might be in order of strength in comparison to the other samples.

- Once you are sure of the sample order, copy down the codes onto the lines below in order of increasing strength: the weakest goes first. This saves lots of time and scribbling out.

- Remember to rinse your mouth with water between each sample.

(Sample codes 397, 216, 408, 965, 714)

Least taste/ _____
weakest

_____

_____

_____

Most taste/ _____
strongest

How would you describe the taste of these solutions?

_____

Figure 5.18 Example questionnaire for ranking of basic tastes.

In the presentation to the panellists at the beginning of the screening test, it can be helpful to describe this test as having to describe the product to a friend over the telephone so that their friend can see it and smell it and taste it (or whatever modalities you will be using for your products). An example question is shown in Figure 5.20: you will need to give each applicant several lines to write their descriptions – an A4 sheet should suffice. If you have a specific interest in the measurement of aftertaste of certain attributes, you may wish to adjust this test to take this into consideration. You could specify that the applicant also writes a description of the aftertaste over a

You have been provided with two standard solutions A and B.

Please assess them in the order presented (A and then B) remembering to rinse your mouth between assessments with the water provided.

Note the bitterness intensity values of these two samples on the scale below.

Once you have completed this step, please ask for the next samples. There will be a slight pause before we give you the samples.

Please assess the coded samples in the order presented to you (from left to right).

As you assess each sample, place a mark on the line to indicate the intensity of the bitterness.

Remember to rinse your mouth between assessments with the water provided.

Remember to write the sample code above or below the line you have made or we won't know which sample was what.

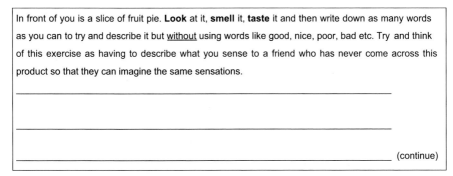

Figure 5.19 Example scaling test with two standards.

In front of you is a slice of fruit pie. **Look** at it, **smell** it, **taste** it and then write down as many words as you can to try and describe it but <u>without</u> using words like good, nice, poor, bad etc. Try and think of this exercise as having to describe what you sense to a friend who has never come across this product so that they can imagine the same sensations.

_____

_____

_____ (continue)

Figure 5.20 Example question for a descriptive ability screening test.

particular time period. When you are marking this part of the screening test, consider which elements are important to you. For example, in the mark scheme for the fruit pie test, the interviewer was interested to see how many people described the temperature of the product (only two out of 92 people screened).

You could also extend this test to include a comparison to another product. This can be useful to help you understand how many similarities and differences the applicant might find in two similar products. If you are going to do this, it can be better to give them one product to begin with and ask them to describe it. Then give them the second product and ask them to compare it to the first. Otherwise some panellists can get quite confused and overloaded and miss certain modalities: but it can also be a good test of their reactions under stress!

One time when I was running this test many years ago, we had some interesting descriptions for a cup of soup. One panellist described the croutons as 'little islands, moving slowly over the red surface, telling my fortune like tea leaves are said to do'. Another described the soup as 'red' (that was it – nothing about the croutons or little pieces of vegetables, nothing about the flavour or aroma). Needless to say, neither of these applicants were invited to join the training programme.

5.4.5.12   RM5 Section 9: Running a mini panel session

If you will use your sensory panel to create product profiles or use any of the tests that involve some form of discussion or consensus, it can be useful to run a short panel session to assess how the potential panellists behave as a group. If you are recruiting externally, this part can be difficult to combine with the screening tests and interviews but is well worth the effort, particularly if you are moving directly from screening to employment.

The test is best carried out with your company's products, but sometimes this can be difficult to organise with home and personal care products. You may need to think of some alternative approaches. For example, if the panel are to be assessing their own hair in booth showers on site or at home, you could prepare some switches or mannequin heads for assessment. If the panel will be working assessing the efficacy of deodorants and subjects are not available, model malodours in jars could be used as an alternative. Remember to instruct the panellists how to assess odours before they begin (see Figure 7.1).

Arrange the session in a similar way to how you would conduct a standard test, as this will give you the best feel for how the panel will perform. If you are recruiting new panellists to join an existing panel, it can be helpful to invite an existing panellist to help screen the potential panellists. Do not expect the panellists to act exactly how they might once you recruit them: they may well be a little quieter as they get used to the different people and situation. As you run the session note their reactions to the samples, to other panellists and to you and consider if you would like to work with them. For example, if one panellist mentions something in the product, note who is confident enough to disagree and the manner in which they disagree.

5.4.5.13   RM5 Section 10: Testing memory

In many ways the use of the odour vials is a good test of memory as generally, when we assess and describe odours, the food or product is in front of us. But you may wish to include a memory test with the panel by using the simple memory tray game you may have played as a child. This can be carried out at the end of the mini panel session or in the meeting room after you have finished describing the screening process.

Collect together around 15 to 20 items from around the office (for example, a paper clip, a pen, a birthday card, a postcard, an empty drinks bottle, a pen, a key, a notebook, etc.) and place them on a tray which is covered by a cloth. Explain the process to the panellists and then bring in the tray covered with the cloth and put it where everyone can see it easily. Set the timer for 30 seconds and remove the cloth. The panellists are not allowed to write anything down and should not really talk either – they are more likely to remember an item on the tray if someone comments about an object. Once the 30 seconds are up, cover the tray again and give the panellists pens and paper and see how many items they can remember. Ask them to write down how many items they remembered on their sheet. Then remove the cloth again and let them chat for a few minutes about the process.

Collect in the sheets and check the number of items each person remembered. If there were panellists who remembered less than 20% of the items (i.e., only remembered four items if there were 20 on the tray) consider running the test again. An adaptation of the method is to show the tray for 30 seconds, cover it and remove it from the room. Remove one or two items from the tray. Then bring the tray back in and ask the panellists to write down which item(s) you removed.

You can also do this with words. Type up a list of words and project them in the room or write them (neatly!) on the flip chart. You can write some simple nouns such as piano, camera, book and cat but make sure that the words are not especially connected in any way. The trouble with this option is that it is rather dry and boring so you could opt for pictures if you are using presentation equipment or if you are a good artist to liven it up.

There are also many memory tests online that you could use. Simply search 'memory test' and you will find several hundred. You could use the principle of these tests to build something similar in your sensory software. For example, some of the tests involve matching the patterns of coloured tiles which would be easy to reproduce via a ranking test, for example. If you search 'sensory memory test' you will also find several useful videos that you could use or adapt for a memory test.

5.4.6   Recruitment Module 6 (RM6): Interviewing

5.4.6.1   RM6 Section 1: Introduction

To the best of my knowledge there is no literature or published advice regarding the interviewing of sensory panellists and therefore I have tried to collate information from various sources and apply it to the recruitment of sensory panellists. If you are recruiting an external analytical panel, the recruitment phase will need to incorporate a personal interview, as well as the sensory tests. If you are recruiting internally you can probably skip this step as the person will have already been through the interview process. However, you might find it useful to talk to the person on a one-to-one basis to find out their level of interest in the role (see Section 5.4.6.3). If you are recruiting a consumer sensory panel for a single test you will not need to conduct an interview, however, if you are recruiting for a long-term consumer discussion group, this module should prove to be very useful.

If you can stagger the screening assessments you may only need to interview those people who pass the screening stage. Sometimes this is difficult to arrange, but the interview need only take a few minutes. Your aim is to determine the level of interest the person has in the role and whether they would be a good fit for the type of panel sessions you are planning. Some of your job will have been completed for you by your colleagues and existing panellists (if you have them) who have also been assessing the panellists throughout the screening assessments.

5.4.6.2   RM6 Section 2: Preparing for the interview

First things first: check with your HR department. They may have specific rules and paperwork to complete. They might also be able to offer you interview training or suggest courses for you to attend. Attending an interviewing course can be really helpful, particularly if you have never interviewed before. Choose a course that allows you to practice interviewing rather than one based just on the theory. This way you get the chance to practice your new skills and get feedback on what you can do to improve. This can give you more confidence and help the sensory panel interviews run smoothly.

One of the best things you can do before the screening stage and personal interview is not only to create a job description or specification for your panellist (see Section 2.5) but also to imagine the type of people you would like for the role. You can use Figure 5.2: Attributes of a good panellist and Table 6.2: Panellist dos and don'ts, as a starting point and write a short description of your ideal panellist. You might even like to involve your existing panellists to help you write this description. It's quite interesting to hear their requirements for the new people and for them to perhaps realise that they do not quite fit the descriptions!

Choose somewhere for the interview that is comfortable and informal: if the potential panellist is relaxed they will be able to give their best. You might like to choose an office near to the screening location with a small table so that you can sit side-by-side, for example. Make sure the chairs are the same height: there is no better way to intimidate someone than have the interviewee sit on a very low chair while you tower over them!

You will also need to decide who will conduct the interviews. You could do it alone but having a colleague with you can be very beneficial, as they can give you feedback on the person as well. If you are not going to be the person actually running the panel sessions, then do invite the person who will be. You might also choose to invite an existing panellist (but if you do, choose several different panellists to help) or your sensory technician or another sensory scientist. Share out the questions so you both get the time to consider and make notes about the interviewee.

You will need to prepare a list of questions for the panellists to answer. These do not need to be onerous, in fact just a couple may well be enough. You will also need to check the application form for any additional queries you need to ask. For example, if the panellist has written on their application form that they wear tinted glasses, you may need to check if they have an untinted pair they can wear for panel sessions.

5.4.6.3   RM6 Section 3: The interview questions

The questions you ask should be targeted at your aims of the interview: finding out about the interviewee's personality and potential fit to the work of a panellist. Some suggested questions are given in Table 5.6. In asking the applicant if they have any questions, you hope to get some questions about the role as this will show that the applicant is interested.

Table 5.6 Example interview questions

Question	Objective for asking
What hobbies do you enjoy?	Understanding what people like to do in their free time can give some good indications about their personality. Also, if you are recruiting a panel working on food flavour, for example, if the interviewee expresses an interest in cooking, or growing vegetables, this might be relevant to their application.
If your friends had to describe you with three words, what three words might they choose?	The words the interviewee chooses can be very enlightening. Also, if they seem to be very hesitant, they might be thinking of words they expect you want you to hear, rather than the actual truth…
Do you prefer to work alone or in a team?	This answer can give you valuable information about how the interviewee might work as a panellist.
If you were taking part in a group discussion and someone else disagreed with you, what would you do?	Hopefully the answer will include something about finding a win-win situation or finding a compromise position. You can add additional questions to delve deeper if you wish.
Application form related questions	You might need to check some aspects of the application form: for example, more details about their health answers or availability.
Do you have any questions for us?	Always a good question and can be interesting to hear good questions about the role or the company.

5.4.6.4   RM6 Section 4: Conducting the interview

Make sure the interviewee feels at ease. Walk into the interview room together or stand up when they arrive and make them feel welcome. Imagine yourself in their shoes – how would you want to be greeted? Be polite and respectful and smile! Introduce your colleague if you opted to have someone else conduct the interviews with you. You might like to kick off with a discussion about their travel to the site or a comment about the weather, to help you both feel at ease. Explain what this part of the screening process is about and then begin with your questions. Make some notes but also look at the interviewee whilst they speak to gauge their body language and expressions. Remember that while you are assessing the interviewee, they will be assessing you and the company and deciding if they would like to work with you – it works both ways.

During the interview you are assessing the panellist for the following points:

• How well the person communicates;
• How they will fit working in a group situation;
• Someone who is interested in the role.

Remember though, that if the interviewee seems a little quiet, they may be a little overawed by the whole process, however, if they are overly loud and dominant it is likely they could take over in panel discussions and prevent everyone from contributing.

If you do not think the person is a suitable panellist, you should still continue the interview and finish in an unhurried and polite manner. The interviewee deserves respect and you are also representing your company. If you think you have found a good potential panellist, make sure you have all the information you need and check you have given them the information about the next steps and dates. It can be helpful to have the information relating to the training plan prepared in advance so that this can be given to all candidates on leaving.

5.4.7   Recruitment Module 7: Panellist selection

If you are conducting the screening tests and interview to determine who to employ for the panel, the selection will need to be done very carefully. If you have planned to recruit more panellists than you need to allow for further selection after training, as recommended by BS ISO 8586 (BS ISO, 2012), you can be a little more flexible about your choice of panellist as you have the option to deselect them later in the process. If you are recruiting an internal panel, this part may well be much easier, as you probably only have to consider the panellist's sensory ability.

From each of your application and screening forms create a 'mark scheme' to enable you to assess each applicant in the same manner. Example mark schemes are given for a typical externally recruited food and drink panel and a home and personal care panel assessing toothpaste in Figures 5.21 and 5.22, respectively. Collate the comments from the personal interview and comments from the mini profile sessions. These may be in the form of: 'yes', 'maybe' or 'no' from each person involved, or might be more detailed. Your job is to collate all the information and make the final decision.

You need to assess whether the panellists have the abilities you need, with no impairments that might impact your requirements. You also need to know that they can discriminate between samples, can consistently rank samples in order for key attributes and can follow instructions. You also need to assess their motivation and interest in the role and how they fit into the panel. If a person is available to attend less than 80% of the training dates, you might need to reconsider whether to recruit them or not.

It is better to recruit someone who has the potential to be a good panellist, with an interest in being part of your panel and the work you conduct, than someone who has excellent sensory skills but appears disinterested or unmotivated, or someone who you are not sure has the right personality for the panel. Discuss each person with everyone in the team and come to an agreement about who gets recruited. If you are recruiting with a probationary training period, you may have room to take on people who you are a little unsure about. You can always check out their personality, ability to detect aromas and ability to scale, for example, during the training period. Recruiting in this way, if you are able to, relieves the stress in the recruitment procedure.

They key message here is that the panellists you recruit will have a major impact on the success of the sensory team's work: it is better to do it right first time.

TEST 1 – BASIC TASTES

In front of you are solutions which represent the basic taste sensations. One or more of these may be a blank (water) or some solutions may be repeats. Please taste each one in turn and identify the dominant taste in each sample. Please describe the sample below as SWEET, SALTY, SOUR, BITTER or BLANK (plain water) against the appropriate code number. Ensure that you cleanse your palate thoroughly between each sample.

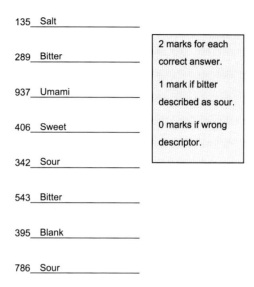

135 Salt _____

289 Bitter _____

937 Umami _____

406 Sweet _____

342 Sour _____

543 Bitter _____

395 Blank _____

786 Sour _____

> 2 marks for each correct answer.
>
> 1 mark if bitter described as sour.
>
> 0 marks if wrong descriptor.

TEST 2 - ODOUR RECOGNITION

In front of you there are three coded screw-capped bottles containing odour samples. Please smell each one in turn (replacing the cap before proceeding to the next sample). Describe the odour in the space next to the appropriate code, either with what you think it is or what it smells like. The odours may be described by several words if you feel it is necessary.

319 _____ Lemon _____

962 _____ Orange _____

118 _____ Strawberry _____

> 5 marks available for each answer – mark right answers and good descriptors – most important thing is to have some description. For example for 319 you might give 5 marks for lemon, 4 marks for citrus or lime, 3 for washing-up liquid, etc. If you have existing panellists they can be used to develop a mark scheme. Descriptions such as alcohol, chemical, weak or blank may indicate anosmia.

Figure 5.21 Mark scheme for a typical food and drink panel recruitment screening test.

TEST 3 - INTENSITY RANKING

In front of you there are five solutions of different strengths. Please taste each one and then put them in order of increasing strength of taste. Remember to cleanse your palate thoroughly between each sample.

Least taste/   863 _(water)_____ | 2 marks for each
Weakest | solution if in correct
                                                       | order. 1 mark for
                      214 _(10.0g per litre)__ | each solution if
                                                       | swapped around by
                      629 _(20.0g per litre)__ | one place. No marks
                                                       | for wrong place.
                      983 _(50.0g per litre)__

Most taste/   161 _(100.0g per litre)_
strongest

How would you describe the taste of these solutions?

_____sweet_____

| 2 marks if correct (sweet), 1 mark if in right area, 0 marks if wrong |

TEST 4 - ODOUR RECOGNITION

In front of you there are three coded screw-capped bottles containing odour samples. Please smell each one in turn (replacing the cap before proceeding to the next sample). Describe the odour in the space next to the appropriate code, either with what you think it is or what it smells like. The odours may be described by several words if you feel it is necessary.

133_____Cola_____

519_____Vanilla_____

688_____Peppermint_____

5 marks available for each answer – mark right answers and good descriptors – most
important thing is to have some description. For example for 519 you might give 5 marks for
vanilla, 4 marks for sweet, custard, etc. Descriptions such as alcohol, chemical, weak or
blank may indicate anosmia.

Figure 5.21 *(Continued)* Mark scheme for a typical food and drink panel recruitment screening test.

TEST 5 - INTENSITY RANKING

In front of you there are five solutions of different strengths. Please taste each one and then put them in order of increasing strength of taste. Remember to cleanse your palate thoroughly between each sample.

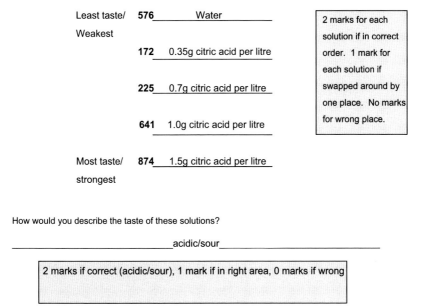

Figure 5.21 *(Continued)* Mark scheme for a typical food and drink panel recruitment screening test.

TEST 6 - TEXTURE DIFFERENCES

Please assess the two drinks in front of you and describe their textures in the spaces below.

Following this, please describe any differences between the texture of each drink.

Code 236 - Texture descriptions   Code 414 - Texture descriptions

(Flat cola)*   **(Fizzy cola)***

_____   _____

_____   _____

_____   _____

Differences in texture between the drinks:

_____

_____

Each descriptor given 1 mark, extra marks given for recognition and use of the five senses. For example if they describe the bubbles visually as well as how the bubbles felt in the mouth, or if they dipped fingers in drink and described it as sticky. Each comparison descriptor given 1 mark. No maximum.

*Note: these descriptors would not be on the panellists' worksheet.

Figure 5.21 *(Continued)* Mark scheme for a typical food and drink panel recruitment screening test.

TEST 7 - INTENSITY RANKING

In front of you there are five solutions of different strengths. Please taste each one and then put them in order of increasing strength of taste. Remember to cleanse your palate thoroughly between each sample.

Least taste/   **459** ___Water_____

Weakest

                   316 _0.2g caffeine per litre water_

                   584 _0.5g caffeine per litre water_

                   693 _1.4g caffeine per litre water_

Most taste/   **751** _2.6g caffeine per litre water_

strongest

2 marks for each solution if in correct order. 1 mark for each solution if swapped around by one place. No marks for wrong place.

How would you describe the taste of these solutions?

_____bitter_____

_____

2 marks if correct (bitter), 1 mark if in right area, 0 marks if wrong

Figure 5.21 *(Continued)* Mark scheme for a typical food and drink panel recruitment screening test.

TEST 8: DUO-TRIO TEST

You have been provided with an identified reference sample and two coded samples.

Taste the reference sample and then the two coded samples from left to right.

You must rinse your mouth between each sample.

Circle the sample you identify to be the **same** as the reference.

Explain why the other sample is NOT a match.

If you cannot determine which the matched sample is, please make a guess.

| Reference | 701 | 465 |

Can you describe the difference between the samples?

_____

_____

Digestive biscuits

For variant leave out of packet to stale for two hours. Reference is fresh sample.

5 Marks for identification of matched sample.

5 Marks for correct response in terms of other sample being softer.

Figure 5.21 *(Continued)* Mark scheme for a typical food and drink panel recruitment screening test.

TEST 9 - VOCABULARY GENERATION

In front of you is a slice of fruit pie. **Look** at it, **smell** it, **taste** it and then write down as many words as you can to try and describe it but <u>without</u> using words like good, nice, poor, bad etc. Try and think of this exercise as having to describe what you sense to a friend who has never come across this product before so that they can imagine the same sensations.

Layout (use of titles for appearance, aroma etc.): max. 5 marks

Handwriting: max 5 marks (if relevant)

Each descriptor 1 mark

English fluency (use of sentences etc.): max 5 marks.

Extra 5 marks if temperature of pie mentioned.

THIS IS THE END OF THE TEST. THANK YOU FOR PARTICIPATING

Figure 5.21 *(Continued)* Mark scheme for a typical food and drink panel recruitment screening test.

The next figure is a mark scheme for the assessment of panellists to join a tooth-paste panel.

TEST 1 – BASIC TASTES

In front of you are solutions which represent the basic taste sensations. One or more of these may be a blank (water) or some solutions may be repeats. Please taste each one in turn and identify the dominant taste in each sample. Please describe the sample below as SWEET, SALTY, SOUR, BITTER or BLANK (plain water) against the appropriate code number. Ensure that you cleanse your palate thoroughly between each sample.

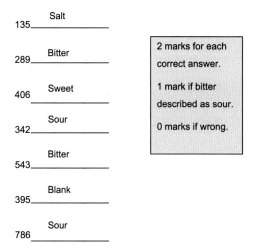

TEST 2 - ODOUR RECOGNITION

In front of you there are three coded screw-capped bottles containing odour samples. Please smell each one in turn (replacing the cap before proceeding to the next sample). Describe the odour in the space next to the appropriate code, either with what you think it is or what it smells like. The odours may be described by several words if you feel it is necessary.

173_____ Cinnamon _____

529_____ Spearmint _____

468_____ Peppermint _____

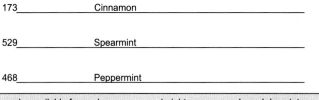

Figure 5.22 Typical mark scheme for a home and personal care panel.

TEST 3 - ODOUR RANKING

In front of you there are three coded screw-capped bottles containing an odour. Please smell each one in turn (replacing the cap before proceeding to the next sample). Please put them in order of increasing strength of odour.

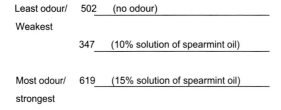

Least odour/   502   (no odour)
Weakest

                347   (10% solution of spearmint oil)

Most odour/   619   (15% solution of spearmint oil)
strongest

---

2 marks for each solution if in correct order. 1 mark for each solution if swapped around by one place. No marks for wrong place.

TEST 4 - TEXTURE

In front of you there are two different products (you will need to let us know by raising your hand when you need the third product). Please taste each one and then describe the texture of each in the mouth. Remember to cleanse your palate thoroughly between each sample.

301 Digestive biscuit: gritty, sharp edges, crumbles, mixes to a paste in the mouth etc.

972 Sweetened condensed milk: dry, powdery, liquid, runs over tongue etc.

584 Whipped cream from a spray canister: foamy, airy, light, creamy, disappears etc.

---

5 marks available for each answer – mark right answers and good descriptors – most important thing is to have some good textural descriptive words.

Figure 5.22 *(Continued)* Typical mark scheme for a home and personal care panel.

TEST 5 - INTENSITY RANKING

In front of you there are four solutions of different strengths. Please taste each one and then put them in order of increasing strength of taste. Remember to cleanse your palate thoroughly between each sample.

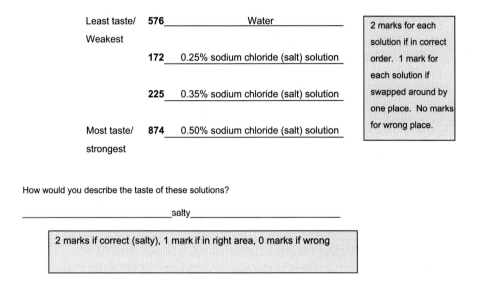

Least taste/ Weakest	**576** _____ Water _____	2 marks for each solution if in correct order. 1 mark for each solution if swapped around by one place. No marks for wrong place.
	172 ____ 0.25% sodium chloride (salt) solution	
	225 ____ 0.35% sodium chloride (salt) solution	
Most taste/ strongest	**874** ____ 0.50% sodium chloride (salt) solution	

How would you describe the taste of these solutions?

_____salty_____

> 2 marks if correct (salty), 1 mark if in right area, 0 marks if wrong

TEST 6: INTENSITY TEST (2-AFC)

You are presented with two coded samples.

Please smell the samples in the order given (from left to the right)

Indicate which sample has the more intense menthol aroma.

930		751

Code of the sample with more menthol aroma:

_____

> Toothpaste samples that look identical but have different levels of menthol.
>
> 5 Marks for identification of most intense sample.

Figure 5.22 *(Continued)* Typical mark scheme for a home and personal care panel.

TEST 7 - VOCABULARY GENERATION

In front of you is a sample of yoghurt. Look at it, taste it and feel the texture in your mouth and write down as many words as you can to try and describe it but <u>without</u> using words like good, nice, poor, bad etc. Try and think of this exercise as having to describe what you sense to a friend who has never come across this product so that they can imagine the same sensations.

Layout (use of titles for appearance, aroma etc.): max. 5 marks

Handwriting: max 5 marks

Each descriptor 1 mark

English fluency (use of sentences etc.): max 5 marks.

THIS IS THE END OF THE TEST. THANK YOU FOR PARTICIPATING

Figure 5.22 *(Continued)* Typical mark scheme for a home and personal care panel.

5.5   Introduction to the recruitment of consumer sensory panels

Generally, consumer panels for hedonic tests or focus groups require no *sensory* screening prior to taking part in a test (Resurreccion, 1998). The screening that is generally required is related to specific demographics, product usage or geographical location. For example, you may wish to recruit 300 consumers for an affective (liking/hedonics) test with 150 being male and 150 female. As part of this you may also wish to recruit users of your brand and users of a competitor's brand in a 50:50 ratio, say. You may also wish to recruit people from different races, ages, household income and size and so on. Sometimes the demographics can be quite complicated and it can be difficult to find people who fit the criteria you need: that means it will take time to recruit the right consumers. And remember that choosing the right consumers is the most important step in consumer research. There is no point testing the products with the wrong people! Whether your product is liked or disliked, it is not as a result of the product itself: it is the interaction with the consumer that is critical. A product on its own, sitting on a shelf, is neither liked nor disliked until a consumer picks it up and begins to consume or use it.

There are several statistical software packages to allow you to determine how many consumers you might need for your study. The calculations are generally built around the risks involved (for example, the alpha and beta risks or Type I and Type II errors as they are also known), the difference in means that you wish to find and the standard error (a measure of the data spread). If you would like to understand more about the number of consumers needed for an acceptability test, the papers by Hough (2005) and Hough et al. (2006a,b) give a very readable and usable explanation.

Occasionally the interviewer may wish to conduct tests to check for sensory acuity if the applicants are going to take part in discussions that require the sensory assessment of products or detailed discussions about flavour, texture or aroma, for example. When this is the case, modules from Section 5.4, such as the application process, sight acuity tests, basic taste acuity tests, olfactory screening, discrimination tests and ranking tests might be useful if the group work will be discriminative in nature. Modules to test for descriptive ability, ranking and scaling tests might also be useful.

If the consumers are to meet regularly and take part in group discussions relating to cocreation, product design or product feedback, one-to-one interviews would be advisable and a mock group session might help weed out those who will be unsuitable for the role (see Sections 5.4.6 and 5.4.5.12 for more information).

The use of consumers to give sensory scientists analytical sensory information about products has grown, and this growth has been fuelled by the rise in rapid methods which can be performed by naïve consumers as well as by semi-trained or even highly trained panellists (Valentin et al., 2012). Many publications and presentations have shown that consumers can produce reliable analytical-type results; in fact we have known this since free choice profiling (FCP, Williams and Langron, 1984) first started being used. Often, when people recruit consumers for discrimination testing or rapid methods, they do not conduct sensory screening as they are 'recruiting consumers'. However, as a proportion of consumers are unable to detect quite large changes in even their favourite products, it might be advisable to screen prior to the test so that you know who you are dealing with (Stone, 2015).

An excellent text for further reading about consumer testing is Jaeger and MacFie (2010) as it includes many industrial applications and innovative approaches.

5.5.1   Recruitment Module 8: developing a screener for a consumer study

Once you have developed one screener for a consumer study you can adapt and edit to create the next one, so it's worth taking the time to create a template screener with all the information that might be needed: you could even include handy tips and examples. Screeners are critical because they will help you, or your agency, recruit the right consumers for the test.

You will need to consider the test method first to determine the number and type of consumers to recruit. For example, if you are recruiting for a focus group you will need a smaller number of people and they will need to be able to communicate with you or the focus group moderator.

You will also need to consider the demographics of the people you need to recruit, including their purchase habits for the category, the brands they purchase or do not purchase and many other aspects such as age, gender, household income, etc. The test location also needs to be part of the decision-making process. Not just the geographical location but actually where the test is being conducted: in a local hall, at home, in a bar, etc. Once you have this information you will be able to consider how the person is to be recruited. For example, you need to decide if you are recruiting from a database/telephone, intercept (on street generally in a shopping area) or via online advertising. Recruitment can also be carried out via mobile phones through email, text messaging or any messaging app (for example, WhatsApp). An example screening questionnaire for a typical product liking test is given in Recruitment Module 11: screener questionnaire. If the number of people who consume/use your product is low, you might like to bring some of the product related questions to the beginning of the questionnaire.

5.5.2   Recruitment Module 9: developing a respondent information sheet

The respondent information sheet below can be used as template for your study. You will need to complete the areas in the square brackets [insert information]. For example, if you are running a test regarding fragrances you may wish to add in information about asthma, respiratory issues and hay fever. If the test is involving the use of washing-up liquid or shampoo, for example, you might wish to include restrictions regarding eczema, psoriasis or broken skin on the hands.

Respondent Information Sheet

Project leader:

Contact details:

Phone number:

Project code and title:

Introduction
You have been invited to participate in a product assessment. It is important for you to understand why the assessments are being done and what they will involve. Please take time to read the following information carefully. Ask the person in charge if there is anything that is not clear. Take time to decide whether or not you wish to participate in this sensory assessment.

You may stop taking part in the study at any time without giving a reason.

You have the right to ask the team any questions concerning any aspect of this product assessment at any time.

Guidelines
We would like to remind you of the following guidelines which are necessary when participating in a product assessment:

- Please do not attend if you are unwell, for example, if you have a cold or a bad headache, but do contact us to let us know that you are no longer able to take part.
- Please do not wear any perfume or aftershave before coming to the session.
- Please do not consume any alcohol prior to attending the session.
- Please do not smoke in the last hour prior to the session.
- Please do not brush your teeth in the last hour prior to the session.
- Please do not eat or drink 30 minutes prior to the session, this includes coffee, mints, sweets and chewing gum. You may drink water.
- Please arrive on time.
- Please do not bring your children on site.

This study is being conducted by [insert company name].
What is the Purpose of this Study?
[Insert purpose.]
Who can participate?
[Insert information.]

You will be eligible to participate if you:

[Insert information.]

You must not take part in the study if you:

[Insert information.]

What do I have to do?

[Insert information.]

If you agree to participate we will ask you to sign the consent form and you may keep the form.

What are the possible disadvantages and risks in taking part?

[Insert information.]

What are the possible benefits of taking part?

[Insert information.]

This study is for research purposes only. There is no direct benefit to you from your participation in the study.

Will I receive compensation for my participation?

[Insert information.]

What if something goes wrong?

In the event that you do experience any unusual symptoms please report this as soon as possible to [Insert information.]

[Insert legal requirements.]

Will my taking part in this study be kept confidential?

Yes. Your data will remain confidential in accordance with applicable laws and corporate policies except when sharing the information is required by law or as described in the additional informed consent form.

5.5.3   Recruitment Module 10: developing an informed consent form for a consumer study

The informed consent sheet below can be used as template for your study. You will need to complete the areas in the square brackets [insert information]. This part of the form is usually printed as part of the respondent information sheet so that all the information is kept together.

INFORMED CONSENT

Your decision to participate in this product assessment is voluntary.

The project manager can stop your participation at any time without your consent for the following reasons:

- If you do not follow the directions for participating in the product assessment;
- If it is discovered that you do not meet the product assessment requirements;
- If the product assessment is cancelled; or
- For administrative reasons.

List of Active Ingredients

[Insert information.]

CONSENT FORM [insert reference number/project code]

- I confirm that I have read and understood this informed consent for my participation in this sensory study. I have had enough time to allow me to consider the information, ask questions and I have had any questions answered satisfactorily.

- I have seen and read the ingredient list for these products. As far as I know, I confirm that I am not allergic to nor intolerant to any of these ingredients.

- I understand my participation is voluntary and that I am free to withdraw at any time without giving any reason. I understand that I am under no obligation to take part in this testing.

- I have received both a written and oral explanation of the following:

  - the purpose of these assessments

  - what I will be asked to do

  - the safety procedures supporting this testing

  - the action I should take in the unlikely event that I have an unusual reaction to any of the test products or procedures

- I will inform the principal investigator or study coordinator as soon as possible of any unusual experiences related to my participation in the study.

- I understand that any data that I provide will only be used for the purpose of this research as defined by the Data Protection Act/Market Research Society Code of Conductor [Insert relevant information].

- I understand that this study, the test products and information about them are confidential and I agree not to disclose or discuss any information concerning this study to anyone other than the study personnel. I understand that all information is and will remain confidential.

- I give my consent to participate in the consumer study.

Name of respondent:	Date:	Signature:
Respondent number:		
Name of project manager:	Date:	Signature:

{Give one copy to the respondent and keep one copy for the file.}

5.5.4   Recruitment Module 11: screener questionnaire

Typical screening questionnaire for a central location test (CLT) for a liking or preference study

Key

[...] Edit prior to survey

Italics: to fill in by interviewer

RESPONDENT ID Number *(To be filled in by interviewer)*	

SUPERVISOR ID **(TO BE FILLED IN BY** **SUPERVISOR)**		PROJECT NUMBER **(TO BE FILLED IN BY** **SUPERVISOR)**	

Interview date and time *(To be filled in by interviewer)*	RESPONDENT NAME (first name and then second name): *(To be filled in by interviewer)*
		RESPONDENT CONTACT NUMBER: *(To be* *filled by interviewer)*

[Insert main location here e.g. Spain]	
1	*[insert specific locations here e.g. Madrid]*
2	
3	

INTERVIEWER TO SAY:

"Good morning/afternoon/evening. My name is [insert interview name] and I am an interviewer from [insert agency name], a market research company.
We are currently carrying out research about [insert product name].
Your opinions and responses are very important to us.
Would you be able to spare us [insert amount of time] to complete a survey? Thank you!"

[Insert relevant information to give to the respondent about data protection act and assurances about how their contact information will be stored and used. This can also be given as a card to the respondent so that if they have any questions in the future they can contact the agency and ask.]

The respondent

Q1. Ring code of respondent's gender:

1	Male	[Put quota here for reference]
2	Female	
3	Prefer not to say	

Q2. We are looking for people of a particular age to participate in this survey. What is your exact age?

[Edit this question as required for demographic requirements]

Enter exact age
Then code age range

EXACT AGE:			
[Insert main location]		**[INSERT QUOTA HERE]**	
1	Less than 18 years old		**TERMINATE**
2	18 to 34 years old		
3	35 to 49 years old		
4	50 or over		**TERMINATE**

SHOW CARD for Q3

Q3. Which of the following currently applies to you?

1	Currently pregnant or breastfeeding	
2	Suffering from cold/blocked nose/cough	
3	Suffering from a headache	
4	Suffering from a fever	**TERMINATE**
5	Health issues that prevent you from [consuming/using] [product]	
5	Have eaten spicy food in the past [insert timing] [remove question if prerecruit or not relevant]	
6:	None of the above	

Q4. Do you have any food sensitivities, allergies or follow certain guidelines due to medical reasons?

1	Yes	**TERMINATE**
2	No	**CONTINUE**

SHOW CARD for Q6

Q6. We are looking for certain people to participate in this survey. Are you, or any member of your household, employed in ANY of the following industries? *Code all mentioned.*

1:	Advertising	TERMINATE
2:	Marketing/market research company	
3:	Media (newspaper, TV, radio), public relations	
4:	[Insert as required]	
5:	[Insert as required]	
6:	[Insert as required]	
7:	Finance	CONTINUE
8:	None of the above	

Q7. Have you participated in any market research related to [insert product type or make question generic depending on project specifications] in the last 3 months?

1:	Yes	TERMINATE
2:	No	CONTINUE

QD. Ring correct code for each demographic below.

A	What is your home address?	
B	What is your telephone number	
	What is the highest level of education you have completed?	1. Primary 2. Secondary 3. Vocational 4. University
C	Are you currently:	1. Employed 2. Self-employed 3. Full time student 4. Retired 5. Unemployed 6. Other: specify_____
D	Which of the following best describes your current job level?	[Insert options]
E	Which of these best describes your current status?	1. Married/living as married 2. Single 3. Widow 4. Divorced/separated
F	In which of the following bands is your family monthly net income included?	[Insert options]
G	How many people currently live in your household?	
H	Which of the following best describes your current occupation?	[Insert list here]

The respondent and the product
SHOW CARD for Q8

Q8. Which, if any, of the following [insert product types] have you [consumed/used] in the last three months?

Code all mentioned

1	[Insert product types required here – add new rows as required]	**CONTINUE**
2		
3		
4		
5		
6		
7		
8	None of these	**TERMINATE**

IF [product type] MENTIONED CONTINUE otherwise TERMINATE

Q9. How often have you [consumed/used product] over the last three months?

	[edit as required]	**MALES**	**FEMALES [remove column if not relevant to split by gender]**
1	At least once a day	1	1
2	4–6 times a week	2	2
3	2–3 times a week	3	3
4	Once a week	4	4
5	2–3 times a month	[edit termination as required]	5
6	Once a month	**6 – TERMINATE**	**6 – TERMINATE**
7	Once every 2 months	**7 – TERMINATE**	**7 – TERMINATE**
8	Only once or twice	**8 – TERMINATE**	**8 – TERMINATE**

[Insert required frequency here and codes in brackets (e.g. CODES 1-4)]

Q10. [insert any additional frequency of consumption/use questions here] [Insert criteria for recruitment here]

SHOW BRAND LISTING CARD

Q11. Which of these have you [consumed/used] in the last month?
Code all mentioned in table below.

Q12. Which brand do you [consume/use] *most* often?
SHOW ALL ANSWERS FROM BRAND LISTING CARD.
One code only in table below.

Q13. And which of these brands, if any, would you never consider [consuming/ using]?

SHOW ALL BRANDS EXCEPT THE ONE SELECTED AT Q12.

Code all mentioned or select 'None of these' if the respondent is happy to consider all brands. [Edit table to show action e.g. if we require half respondents to have consumed/used brand 7, leave cell white and add quota to Q11 and Q12 column. If we do not wish to recruit people who would never consume brand 2, grey out the cell and add terminate.]

BRAND LIST	Q11 – [Consumed/ used] IN LAST MONTH *(Code all mentioned)*	Q12 – [Consumed/ used] MOST OFTEN *(Single code)*	Q13 – WOULD NEVER [Consume/ use] *(Code all mentioned)*
[List brands here – insert new rows as required]	1	1	1
	2	2	2
	3	3	3
	4	4	4
	5	5	5
	6	6	6
	7	7	7
Other	15	15	15
None of these	16	16	16

[Insert 'must consume/use' codes here for the respondent to continue.]
[Insert terminate details here.]

INTERVIEWER TO SAY: "Thank you for taking part in this survey".

If continue: "As I mentioned, we are carrying out some research [insert where, when and how long and any other relevant information (e.g. payment]".

Q14. Are you willing and available to participate in this test?

1:	Yes	**CONTINUE**
2:	No	**THANK AND CLOSE**

If terminate:
INTERVIEWER SAY: I'm sorry you don't meet the requirements of our test today. Thanks again for your time.

[Insert next steps here]

DISCLAIMER (if required)

I can confirm:

I am over 18 years old.

I do not have any known allergies to [insert product].

I do not have any medical reasons or I have not been advised by a doctor to refrain from consuming/using [insert product].

I am not taking any prescriptive or over the counter medication.

I declare that I have read the above and I can confirm that I am eligible to take part in the test.

Name (Print):	
Signature:	
Telephone No:	
Date:	

Part two

Training of sensory panels

Post-screening/initial/introductory training of sensory panels

6.1 Introduction

The training required for each type of panel will be very different. This chapter, as well as Chapters 7 and 8, includes ideas that you can select from to develop your training programme. The objectives of the training are simply to convert the consumer to a panellist: someone who produces reliable and valid results. Your aims are to help them improve their ability to recognise sensory attributes in complex products, gain experience in the product(s) and test methods, and to become a trained panellist who is a consistent and valuable member of the sensory team. The training can include:

1. Teaching the panellists about how their senses work and interact, about potential biases and how to avoid them, and explaining how their data will be used
2. Training the panellists to be objective
3. Training them to work as an effective team (where necessary)
4. Training to ensure the panellists understand the task they have been asked to do
5. Training them in the use of the data collection device.

There are different levels of training. A 'trained panel' mentioned in the literature may in fact not be a trained panel at all, but simply a group of consumers who have been screened and introduced to the method to be used. This is not necessarily a bad thing (except in that the nomenclature makes it confusing to determine the level of training) as often a panel does not need the same level of training as another panel: it will depend on the objective. Some panellists that have been trained in a particular method may have several hundred hours of experience but still be referred to as a trained panel. It would be useful to have some form of naming that makes it easier to determine the difference in training level. Naïve, informed, semi-trained, trained, highly trained and very experienced might be a good starting point.

There are various ideas for training included in this chapter, again in modular form, so that you can pick and choose those elements that will be useful for your panel. Section 6.3 gives examples of the complete training plans for a panel working on discrimination and profiling tests. An additional section for panellists requiring training in temporal methods is given in Section 7.5. The complete training plan in Section 6.3 can be easily adapted for any type of products by selecting only the relevant aspects. Method training for discrimination testing, descriptive profiling and various 'rapid' profiling methods such as ultra-flash profiling are given in Chapter 7. Obviously you do not need to include all the various sections for your panel, you might like to skip certain sections, however, taking the time to build and train your sensory panel can be very beneficial if you would like to recruit and train an excellent panel that are highly motivated and will show commitment to the role.

The topics you will need to include for the panel training sessions will vary depending on whether the panel is internal or external, if they will work from home or on-site and the type of products the panel will be assessing. There is a BSI standard that covers the recruitment, training and monitoring of panellists (BS EN ISO 8586:2014) and also an ASTM publication: STP758, which although it is quite old, includes some very useful information (ASTM, 1981). Lyon (2002) states that there are two steps in panel training: (1) an introduction to sensory analysis, sensory tests, panellists' rules and best practice and (2) practical training on the methods and products that the panel will be working on. This chapter covers the majority of the first step and the next chapter covers the second step.

If you are recruiting new panellists to join an existing panel, either internal or external, it can be advisable to train them separately from the existing panel to prevent them from becoming demotivated when they witness the current panel's abilities and skills. It's also advisable to recruit several panellists at the same time, otherwise there will be a lot of resource investment for one or two panellists and it will be more difficult to train on teamwork. For more information about introducing new panellists to an existing panel please see Section 6.6.

If an external panel is to be working on the site, a tour of the areas that are important to them will be useful. Most panels will benefit from an introduction to sensory science and why it is important to the business. The panellists start off as consumers and gradually develop during training to become trained sensory panellists, and therefore they will probably be completely unaware of sensory as a science so will need to be told. And instructing them early on can be really helpful as then they can understand why the products need to be coded, why they need to follow instructions to the letter and how the methods are based in scientific practice.

It can be helpful if a senior manager can reinforce the importance of sensory science to the business during the introductory training and again at regular intervals to help enhance motivation. It can be very beneficial to include ways of working at the outset. This can also include rules that the panellists need to abide by as well as what you expect from them and why. See Sections 6.3.3 and 6.3.4 for more information.

Including information about the senses, how the senses interact and how to do the assessments will be critical for most panels, however, do not feel you have to include all the senses if they are not relevant to your panel. Some information about psychological and physiological biases and how the panellists' responses can be affected by environmental and other issues can also be helpful. Kemp et al. (2009) have a very useful section relating to factors affecting sensory measurements and you could use this to give the panellists a short introduction. Teaching the panellists about the psychological and physiological biases can be incredibly helpful so that they understand the reasons behind not drinking strong coffee just before a session or why you have asked them not to wear lipstick or fragranced deodorants.

If the panel is going to be working as a team, for example, in developing product profiles, you might like to include elements of communication and negotiation into your training sessions.

Your aim in the training is to build up the panellists' confidence and knowledge, but remember, not everyone will feel the same way: some people might take longer to learn particular aspects or feel happy working in the panel environment. The panellists will need to learn that the tests will take time and that they need to concentrate to get good results. This message is particularly important to instil in internal panellists who may be in a rush to get back to their 'day job'.

Learn about each panellist's strengths and weaknesses so that you can tailor your training plan to meet their needs as you move through each session. Consider the objectives of each session after the event: did you meet the objectives? How well did you meet the objectives? What additional training might be required?

Each of the training aspects is discussed in more detail in the following sections.

6.2 Ideas to help your training have maximum impact

The first item in this list is related to the training objectives. If you decide up front what you would like the panellists to be able to do by the end of each training session and by the end of the training itself, you will be half way to developing your plan. And developing a plan for the training is essential. You will need to plan out all the items you hope to cover, when each task will be conducted, what needs to be prepared beforehand and by whom, what is required on the day, the presentations you will need, stationery requirements, etc. An overview plan for the training of a food and drink panel for discrimination and profile tests is given in Figure 6.1.

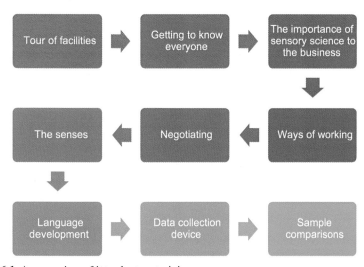

Figure 6.1 An overview of introductory training.

The next critical item is to decide where you will conduct your training: what rooms will be needed, the layout of the room as well as what you will need in the room. The room needs to be large enough to fit in all the exercises and group work you will be planning but with easy access to the booths when you wish the panellists to use them for assessments. Not all assessments need to be done in the booths, many can be done in a conference room environment, but as soon as you want to start demonstrating the correct approach to the panellists or your products require specific equipment (for example, ironing boards or rapid air changes), it's a good idea to do the assessments in the booths.

The training room also needs to be comfortable, with projection equipment, enough tables and chairs for the trainee panellists and perhaps other equipment such as sinks or sample presentation devices (for example, mannequin supports for hair assessments). Consider how the tables will be arranged for each session depending on whether there is group work or activities that require a particular layout. Think about the first session particularly – if you lay out the chairs and tables like a schoolroom, the interaction between you and the panellists may be less than if you arrange the chairs around the screen in a curve like a cinema. And it can be useful to get panellists to sit in different seats each time so that they get to work with different people. This can also help for future sessions to prevent panellists getting into the habit of sitting in the same chair or same booth.

Next, you will need to decide how you will conduct the training. It's a good idea to look through your list of items that you need to train the panellists in and try to mix and match various training methods. For example, if the first session involves presentations and quizzes, you might like to arrange several practicals for the next session. Alternate between videos, games, group discussions, questionnaires, presentations, practicals and demonstrations to make the sessions interesting. Discussions in pairs or in groups can help the panellists learn more efficiently and can help your session run more efficiently too.

Whilst presenting to the panellists, for example, on the senses, make sure you start with something to attract their attention: maybe something funny (related to the senses in this example) or something they will be asked to do at the end of the presentation so that helps them to pay attention. Use plenty of examples and practicals to demonstrate the topics: for example, do not just drone on about retronasal olfaction: make it come to life with an example! (See Section 6.3.6 for one such example.) Tell the panellists that they can ask questions or make comments as you talk. This way it becomes more of a conversation. Do not be afraid to ask them questions too. For example, if you are talking about how the eyes work ask them, 'Does anyone know how our eyes work?' You never know, someone may well be able to describe this much better than you, or a panellist may have worked in an opticians and be able to give some really nice examples.

You also need to consider the person who is going to conduct the training. This might be you, a colleague or an external consultant hired for the role. To be a successful trainer they must have the following qualities:

- The ability to communicate effectively: this is a critical quality for a good trainer. If the training is delivered by someone who speaks in a monotone with no eye contact or interaction with the audience, this can render the time spent completely worthless. Telling the panellists what you expect them to learn or what you expect them to be able to do at the end of the training session can help them understand the point of the training. In fact explaining *why* the knowledge is needed is very important; but do not give it all away before you start. For example, if you are going to demonstrate retronasal olfaction do not start by saying, 'You are going to be *so amazed* by this – when you take your hand from your nose and the aroma volatiles hit you, you will be *so impressed*', as you can pretty much guarantee that they will not be – you have built their expectations up and told them the story plot. But you can tell them that they will be learning how their senses work for example.

- Annoying mannerisms: it will be worthwhile checking the trainer does not have any annoying mannerisms such as jangling their keys in their pockets, pacing backwards and forwards or waving their arms like a windmill (the last one is mine, kindly demonstrated to me by a recording of a presentation by a colleague. I don't do it any more…).
- Excellent organisational skills: the training needs to be planned out in advance with everything organised and ready to go. In the sessions themselves the trainer needs to have visualised the steps for the various objectives so that items for each and every stage have been thought out and prepared. They also need to be flexible enough to reorganise the plan if the panel needs longer for particular tasks.
- Knowledge, interest and enthusiasm about the subject matter so that they can pass on the right information to the panellists in the right way, helping to motivate them, answer their questions and develop a good relationship with the trainees. The idea is to get the panellists excited, curious and fascinated as this will stand you in good stead for future motivation and interest in the role. A motivated panellist gives better data.
- Creativity in teaching methods: we are lucky to be training the panellists in sensory science as we have the option to include interesting and exciting examples of the various training elements, but do consider other teaching methods such as handouts, videos, group discussions, internet searches, learning summaries, votes, quizzes, homework, shopping trips, posters, reading, post-it question sessions and question collections.
- Consideration for the panellists: the trainer needs to be aware of the panellists' personal needs as well as their training needs. Many of the panellists may not have been in a learning environment for some time and might be fazed by the whole concept of learning something new. Something as simple as referring to a triangle *test* might make them panic with memories of maths tests at school. All that is required is a short introduction to tell them that all the discrimination methods tend to be referred to as 'tests' and this does not mean that they are being 'tested'. In fact you could fall back on a saying that I think really helps: 'the panellist is always right'. After all they can only choose, rate and select what they perceive and if it's different to everyone else it may well be related to their physiology, not their ability.

6.3   Preliminary training plan for sensory panel

The information below gives suggestions for a complete introductory training plan for a panel working on discrimination and profiling tests. The additional method training sections are given in Chapter 7. You can pick and choose which of the sections below you need. Some sections will be more applicable to a food or drink panel and some to a home and personal care panel. The earlier sections give the detail for each of the sections of training, whereas the later sections simply present a list of things that could be covered. The detail for the list headers can then be found in the method or product training sections in Chapters 7 and 8, respectively, as well as Chapter 10 which has some more advanced training topics. Topics included in the introductory training are shown in Figure 6.1.

6.3.1   Tour of the facilities

An external panel will need to be given a tour of the areas of the site that are relevant to their daily working routine, as well as any elements that are related to health and safety requirements, e.g., the fire evacuation routes. All panellists, external and internal, will need to do a tour of the sensory facilities, as working in a booth can be quite alien to most people. It's a good plan to show external panellists where they can wait prior to a panel session and what the rules might be for the waiting area. For example, if senior staff often see and perhaps more importantly hear the panellists while they wait and they appear not to be working, this may well cause you issues.

6.3.2   Getting to know you

It's a good idea to include some time for the panellists to get to know you, the team and the other panellists. A fun way to do this is to give each person a quiz sheet like the one shown in Figure 6.2. While the panellists complete the quiz you will notice how some people will try and complete all the questions, while others, having met someone they have a lot in common with, will still be talking to them by the time the 15 minutes you gave them to complete the quiz is up.

There are several other options for a getting to know you exercise and these are listed in Table 6.1. The panel quiz activity above can also be carried out in the form of 'human bingo'. Simply list the people headings, e.g., their first name begins with L (these are usually the numbers in Bingo) in a four by four grid (if you have 16 people headings) with space under each heading to write in the names. Other fun headings can include: can ride a horse, is a musician, likes to watch [insert TV programme name].

Getting each person in a pair to introduce the other person to the group can work well, but sometimes, if the pair runs out of time, only information from one of the pair gets divulged. If you are going to use this method, it can work better if you call time half way through so that the questioner becomes the questioned.

Panel Quiz

Find someone in the room ....

1. Whose name begins with M_____

2. Whose star sign is Aquarius_____

3. Who has two children_____

4. Who lives in [insert suitable location] _____

5. Who was born in the month of May_____

6. Whose star sign is Libra_____

7. Who supports Manchester United_____

8. Whose surname has nine letters_____

9. Who has a pet dog_____

10. Who travelled here by car_____

Figure 6.2 Example 'getting to know you' quiz.

Table 6.1 Ideas for 'getting to know you' activities

Panel quiz	Human bingo
Paired introductions	Five things I'm not
Famous partners	Chinese portrait

'Five things I'm not' is another fun way to get panellists to know each other. Each person gets two minutes to consider what they are not and then shares it with the rest of the panel. For example, they may say, 'I'm not a football fan, I'm not married, I'm not a parent, I'm not confident and I'm not very good at maths'.

'Famous Partners' can also be fun if you have good (sticky!) post-its: write pairs of famous people on post-its and stick one on each panellist's back. For example, Romeo on one post-it and Juliet on another. Other ones that work well are: Tom and Jerry, Adam and Eve, Batman and Robin, Beauty and the Beast, salt and pepper. The panellists then need to ask questions about 'themselves' to find their famous partner. This is a fun ice-breaker if you would rather not get the panellists talking about themselves.

The 'Chinese Portrait' is another 'get to know you' idea that works well. The panellists get a couple of minutes to think about what they would be if they were a tree or an animal or a song or a famous picture (or anything else you would like to ask them to consider – just pick one at a time though!). They could also tell the group why they chose that particular item.

6.3.3   The importance of sensory science to the business

If you have a video of your company history or a public relations film, show this to an external panel as it will increase their sense of belonging to the company, can be motivating and make them feel part of the team. If you do not have a video, some information about the company via a presentation can work just as well.

Giving the sensory panellists a quick overview of how their work might impact the business can be helpful in motivating the panellists and also in giving them an idea of the work they might be doing. For example, if the panel is going to be producing quantitative descriptive profiles, show them an example of a finished profile and how the results might be used. Be careful about the information that you share so it does not bias the panellists. If all the panel's work is going to be related to cost saving projects, it's probably a good idea to include some other applications as well, such as new product development, shelf life and competitor analysis, or the panel may well give you results they expect (for example, 'less flavour' or 'thinner foam') rather than describe the product correctly. Including an introductory welcome and hello from a member of the senior management team can be really useful in demonstrating how important the panel is to the whole company.

6.3.4   Introducing ways of working

If you are recruiting externally you will need to inform the panellists about health and safety on site and how to deal with paperwork (e.g., booking a holiday, reporting sickness). Give each panellist a business card that they can easily keep with them with the relevant contact details in case they are ill, delayed or have a query about a product they have assessed. A talk from the human resources team can also be useful to explain the information about holiday entitlement, pensions and anything else that may be relevant for an employed external panel. You may have other forms for the panellists to sign such as their employment contract, informed consent and confidentiality.

At this stage, for internal and external panellists, it can be useful to give an introduction to the panellists' rules (see Figure 6.3 and Table 6.2), as well as how to work in the booths and how to report illnesses or the use of medication. Meilgaard et al. (2016, pp. 194–198) have an excellent 'Panel Guidelines' that could also be helpful in developing your panellists' rules.

If your panel involves teamwork it can be helpful to create a video of a sensory panel in action and to show this in one of the introductory sessions. Include examples of the panel behaving well so that there are clear examples of expected behaviour. For example, demonstrate panellists listening to each other, discussing ideas but not

Be punctual for sessions

One person speaking at a time (although the panel leader may need to interrupt to help us achieve our session goals)

Listen when others are speaking – each person's suggestions are equally valued

Be considerate and polite – discussions and disagreement will be about the topic not about the person

Do not be stubborn – listen to others' opinions and be willing to compromise

Be keen to improve your skills and performance – feel free to ask for help from your colleagues or the sensory team

Speak up if you think you have the answer

Use your list of objectives and KPIs to become the best panellist ever

Figure 6.3 Helping the panel sessions run smoothly and efficiently.

Table 6.2 Panellist dos and don'ts

Do	Don't
Listen	Use powerful fragrances, soaps, deodorants, etc.
Maintain good hygiene	
Rest your senses between samples	Smoke before a session
Cleanse the palate or test site properly where applicable	Talk over other panellists
	Eat or drink within 30 minutes of a test
Take sensory testing seriously	Eat or drink any strong flavours within an hour of a test (e.g., mint, chilli)
Do not rush – take enough time when carrying out tests	
Switch off your phone	Do not participate in a test when you cannot smell
Respect and follow test protocols, procedures and instructions	Do not participate in a test if you are unwell
	Do not take part in a session if you have a strong dislike for the type of food/drink
Ask questions if you are unsure	
Tell us if you have any issues	Do not participate if you have too much prior knowledge
Consider others' feelings	

arguing and even panellists compromising where necessary with good grace. Also demonstrate the difficulties created when the panel behaves 'badly' to give clear examples of the behaviour to avoid. Show the outcomes when the panellists do not listen to instructions or each other, when they do not follow the standard rules and also demonstrate the impact of that panellist who is always late. If the video is shown early in the training, you can refer back to it easily if things do start to go wrong in later training sessions. For example, let's say you are in training session six for a panel who are training to do descriptive work, and a panellist begins to talk over another panellist. You can raise your hand and say something like, 'Excuse me folks, remember the video. Why is it important that we only have one person speaking at a time?' Do not ask this question of the two panellists involved, in fact look in the opposite direction for the answer. This saves embarrassment for the panellist who was doing the talking.

6.3.5 *Negotiating*

If the panellists will be developing product profiles, particularly in methods where a consensus is reached about attributes or scores, it can be useful to introduce the concept of negotiation. This can involve a presentation or chat about ways to negotiate as well as a game from the Edward de Bono Mind Pack (de Bono, 1998) which teaches negotiation in an interesting way. Allow about 90 minutes for this training.

Start off by asking the panellists what the difference is between an argument and a discussion. You could get them to discuss in pairs or small groups for a few minutes and then ask a couple of groups for their ideas. You might get comments back about an argument being a more negative statement that tends to be more angry than a discussion for example. Split the panellists into two groups and ask one half to think about a discussion they had recently and the other half to consider an argument. Ask each half to talk about the event in their pairs/groups. Walk around the room listening to the discussions about the differences between discussions and arguments. Pick a pair/group from each half that you think really shows the difference between a discussion and an argument. Let them share their events and then ask each pair/group what they thought the 'values' were in the discussion: for example, what beliefs did they have when they were in the discussion/argument, what did they want from the outcome, why were they discussing/arguing in the first place? It can be interesting to ask various other people in the two halves the same question. You could try listing the values on two halves of a flipchart page as this can neatly show the difference in values between discussions and arguments. There are three types of values: common, separate and key – and these can be demonstrated in the de Bono game later if you opt to include this.

Then you can describe the usual process in a negotiation. In the middle of this description the game takes place. A good example of this is in a negotiation for a pay rise between a union and a company. For example, both parties often take an extreme position so that if they have to give something up, they might actually end up where they wanted to be anyway. The values for each party are quite different. They both want to win. Then the discussion between the union and the company takes place. Sometimes this ends up as an argument. Now is a good time to ask the panellists to discuss in their pairs/groups how arguments are best avoided. It can be good to begin by asking what 'things' increase the likelihood of an argument as this can help prompt the discussion.

Then it's time for the game. Begin by explaining that the game is *competitive and cooperative* and that *everyone can win*. Split the panellists into two teams (note terminology) whose aim is to each build three hexagons from six different coloured triangles. These pieces are available in the de Bono Mind Pack but you can easily create them electronically and print them out. There are 30 blue triangles, 15 red triangles, 15 green triangles and 10 yellow triangles. You will also need an opaque bag/pot/hat to drop them into. Team 1 has to use three blue, two green and one yellow triangle to make each of their three hexagons. Team 2 has to use three blue, two *red* and one yellow triangle to make each of their three hexagons. Tell the panellists how many of each triangle there are and that the yellow triangles are rare! Explain that values are generally classified as common (for example, blue triangles as everyone needs these), separate (green and red triangles as one team requires green and one team requires red) and key (yellow triangles as there are not very many of these). It can be a good idea to put this information up somewhere: for example, on a presentation slide (see Figure 6.4: you can download this from http://www.laurenlrogers.com/sensory-panel-management.html so that it is showing throughout the game. Each team gets 20 points for each COMPLETE hexagon and they need to make three each, so the aim is to get to 60 points, but beware – each unused piece held by each team means losing points! Every additional blue triangle a team holds loses them 1 point, red/green minus 3 points, yellow minus 6 points!

You will also be playing in the game – your name is Mr/Mrs Nobody. You only want to make hexagons from blue triangles.

- In the bag are 30 blue triangles, 15 red triangles, 15 green triangles and 10 yellow triangles (rare!!)
- **Team 1** has to use three blue, two green and one yellow triangle to make each of their three hexagons.
- **Team 2** has to use three blue, two red and one yellow triangle to make each of their three hexagons.
- Get 20 points for each COMPLETE hexagon
- Beware - each unused piece held by each team means losing points!!!
  - Blue minus 1 point, red/green minus 3 points, yellow minus 6 points!!

Team 1:   Team 2:

Figure 6.4 Information for playing the de Bono Negotiation Game.

The game consists of a draw phase where each team (and you) draws (takes) a triangle from the bag. Then there is the negotiation phase – the exciting part! Each team considers if they wish to keep their triangle or swap it with Mrs Nobody or swap it with the other team.

You must play fair: if you swap a yellow triangle for two blue triangles for team 1 then you cannot ask for six blue triangles from team 2! But you can just decide not to swap after certain draw phases so that the negotiation is just between the two teams. The teams do not have to play fair. They can keep the red/green triangles from the other team if they like (spot who suggests this...) or they can swap for one, two, three triangles (or more!) of any colour, and they can refuse to negotiate – it just depends on agreement from the other group.

After the negotiation phase, there is another draw phase and then another negotiation phase and so on. Each team tries to build their multicoloured hexagons and you try to build yours. Anyone can look over to see what the other teams have if they wish. The game ends when there are no triangles left in the bag. The scores are totted up and the extra triangles that do not make a full hexagon are also summed and taken away from the team's total. For example, if team 1 had made two hexagons and had two red triangles, three blue triangles and one yellow triangle left over their score would be $20 + 20 - (2 \times 3) - (3 \times 1) - (1 \times 6) = 25$.

But the whole point of the game is for the panellists to realise that they could simply ask that the bag is emptied onto the table and the triangles shared out so that everyone has the same score. I have played this game with many different groups of panellists and only once did a panellist make this suggestion. Once you point this out at the end of the game panellists realise the point was to negotiate and share. On a couple of occasions I had to deal with some panellists who moaned about the game and said, 'What on earth was the point of that?' or similar. One or two of these ended up as excellent panellists, so this game is not a screening exercise as such, just a demonstration that everyone can win in a discussion within the panel, if the discussion is carried out in the correct way.

6.3.6   Introduction to the senses

You might like to create a presentation for the panellists about each of the senses that they might be using in their assessments. You can find plenty of information in the standard sensory textbooks, however, if you wish to go into more detail yourself then *Foundations of Sensation and Perception* (Mather, 2016) is full of excellent information about each sense, with useful diagrams to help explain how the senses work and how they interact. Once you have finished the presentation you can use it the next time you need to run a panel training session. In fact, some of my clients actually have a complete set of training plans, presentations and documents for the introductory sessions as well as the later method training sessions. This way they save time when they recruit a new panel or need new panellists to join an existing panel, by reusing the plans and paperwork over and over again, updating and improving them each time.

One very useful test to include in your session about the senses is to demonstrate what flavour is by demonstrating retronasal olfaction. This can be done quite easily by using tea. Make some slightly stewed tea by putting three standard teabags into a teapot or Pyrex jug and pouring on boiling water. Leave to stand for three to five minutes and then remove the teabags. Cover and allow to cool. A shortcut is to buy a bottle of iced tea. Once the tea is at around room temperature pour into disposable cups. It is best not to use polystyrene cups as you do not want the panellists to realise the drink is tea – it looks somewhat like weak cola. Check that it is possible to hold the cup and drink from it while holding your nose – some glasses or cups make this close to impossible.

Get the panellists to practice holding their nose so that they are unable to breathe through their nostrils. Remind them to breathe through their mouths! If they are doing it correctly they will sound like they have a really heavy head cold when they chat to their partner. Tell them that you will be asking them questions once they have consumed some of the drink but that they must not release their nose to speak to you. Ask that they hold their noses while you give out the drink – this is especially important if you have bought iced tea with an added flavour like peach.

Once all panellists have a drink and are holding their nose, ask them to take a small mouthful of the drink. Ask one or two to describe what they can taste. They may say bitter for the cold tea or sweet if you have bought the peach iced tea. Then tell them that they can remove their hands from their noses and watch the surprise and listen to the exclamations as they realise it was (iced, peach flavoured) tea.

To check learning from the presentation about the senses a handout or quiz can be used: see Figure 6.5. The answer in this quiz is 'false'.

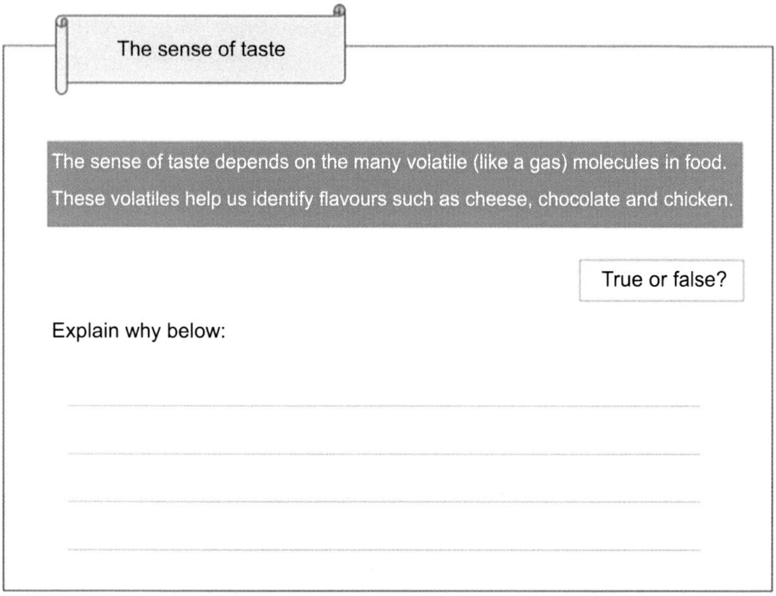

The sense of taste

The sense of taste depends on the many volatile (like a gas) molecules in food. These volatiles help us identify flavours such as cheese, chocolate and chicken.

True or false?

Explain why below:

Figure 6.5 Handout for checking understanding.

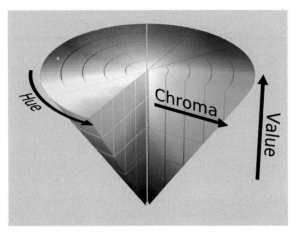

Figure 6.6 What is colour?

It is also useful to train the panellists in the detection and description of colour. Colour can be a difficult topic to discuss, define and reference and therefore it's worthwhile spending some time explaining to the panellists exactly what colour is. Albert Munsell in the 19th century described colour based on three separate dimensions: hue, value and chroma, and these three dimensions can describe any colour (see Figure 6.6). *Hue* describes the actual colour: red, blue, green, etc., and it is the first thing about a colour that the eye detects. This is shown in Figure 6.6 around the outside of the cone. Pure white and pure black are the two extremes of the *value* scale: no other colours can be seen in these two colours. Yellow will tend to be placed high on the value scale as it is generally a light colour and as such is closer to pure white than to pure black. However, yellow can of course be a dark yellow and in that case will be placed lower on the value scale. This is shown in Figure 6.6 by the vertical changes in colour. *Chroma* is all about the colour strength. For example, we might have two colours that are the same hue and the same value but be different colour strength. Let's say we have two colours which are both blue and that these two colours are also similar in lightness (i.e., value), but one might be a very strong blue and the other one a weak, grey-like blue. This is shown in Figure 6.6 by the move from the centre of the circle where the colours are very weak to the outer regions where the colours are very strong. Figure 6.7 shows the chroma changes for blue in the extended segments on the left-hand side. The very strong blue may be like that as shown by the segment numbered 12 and the weak, grey-like blue, by the segment numbered 0.

http://munsell.com/ has some very useful information, handouts and posters that can help make the discussions of how to describe and measure colours much easier. For example, there is a dictionary of colours which is incredibly helpful for use in definitions. There are also several radio and television programmes relating to colour that might be interesting to include in your training sessions.

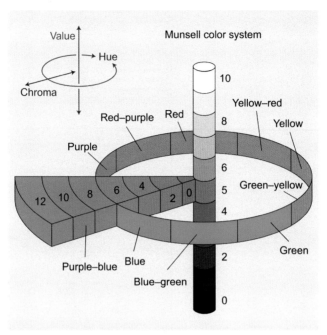

Figure 6.7 The Munsell colour system.

6.3.7 Qualitative and quantitative references

A useful resource for references is the ASTM *Lexicon for Sensory Evaluation* (ASTM, 2011). This is a software package that includes aroma, flavour, texture and appearance descriptors, definitions and references for a wide range of products. The references are particularly useful as the software allows you to search by term and gives reference ideas and recipes for many different product types. An example screenshot is shown below in Figure 6.8 for the term 'floral'.

The software also gives some useful information about lexicon development and describes the process in five steps:

1. Developing the frame of reference: this step involves collecting a wide range of products that span the category or categories that the panel will evaluate. The inclusion of ingredients and prototypes can also be very beneficial at this stage.

   For example, if you were developing the frame of reference for the quality control (QC) of lagers, you might assess a wide range of different lagers (maybe from your country and maybe from other countries if the panel will be assessing these), some lager ingredients such as malt and hops, as well as some commercially available compounds that demonstrate typical lager on- and off-notes. Imagine that you were training people about colour who had never seen or classified colours before. If you simply gave them a series of five cards: white, yellow, blue, red and black and told them this was the colour blue, for example, when giving them the blue

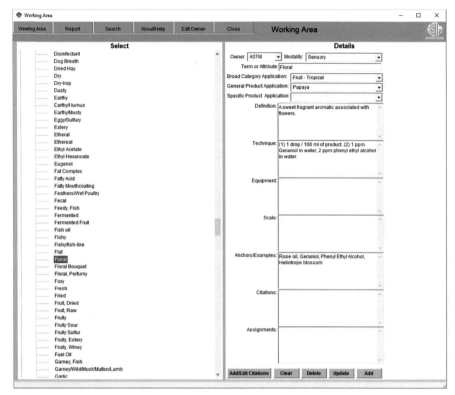

Figure 6.8 ASTM Lexicon for sensory evaluation screenshot for floral.

card, they would find it very difficult to classify navy blue or light blue as blue, as they would be unaware of the spectrum of blue colours. They would ideally need to see a range of blue cards to help them identify each as 'blue' and to be able to assign the correct description (navy, baby-blue, grey-blue, royal blue, sky blue, etc.). When children are small, they are taught in this way, gradually and over time. But no one ever really does this training with odours or flavours until they are lucky enough to become a sensory panellist.

2. The next step is the language development: this involves the assessment of the samples and ingredients by the panel and the generation of a list of words, classified by modality, for the description of the product type.
3. Finding the right reference: a reference is defined by the ASTM as 'a substance (a chemical or a simple substance) that provides a clear and distinct demonstration of the term or characteristic in question; it is important that the characteristic be the predominant trait in the reference'.

It can help to think about a particular example. Let's say that the panellists came up with the term 'coffee' to describe a particular aroma in a bread product. If we look at a lexicon for coffee (Chambers et al., 2016) we can see that there are many attributes describing coffee: woody, spice, honey, nutty, chocolate, etc., and we need to be clear which element of coffee it is that the panellists actually wish to measure: is it woody, spice, honey or nutty? The software could be used to search for coffee related references and these would be created and presented to the panellists in a bid to find the actual coffee element that they were measuring in the bread products.

Chambers et al. (2016) describe how they developed a qualitative reference for the term 'nutty' for use in a coffee lexicon. Originally, they used a blend of almonds and hazelnuts, but the panellists detected a rancid note in the hazelnuts and therefore the blend was changed to almond and walnuts and this was felt by the panellists to better represent the 'nutty' character of coffee.

4. The next step is somewhat counterintuitive as it involves swapping back to find 'examples' of the reference within products. The guide defines an example as 'a product or substance in which the term or characteristic can be readily perceived, although not as singularly as in a reference, enabling the panellists to experience the flavour/texture character and its corresponding term in a product'. As the original sample is an example of the reference, this step seems like an overkill but may be useful in some cases, particularly if you are developing an attribute list for long-term use.

5. The last step is to finalise the list of attributes, definitions and references, checking for attribute overlap and redundancy.

There are several useful publications giving lots of useful information about developing lexicons: Lawless and Civille (2013); Drake and Civille (2003); Muñoz and Civille (1998) and Rainey (1986) to name a few. The message here is to have references that are easily understood by all panellists, simple, specific and not generic and that demonstrate the attribute under discussion. Sometimes the panel will need to try several references until the right one is found.

Some methods require the use of quantitative references. Both Spectrum and QDA profiling methods use qualitative references where required to ensure that the panellists are in agreement about what element of the product they are actually going to measure. However, the Spectrum method also uses quantitative references. These quantitative references mean that all panellists will agree that a particular reference is, say 5.0, on the 15-point scale: i.e., an attribute for a certain product would be scored the same for the same product by all panellists. The intensity values are based on data from several panels over several replicates (Rutledge and Hudson, 1990). Therefore, comparisons across many types of products and panels can be made. For example, you could produce a profile of one product and compare that product to a product profile you created some time back. It also means that the measurements for different attributes are also comparable: a '5' on one scale will have the same intensity meaning as a '5' for a completely unrelated attribute. However, Lawless and Heymann (2010) state that, as there are no published data to support this approach, they are 'somewhat sceptical of this equi-intensity claim'. They also mention, in regard to the calibration by quantitative references, that, 'We are not sure that this level of calibration can be achieved in reality'.

Some other profiling methods also have attributes and definitions already developed that can be used or adapted. An example of this is the Texture Profile (ISO 11036: 1994(E)) which is based on technical and rheological measurements.

It includes the attribute titles (for example, hardness, cohesiveness and springiness), the definition (e.g., cohesiveness: mechanical textural attribute relating to the degree to which a substance can be deformed before it breaks) as well as describing the technique or protocol for the assessment (e.g., cohesiveness: place the sample between the molar teeth, compress it and evaluate the amount of deformation before rupture). The details regarding the Texture Profile method were published in 1963 (Szczesniak, 1963; Szczesniak et al., 1963) and both of these papers are well worth reading as they include useful information about the criteria for reference selection (see later in this section for more information) and the determination of the intensity values (see next paragraph).

The panel that set the texture profiling intensity reference values was very familiar with the Texture Profile method and the definitions for each of the attributes (Szczesniak et al., 1963). The panellists assessed each potential reference individually and then a round table discussion was held to arrive at an 'average rating'. The references were also ranked for each texture attribute: each scale 'encompassed the entire range of parameter intensities encountered in food products'. The ranking and scoring information was used to determine the references that were equidistant from the intensity reference on each side: these were selected for the final scales. The scales for each attribute are different. The gumminess scale runs from 1 to 5 and is based on flour paste mixes. Brittleness (now called fracturability) runs from 1 to 7 and hardness from 1 to 9. Each of the references was also assessed instrumentally and the panel data correlated well to these measurements. Portions of the scales can be selected to measure samples that are more similar in texture, however, new intensity references would be required to mark the points between each documented scale point. Prior to each profiling assessment the references are viewed again either in the booths with the samples to be assessed or a short training session beforehand.

Lyon (2002) state that there are four main methods for calibrating trainee sensory panellists on the use of the scale: the Spectrum method (we could also include the Flavour Profile and Texture Profile) with its absolute scales and extensive quantitative references; the use of quantified references for some attributes or some scale points (a kind of 'hybrid' Spectrum approach) and 'auto-calibration with the samples under investigation'. (The fourth method they mention is qualitative references but as these are not used for calibration on the scale, this method cannot really be included in the list.)

A review of some recent publications relating to lexicon development, suggests that there are a couple of other methods as well. Chambers et al. (2016), Ting et al. (2015) and Corollaro et al. (2013) have developed their own intensity references for their lexicons, mainly for the top and bottom of the scale. Chambers et al. (2016) describe the development of a lexicon for coffee and lists quantitative references for a range of coffee types. The scale used in Chambers et al. (2016) was a 15-point scale with 0.5 point increments and the intensity references were determined by the panellists through reference assessment, discussion and agreement. Several references were assessed by the panellists and the choice of these and their related intensity values were based on group consensus.

Other authors use one of the samples as the intensity reference. For example, Griffin et al. (2017) having gathered data over six different replications for the reference sample, present the sample blind on several occasions to check the panel's calibration. Panellists who did not agree with the reference scores were retrained with the reference again and allowed to discuss their scores and reasoning until a consensus was obtained. The reference sample can then be used at the beginning of each session to recalibrate the panel and inserted as a blind reference to check scaling.

For some products it can be quite difficult to find references that can be used at the beginning or end of the scale. For example, if you were working with a panel on the assessment of hair shampoos, imagine the attribute 'tangly' to describe how tangled the hair is after shampooing. To create the end of the scale you could use a mannequin head, strip the hair with a harsh shampoo and create a nice 'tangly' reference. By making a very tangled head of hair, though are you restricting the panellists' use of the scale to the first part only? And how tangled should the hair be to represent the end of the scale? Will some panellists hair actually be more tangled than this reference? One way to get around this issue is to create a reference that is at the midpoint of the scale. In this way you can create an example that is fairly tangly but not overly tangly and leave room for assessments on either side of the scale.

If you would like to use quantitative (or intensity) references with your panel it will be well worth doing some thorough research before you begin. Read up on scales: Chapter 7 in Lawless and Heymann (2010) and Chapter 3 in Stone et al. (2012) are both excellent resources. Read the various standards and relevant chapters for the method you are considering. For example, if you are planning on running texture profiles you will need to read ISO 11036 Texture Profile (1994-11). This standard has some very useful criteria for references and explains that some of the foods used in its appendix for each of the texture scales may not be available in different countries, may become unavailable in the country of origin, or may change in intensity due to different raw material usage or changes in processing. It suggests that if this is the case, that other products 'should be selected to fill out the scales'. There are several criteria for good references. One of the most important is that the reference product should demonstrate not only the texture attribute being measured (obviously) but also that other attributes of the product should not 'overshadow' the attribute under consideration. The reference product also should be quite stable over shelf life. Other criteria such as easy to store, easy and quick to prepare, easily available and of a constant quality may be more obvious.

If you are trying to decide between different profiling methods, the ASTM standard E1490-11 is an interesting read as it includes an example of a profile for hand cream compiled via Spectrum and QDA methods.

The assessment of references usually occurs in the profiling sessions when the panel are defining each attribute. The panellists produce the list of references they feel would be useful as they are assessing the samples. Well-trained panellists will do this as they describe each sample. There are two options for presenting references. The first involves assessing the qualitative references as required when needed for each discussion. One way to do this is to focus on a particular attribute and ask which references the panellists would like to assess. The panellists assess each reference writing a description of the reference (they could use a different notebook or the back of their current notebook). Once everyone is ready, ask for a show of hands for panellists who think that the reference is suitable for that attribute. If there is little agreement, discuss the descriptions and why the reference was not suitable. Remember that sometimes you may need to try several references until the panellists agree that the note they are measuring is present in the reference as well as the sample(s). The second option is to work through assessing each of the references on the list and to ask the panellists which reference most closely matches the sensations they described. If an attribute is left without a reference, ask the panellists for further suggestions and repeat until the majority of attributes have a reference.

6.3.8   Other training

The next steps in the training of your new panellists involve method training, for example, training in discrimination testing, rapid methods, profiling or temporal methods. Information for this part of the training can be found in Chapter 7.

You may also need to do some training on the data collection device. If panellists have not used a computer or tablet before, this can be quite time consuming, but it's easier to arrange one-to-one sessions with these panellists and go through the different things they need to do, with them 'in the driving seat' so to speak, as just showing them what to do does not mean they will be able to complete the task themselves. For panellists unused to using a mouse, playing solitaire or something similar on the computer can be a fun way of learning. If you are giving panellists a device to use at home you will need to draw up a contract with the panellists: check with your information technology and human resources groups.

Another useful thing to train the panellists to do is to keep a sample comparison chart in their notebooks. One way to do this is to use 'Lauren's bucket method'. Each panellist draws some 'buckets' (essentially three lines to make a bucket, or just draw some circles, no artistic skills required) on a page in their book and they write in the sample numbers that are the most similar, so that those that are different end up in different buckets. They can do this as an overall comparison, by modality and even by attribute or key attributes. This can really help you keep track of the samples and also gives the panellists a 'view' of the sample similarities and differences. See Figure 6.9 for an example. In the example, samples 1 and 6 were found to be quite similar but different from the other four samples. Sample 4 was different to all the other samples, while samples 2, 3 and 5 were found to be similar. Three buckets were needed as there were three different groups of samples, but one bucket may have been enough if all the

Figure 6.9 An example of 'Lauren's buckets' for a six sample descriptive profile.

samples were pretty much the same, or two buckets if five of the samples were similar and one was different. The buckets can also be drawn overlapping if required to show that some samples have some attributes that are similar and some attributes that are not.

Another way to do a sample comparison is to create a simple ranking sheet for the panellists using a table in word processing software or a spreadsheet as shown in Figure 6.10. In this example four columns are needed for the four samples in the test. More rows would be required for all the different modalities and attributes being measured.

Another more detailed way for the panellists to keep a note of sample comparisons is shown in Figure 6.11. This can help give the panellists an idea of how the samples compare in terms of intensity as well as the attributes each sample is described by. Each modality would require more rows: this is just an example to show how each modality is laid out. If the headers do not match the attribute title the panellists can write the correct terminology above the sample numbers as shown for the attribute 'depth of cracks'.

Attribute	Least sample			Most sample
Bitter	4	3		1 and 2
Nutty	3 (none)			1, 2 and 4
Spicy	1, 2 and 4			3
Floral	4	2	3	1

Figure 6.10 Example ranking sheet for recording panellists' initial impressions during the descriptive analysis of four samples.

	Intensity	Not	Slight	Slight to Moderate	Moderate	Moderate to Strong	Strong	Very Strong
APPEARANCE	Shiny	Samples 1, 2, 4 and 5			Sample 3		Sample 6	
	Brown	Sample 6		Sample 3				Sample 1, 2, 4 and 5
	Depth of cracks	None Samples 1, 2, 4 and 5			Moderately deep Sample 6			Very deep Sample 3
AROMA	Overall impact	Sample 6	Sample 3		Sample 5	Sample 2 and 4		Sample 1
FLAVOUR	Overall impact							
TEXTURE								
AFTER TASTE / FEEL								

Project _____ Date _____ Name _____

Figure 6.11 Detailed sample comparison sheet for recording panellists' initial impressions during the descriptive analysis of six samples.

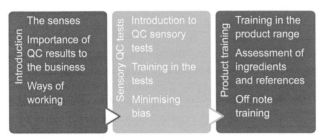

Figure 6.12 Overview of training for a quality control panel.

6.4 Overview of training for a quality control panel

Although QC panels do not often get involved in descriptive work, training a sensory panel for a QC function is similar to that for the training of any other sensory panel, especially if the panel also works on shelf life projects where descriptive work is required. An overview of the training plan is shown in Figure 6.12. Details for the training in each of the sections can be found in Section 6.3. In Chapters 7 and 8 there are some detailed plans for training a sensory panel for a QC function.

If you need to train QC panels across the business, it can be useful to design a training programme that trains the staff at the various sites to run their own training sessions. This is similar to the food hygiene 'train the trainer' ethos and can work very well. You will also need to develop a maintenance and monitoring system to ensure that these panels are not 'out of sight, out of mind'.

6.5 Developing your own training plan

There are several examples of training plans within this book that can be used or adapted for food, beverages, home and personal care products, but you may need to develop your own training plan from scratch if none of these plans quite suit your requirements. Your first step will be to develop the panellists' job description and from there decide what they will need to learn about (see Sections 2.6 and 4.2.3 for more information). Consider how you will validate their learning at the end of each training session as well as at the end of the training period. This might be through a series of discrimination tests, some qualitative descriptive work or via several replications for a quantitative descriptive profile on known samples (see Section 7.7 for more information). Make a list of the information the panellists will need to learn about, but do not forget the items in Sections 6.2 and 6.3, as some of these may prove useful.

Once you have your list of what the panellists need to learn, think about how you might deliver the training to them. This might be in the form of practical sessions, lectures, booth work, quizzes, etc. Write out some ideas on post-its along with a rough idea of when in the training plan the item is required and how long you think it might take. Take some blank A4 sheets, one for each training session (at this stage you can have as many blank A4 sheets/training sessions as you like – you can wheedle them down later) and place the post-its on the 'sessions'. Consider if the tests can be carried out on the same day: there might be limitations on carry over or number of samples that can be assessed. Move the post-its around and juggle the number of training sessions you need. Ask a colleague for help with this as it's very easy to jam far too much into one session which you may not be able to complete.

Once you have a general layout, you can start writing the plan up. A useful way to do this is to create a table for each session with the following headers. It's quite useful to do this in a spreadsheet as each tab can be a session, with the first tab giving the overview plan with dates. More information about each header is given in the paragraphs below.

- training objective – what you hope the panellists will be able to do after this session
- training plan – details of what will be done in the session
- activities – list what you plan to do in that session
- documentation – what you need for the session
- equipment, rooms, products – what you need for the session
- evaluation – how you will judge whether the training was successful
- connections – how this session fits into other sessions.

Let's look at an example. The training objective might be: panellists begin to develop an understanding of discrimination tests. The training plan might be: introduce panellists to discrimination tests, let them experience one type of test three times, discuss how they approached the test, conduct some qualitative descriptions of the pairs of products so that the panellists can really experience the differences and similarities between the pairs. Your activity list may then include something like: recap of training so far (five minutes), time for any questions (5–10 minutes), presentation about discrimination tests (15 minutes), three discrimination tests (30 minutes in total with a five minutes break between each test for any queries before the next test), break (5–10 minutes), group discussion about approach to the discrimination tests, e.g., whether to reassess samples, layout of samples in the booth etc. (20 minutes), qualitative description of each of the pairs of samples from the discrimination tests (45 minutes with a 5–10 minutes break), group discussions about the differences and similarities between the samples (10 minutes), panel discussion about the differences and similarities between the samples (10 minutes), recap and then questions and answers (10 minutes).

Your activities list might then include create presentation, prepare samples for three duo-trio tests and the qualitative assessments and create paperwork. The documents required would then be the presentation, the paper sheets for the duo-trio tests and the qualitative description sheet. It's best to do the tests on paper during training sessions so that the panellists can see their own completed forms and can make their own comparisons between the discrimination test results and their qualitative descriptions. The rooms, equipment and products list will give details of the rooms required, including the booths for the discrimination tests, the products for the duo-trio and qualitative assessments, as well as how the products will be presented (plates, trays, tress holders and number of subjects needed).

How you will evaluate the training is slightly more difficult to document. You might like to record how many panellists matched the correct sample in the duo-trio tests, or maybe just the third test (see Section 7.3). You might like to write down how you felt the group discussions went: was everyone contributing, was everyone enthusiastic and enjoying the session? Or maybe collect and read through the qualitative descriptions to see who gave the best descriptions of the differences between the pairs. Or maybe a mixture of all three.

Once you have your plan written up, check that each item flows into each session. For example, in the session described above, it would be important for the panellists to have learnt how to write qualitative descriptions and how to work in a group discussion before taking part in this session. There are some useful project management tools that can help with this part, but if you have written out your post-its in a logical fashion first you might find using a simple spreadsheet will work just as well. The final thing to complete on the plan is who will do what and when. For example, if you have samples to prepare for session 7 that require treatment of some sort for one week before assessment, you will need to add the date this needs to start and who will be responsible.

6.6   Introducing new panellists to an existing panel

You may have been recruiting internal or external panellists to make up numbers for an existing analytical sensory panel. Or maybe the panellists are to join an existing consumer panel working on the development of new creative product ideas. Either way, the best approach is to train the new panellists separately to the existing panel. This way they would not be overawed by your current panel's ability and lose confidence in their own ability.

It is not an easy task in my experience and there is often some jostling for position and general niggles. It is not always the new panellists who might lose confidence: sometimes the existing panellists may feel that they have forgotten their training or that you are hoping to replace them with new 'better' panellists, so consider both groups of people. Hopefully you will have been able to involve your existing panellists in the recruitment of the new panellists, as this is very motivating for the existing panellists: they will feel they have some say in who gets recruited.

If you can organise it, plan your sessions so that the new and existing panellists get the chance to meet while the new panellists are still training: maybe during break time or arrange a tea and cakes session. This starts the team building off well, especially if a special session has been arranged! Another approach is to run some of the training sessions with the existing panel as well: perhaps introduce a new method you were hoping to use, as this way everyone is learning together and no one is the 'newbie'. You could also invite two or three of the existing panellists along to some of the new panellist training sessions to help with the training. This way they get to meet up and share their experiences.

Asking the existing panellists to volunteer to be a mentor to the new panellists can work well, but be careful in your choice of panellist pairs and decide an end point to the system ahead of time, so that the new panellists are no longer new and the team works as one big team.

If you think that any of your existing panellists might 'play up' when the new panellists join, for example, acting the 'know-it-all' or the reverse, speak to them ahead of time and ask them specifically to do the opposite. Consider what might go wrong before it does: it can be a lot easier to prevent problems than to solve.

Once you think the new panellists are ready to join the existing panel, run the first few sessions with an eye on the new panellists and how they seem to be coping. You will need to keep a close eye on them over the next few weeks. It can be useful to keep two separate datasets by labelling the new panellists in your software as a different panel. This way you will be able to monitor both groups' progress. However, in the panel sessions do not treat the two groups any differently. If you ask an existing panellist to help define a difficult attribute, then for the next attribute that needs defining ask a new panellist. This way the 'newbies' will gradually merge into the existing panel and you may find you cannnot quite recall who was new after a few sessions.

Method training for sensory panels 7

7.1 Introduction

This chapter is again split into various sections so that you can look specifically for what you need depending on the panel you have recruited and the work you intend them to be involved in. Instructing the panel about the various different categories of tests in sensory science, for example, discrimination, descriptive and temporal, can be helpful in showing them the big picture and why they need training in these various tests. Demonstrating the results of these tests, for example, showing them what a profile might look like by the time you have analysed the data, can be beneficial as it shows the panellists where they are heading for.

Training your new recruits will take some time, and the approach and the time taken will depend on the methods that you plan to be using. Training is generally not required for consumer panels who are being asked about liking, emotions or preference; however, if you plan to conduct discrimination tests with consumers it's advisable to familiarise them with the test you are asking them to conduct using a dummy test first (see Section 7.3 below for more information). Training panellists to take part in descriptive or time-intensity (TI) work will require more time than training panellists in discrimination tests or quality control tests, for example. It's a good idea to plan your approach to the panel training before you start; however, keep your plan flexible as you may have to change things as you go along. For more information about planning the training sessions, see Section 6.5.

When you are running any training sessions, particularly with an existing panel, it can be a good plan to not divulge the fact that you are running a training session. For example, let's say you have a set of products that you use for validation of new panellists and for regular validation of the existing panellists. You have six products in this set that demonstrate various differences and similarities across the range of attributes your panel assesses for that product type. If you were to tell the panellists that you were about to run a training or validation profile or assessment, they may well pay more attention to the test, recall differences between the samples from last time or generally act in a way to make the whole test a waste of time. Therefore, it is better to just introduce it as a new project: mix up the sample numbers, repeat a sample so that there are seven samples in the set and start the testing process from scratch.

There are several reasons for conducting method training with sensory panellists prior to using any of their data to make business decisions. The data will be more robust as the panellists will understand the task they have been asked to do. The training will be motivating and will encourage the panellists to give you the best data possible. You will understand a huge amount about each individual panellist's capabilities both from a sensory and personality point of view.

Sensory Panel Management. https://doi.org/10.1016/B978-0-08-101001-3.00007-0

7.2   Training the panel how to assess products/how to use an existing lexicon

For some modalities it can be useful to train the panellists so that they can assess each sample successfully without too much fatigue or adaptation. It's important to explain to the panellists that repeated assessments can make the aroma or flavour of the next sample appear weaker or can even prevent them from detecting weaker aspects of the product. For example, in the assessment of odours of any type, the instructions as shown in Figure 7.1 can be useful.

You will also need to explain the importance of deciding, agreeing and following the protocol for the assessments of all types of samples for all methods. The amount of product consumed or used needs to be similar for all panellists and needs to be adequate to make the assessments. Generally, the number of chews or the number of rubs between the fingers also needs to be standardised to enable the measurements to be comparable. It can also be helpful to explain about the importance of cleansing between samples and that the protocol needs to be followed to enable the results to be valid and reliable. If you do all your assessments in the panel discussion room but all your data gathering in the booth room, remember to do a trial run in the booths prior to the actual experiment: the appearance attributes may well change in the booths. This can also give you the chance to try out the sample preparation for delivery to the booths as this can sometimes cause issues for the panel technician or person preparing the samples.

To help a profiling panel develop excellent attributes it can be useful to give them the handout as shown in Figure 7.2.

For a panel that works at home it can be useful to create a pack for them that includes contact details, printouts of the protocols for assessment, the attribute list, photos or videos to demonstrate the assessments and any other handy reminders to keep them on track.

- Remove the cap from the first bottle and gently sniff in the space above the bottle. If you can detect and describe the odour, replace the cap and write down your description in the box next to the code of the sample you assessed.

- If you cannot detect any odour, bring the bottle a little closer to your nose and again, sniff gently. If you can detect and describe the odour, replace the cap and write down your description in the box next to the code of the sample you assessed.

- If you cannot detect any odour, bring the bottle under your nose and sniff gently. If you can detect and describe the odour, replace the cap and write down your description. If you cannot detect any odour, replace the cap and move onto the next bottle.

- DO NOT sniff too hard if you cannot detect an odour, as this may affect your ability to detect the odours in the later bottles.

- Remember to replace the cap on each bottle before moving to the next.

Figure 7.1 How to assess odours.

o   The **attribute** is the word that summarises the description of the appearance, aroma, flavour and texture or aftertaste. For example, white colour, caramel aroma, moistness of internal crumb, butter flavour...

o   The **reference** helps us identify and agree exactly what we are measuring for each attribute. For example, what *type* of orange flavour.

o   The **definition** helps everyone understand what the attribute means – not just us but the person who will be reading the report! For example, caramel aroma: the sweet and toffee-like aroma associated with lightly cooked granulated sugar (reference).

o   The **protocol** describes the actions prior to the assessment of the intensity of the attribute. For example, for the assessment of vegetables: Aroma: assessed from the bowl – cup the bowl in the hands and bring to the nose and take small bunny sniffs. Assess the intensity of the aroma (no cutting – assess the product whole).

o   We also need **anchors** for the ends of the scales such as 'not' and 'very' or 'light' to 'dark' so that we know which direction the scale goes and therefore how to rate the samples.

o   Finally we need to check that the **order** we have listed the attributes will work. Because we wrote up the attributes in the order they appeared, it should all work OK but sometimes we need to shift words around for practical reasons.

Figure 7.2 Panellist handout describing the elements of a sensory profile.

In some profiling methods such as the Texture Profile, you may need to train the panellists how to make the assessments according to each of the defined attributes. Civille and Szczesniak (1973) published guidelines on how to train a texture panel. They state that the initial training for a texture profile panel assessing different product types starts with two weeks of orientation sessions (around two to three hours in length) and then around six months of hourly practice sessions four to five times a week (in excess of 110 hours in total).

If you are training panellists to join an existing panel, you may need to train them on how to assess the attributes that the existing panel have created the definitions and protocols for. You may even have a lexicon from a client or from the literature that you would like to work to. However, remember that if the panellists have generated, defined and referenced the attribute list themselves, they will have more chance of using it successfully (Lyon, 2002).

Start by giving out the lexicon and asking the panellists to read it through first. Then start with the first part of the protocol and ask a panellist to actually read out the instructions. You could, for example, ask each panellist to read an attribute one at a time and discuss as you go through each attribute. This ensures that they have actually read all the text and, in listening to the instructions, it helps other panellists realise that they do not understand exactly what happens when. This can be quite interesting when you realise that the protocol does not always make complete sense to people who were not there when it was devised.

Next step is to give out some products to demonstrate the protocol in action. If you are assessing a home and personal care product, you might actually be able to demonstrate the protocol to the panellists. Using photos and videos from previous panel work where this exists, can be very helpful in training new panellists. Allow the panellists time to try a range of products, ask questions and discuss any disagreements.

Next step is to ask everyone to give the assessment a try and if possible watch them to ensure that they are actually following the protocol. You might be surprised by how many who simply do not, but this step is critical to get good data. Allow the panellists to ask questions and practise the assessments. You can then move through the attribute list, checking, demonstrating and trying each assessment stage and attribute until everyone is following the protocol and understands each of the attributes.

Once all the panellists are happy with the procedures for evaluation and you are happy that they are actually assessing the product in the correct way, you can then start by conducting some profiles of products. The panellists can then learn and practise using the attribute list, the order the attributes will be assessed in, the definitions and details about how much product will be assessed and how. For the Texture Profile method, this is where the panellists develop their own protocol, scales and select references suitable for the product they will be assessing (Civille and Szczesniak, 1973, p. 214). This part of the method is similar to the generation of attributes, methods of assessment, and qualitative references for the Quantitative Descriptive Analysis (QDA) method.

The next step involves training the panel to use the scale for your product and the description in Civille and Szczesniak (1973) is very helpful. It suggests starting by ranking two or three samples on a selection of important attributes (around five or six). Once the panel agree on the rank order of these two or three samples they are introduced to the scale used in the Flavour Profile method which has just five points: 0 not detectable to 3 strong, large (there is an additional point which is called 'just detectable/threshold'). The same samples are now rated using that scale. Once the panel are in agreement and can rate the samples consistently, they can move to an expanded scale and will probably request to do so. The scale still runs from 0 to 3 but has the added points:

)(–1 just detectable to slight
1–2 slight to moderate
2–3 moderate to large/strong.

Once the panel has practiced using this scale they can then move on to the 14-point scale, which still goes from 0 to 3 but includes steps such as 'between moderate to strong but closer to moderate'. Civille and Szczesniak (1973) do not mention if more products are brought in at this time, but I would assume this would be the case. This 14-point scale is not specifically mentioned in the ISO standard for the Texture Profile method (ISO 11036, 1994). The standard only refers to the scales developed for basic texture profile method (e.g., the hardness scale from 1 to 9) and refers to scales 'developed for a specific product including varieties of that product' (see 9.3 p. 9 of the standard).

For more information about references see Section 6.3.7 and Section 7.4.2, and for information about descriptive profiling methods see Section 7.4 and Chapter 8.

7.3   Training assessors for sensory discrimination tests

There are two main elements to the training of sensory panellists to take part in discrimination tests: training in the various discrimination tests themselves and giving the panellists experience in discriminating so they can develop their own approach to the assessment procedure. The first aspect is relatively easy: the panellist is taken through the steps in each test and given the chance to ask questions and then experience the test for themselves. The second aspect is not so easy to teach as such, as it is all about experience in the tests and the panellist's personality, but advice can be given. For example, some panellists will tend to assess the samples presented once only and decide on their answer fairly quickly. Others may assess the samples two or three times (where this is allowed) prior to making their judgement. Let's start by describing the training for some new panellists who will be taking part in a wide range of sensory discrimination tests.

There are 20 or more discrimination tests in common use, but many of them follow a similar pattern and therefore training in a selection of these tests will give the panellists a good grounding and will allow them to develop their own approach to discriminating. Table 7.1 (Rogers, 2017) gives an overview of sensory discrimination tests. One of the main differences between the sensory discrimination tests is related to whether the test has a specified attribute (e.g., sweetness, smoothness) or not. For example, in the 4-alternative-forced-choice test (4-AFC, see Figure 7.3) the panellist might be asked which of the four samples is the most *bitter*, for example, (the attribute of interest, *bitter*, is specified) while the dual-pair (see Figure 7.3) is an unspecified test and hence asks which pair contains the different pair of samples. In the tetrad, the assessor is asked to sort the samples into two similar pairs in the unspecified version, while the specified version of the same test will ask the assessor to group samples based on a specific attribute. Having specified and unspecified versions of most tests takes us to more than 40 different named tests, but many tests are very similar as they are based on the same principle. All that really varies for the assessor is the number of samples, their task and whether or not a reference sample is identified; therefore, discrimination tests can be grouped in a number of different ways. For example

- Type: whether they involve a *specified* attribute such as sweetness or if they are *unspecified*. Unspecified tests are also known as 'overall discrimination tests';
- Reference: if a reference or control sample is identified in the test;
- Task/action: the manner in which the assessor makes the judgement: answering **yes/no**, e.g., same-different test and A-not-A, **matching**, e.g., to a reference, e.g., duo-trio; **oddity**, e.g., picking the different or odd sample, e.g., triangle; **choosing**, e.g., the most intense sample or the different pair; or **sorting**, e.g., putting samples into groups (Gridgeman, 1959);
- The number of samples presented: from 1 to 12, e.g., 1 sample in the A-not-A test through to 12 samples in the six-out-of-twelve test;
- The number of products presented: the majority of tests involve two products, however, tests such as ranking, difference from control and polyhedral tests can contain any number (within reason);

Table 7.1 An overview of sensory discrimination tests sorted by type. The number after each test name gives an idea of the panellist task but should not be used to develop the panellist questionnaire: check the relevant literature for the exact wording for each method. The ellipses (…) indicate that the sequence can be continued where relevant for the product type and experimental objectives

m-AFC (all specified)	x out of y (can be specified or unspecified) x can equal y, i.e., symmetrical samples (AABB) or the samples can be asymmetrical (AAAAAB)	'Reference' (all unspecified)	Response bias (all unspecified)
		A-not-A (5)	A-not-A (5)
		A-not-A-R[a] (5)	A-not-A-R (5)
2-AFC (1)			Same–different (7)
3-AFC (1)	1 out of 3/triangle (2)	ABX[b] (6)	
		Duo-trio/2-AFC-R (6)	
		Reference plus 3 (6)	
		Reference plus 4 (6)	
		Reference plus 5 (6)	
		…	
4-AFC (1)	1 out of 4 (2)	Dual standard (6)	
5-AFC (1)	1 out of 5 (2)	2 References plus 3 (6)	
6-AFC (1)	1 out of 6 (2)	2 References plus 4 (6)	
7-AFC (1)	1 out of 7 (2)	2 References plus 5 (6)	
8-AFC (1)	…	…	
9-AFC (1)			
10-AFC (1)			
…			
	Multiple standards (3)		
	2 out of 4/tetrad (3)		
	2 out of 5 (3)		
	3 out of 6/hexagon (3)		
	3 out of 7 (3)		
	4 out of 8/octad/double tetrad/Harris–Kalmus (3)		
	…		
	Dual-pair/4 interval AX (4)		
		Difference from control[c]	
Example panellist question[d]: (1) which sample is the most bitter?	Example panellist question: (2) which sample is the odd one? (3) sort into x groups of y (4) which pair contains the different pair of samples?	Example panellist question: (5) Is the sample A or not A? (6) which sample matches the reference(s)?	Example panellist question: (7) are the samples the same or different?

[a]When the reference or reminder is present in the test.
[b]No labelled reference is provided – the two initial-coded samples serve as blind references.
[c]Generally unspecified but can be specified by attribute or by modality.
[d]These questions simply summarise the panellist task: they should not be used to develop panellist questionnaires. Please check the relevant chapter or literature for the exact wording for each method.

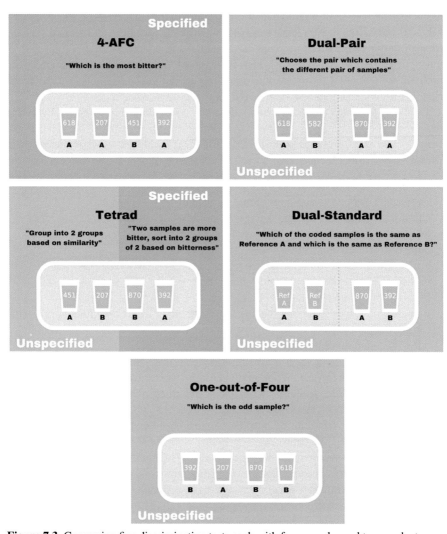

Figure 7.3 Comparing five discrimination tests each with four samples and two products.

- Whether or not there is a response bias (the differences between panellists in terms of their criterion for stating whether there was a difference or not) associated with the test, e.g., same-different test and A-not-A;
- Whether some form of rating scale is included as part of the methodology.

For example, all five tests in Figure 7.3 involve the comparison of two products with the use of four samples. The dual-standard, in contrast to the other four tests, is quite different, as it is the only test to contain any identified references. The assessor's task in the dual-standard is to *match* each of the two coded samples to a different reference sample. In the tetrad, which can be specified and unspecified, the assessor's task is to *group* the samples into two similar groups, while in the 4-AFC the assessor's task is to *choose* the most intense sample for a specified attribute. In the dual-pair the assessor is presented with two pairs of samples, both coded: a matched pair and an unmatched pair, and the assessor's task is to *choose* the pair that is unmatched. And finally in the 1 out of 4 the assessor is asked to pick out the *odd* sample (similar to the triangle test, which could also be referred to as a 1 out of 3 test).

Training the panellists in the various tests can be conducted quite easily by giving them an idea of the task for each test and allowing them the time to experience and practice. The use of samples with known differences is the best approach as you will be able to give direct feedback after each test. For example, (see also Table 5.5):

- combinations of basic tastes: for example, sucrose solutions with added artificial sweeteners or mixtures of sucrose and citric acid
- soft drinks where more sugar, acid or flavour has been added
- face creams: night and day versions of the same cream that look the same but feel quite different on application
- drinks which have been diluted with water or another flavoured drink: for example, orange juice with 10% apple juice added, lager diluted with water
- biscuits where one product has been left to stale for two to three hours
- tissues: more expensive brand versus a cheaper brand where they look very similar (or get the panellists to make the assessments behind a screen) or the same brand but with added ingredients (e.g., balsam)
- food products where there is a low fat, low sugar, low salt alternative
- food mixes where the production method can be varied: for example, custard made from powder can be made with full fat or skimmed milk, gravy powder can be made with 10% more/less water
- products where a taint can be introduced via storage or the addition of taints via commercially prepared capsules
- products where a supermarket or store brand looks identical to the branded product
- products produced specifically for discrimination test training, e.g., products prepared plus or minus a particular ingredient, products cooked or processed for slightly longer than standard, and products prepared to a slightly different formulation to standard.

Once the panellists have had the chance to try a particular discrimination test two or three times, they will have been able to understand the task at hand and begin to develop their own technique to get the best results for these tests.

Coloured lighting can sometimes be used to hide differences in appearance between the products in a discrimination test, but be careful that the lighting does not create some other clue for the panellists to pick up on. One approach can be to set up one or two booths with the test you are planning and ask members of staff (who are have no issues with colour assessments) if they can complete the test by appearance alone. For example, if they can repeatedly pick the correct odd sample in a triangle test or match the right coded sample to the reference in a duo-trio, under white light by appearance alone, you can change the presentation order and ask them again under each of the coloured lights you have access to. Many facilities have white light, red light and blue light. Green light can also be useful.

7.4 Training assessors for sensory descriptive profiling techniques

7.4.1 An introduction to the various sensory descriptive profiling methods

Descriptive methods qualitatively describe the various sensory attributes of a product and can also quantitatively measure the intensity of these attributes, either at the time of assessment or over time when temporal methods are being used. A small number of (often) screened and (often) trained panellists develop or use a vocabulary to measure certain aspects of the product being assessed: for example, the greasiness of a lotion or the sweetness of a fruit pie. The scales used by the panellists are actually based on a psychophysical model and we make the assumption that panellists can use scales to tell us about their experience in a quantitative way. Some element in the product (often called the stimulus) is perceived by the panellist. This is the 'psychophysical process': *psycho* – the mind or mental processes, and *physical* – the physical thing being assessed (soft drink, lotion, fruit pie, etc.). The panellist considers (thinks about) this sensation prior to scaling the intensity. So it is really a two-step process. The panellists do not measure the amount of sugar in a solution: they measure the sensations they perceive. This is why the manner in which the perception is measured can change the measurement: just think about the recent publications on the theme of cup or plate colour and size, room lighting and study context and their impact on the resulting sensory measurements.

There are many applications of descriptive techniques but perhaps the main applications are in new product development, product reformulations, sensory research, monitoring the competition, shelf life assessments and quality control. The methods are often referred to as profiling, descriptive analysis (or DA) or by the published method name. The main published methods are shown in Figure 7.4 in date order. Many sensory teams use their own version of 'profiling', sometimes adding in elements of different methods, so it can be difficult to fully understand how each profile was created. Often in publications the descriptive analysis is not described in a manner which would allow you to replicate the method.

Flavour Profile (Cairncross & Sjostrom, 1950)

Texture Profile (Brandt et al., 1963)

Spectrum Descriptive Analysis Method (method developed in the 1970s, formally named Spectrum Descriptive Analysis Method in 1986) (Munoz, Civille and Carr, 2016)

Quantitative Descriptive Analysis (QDA, Stone and Sidel 1974)

Free Choice Profiling (Williams and Langron, 1984)

Rapid Methods (Napping, Sorting, Flash...)

Figure 7.4 Published descriptive methods.

For example session plans for running qualitative and quantitative descriptive profiles please see Chapter 8.

Each of the descriptive techniques has its own approach to creating the product profile and they are briefly described below. A selection of the rapid methods is also described. Note that although these methods are often referred to as rapid descriptive methods, not all of them are actually describing the products, but they do allow a comparison across products in various ways. Also, although the methods are often classified as rapid in relation to the standard sensory profiling methods such as QDA and Spectrum, they are not necessarily more rapid than conducting a profile with a ready trained and experienced profiling panel (Stone, 2015). A great review of these alternate methods is given by Valentin et al. (2012).

If you require more information about the various methods please consult the standard sensory textbooks or the *ASTM Manual on Descriptive Analysis Tests for Sensory Evaluation* (Hootman, 1992) for the first four methods in Figure 7.4.

7.4.1.1   Flavor profile (Cairncross and Sjostrom, 1950; Hootman, 1992)

This was the first published method for the creation of food product profiles. Around six to eight (a minimum of four) very experienced panellists with around six months of training, prepare a consensus profile for a product for aroma, flavour and aftertaste attributes. The panellists are screened using basic taste tests, odour identifications, ranking and a personal interview. The panel create the list of attributes and select references that demonstrate particular aroma and flavour characteristics. The panel leader also takes part in the assessments. The scale used has seven points and ranges from 0 (not present) to 3 (strong) with half points. The amplitude or overall impression of the aroma and flavour is also measured to take into account the blend/balance and body/fullness of the flavour. Panellists work individually to measure the attributes and over a period of three to five sessions refine the information until a final agreement on the profile, or consensus, of the product is produced. An adaptation of the Flavor Profile method is called Profile Attribute Analysis which allows the assessment of several samples and statistical analysis of the data (Gacula, 1997).

7.4.1.2   Texture profile (Brandt et al., 1963; ISO 11036, 1994; Hootman, 1992)

This method is quite similar to the Flavor Profile method, but the focus is on the textural attributes of food products such as the mechanical (i.e., the response of the product to stress, e.g., hardness, chewiness), geometric (i.e., the size, shape and particle composition, e.g., crumbliness, grittiness, flakiness) and mouthfeel (i.e., surface attributes, e.g., oiliness, greasiness, moistness) characteristics. The texture of the food is measured from first bite through to finish. These characteristics can be measured by appearance, touch (e.g., by hands or lips), in the mouth or after swallowing dependent on the actual product being assessed. These aspects are measured by either kinaesthesis (e.g., through muscle and nerve feedback) or somaesthesis (e.g., through touch). Originally the scale was similar to the Flavor Profile scale but is more often a 10- or

15-point scale. The attributes can be taken from a predefined list or created/edited by the panellists. Between six and 10 experienced panellists (around four to six months of training is required) measure chosen texture attributes from a ready-prepared list on standard reference scales that cover the entire range for a particular measurement. For example, the standard hardness scale goes from 1.0 (cream cheese) through 7.0 (frankfurters) to 14.5 (hard candy). These reference products were checked via instrumental analysis to place them on the scale.

Panellists are screened based on their ability to discriminate textural attributes, as well as a personal interview. It's important the panellists are all trained in the same way so that the frame of reference (both qualitatively and quantitatively) is the same for all panellists. The panel leader also takes part in the assessments as long as they are unaware of the project objective and the identification of the samples in the test. In the original version of the method, consensus scores were gathered after several sessions, however, nowadays individual scores are collected from the panellists and subjected to statistical analysis.

7.4.1.3   Spectrum descriptive analysis (Meilgaard et al., 2016)

This method was developed by Gail Vance Civille in the 1970s based on her experience with the previous two profiling methods and hence has many similarities to these two methods, which are all quite different in principle from QDA. Eight to 12 highly trained panellists take part in creating Spectrum profiles of both food and any other products (e.g., home and personal care products, sounds of cars, environmental odours). The panellists are screened using basic tastes and odour identification, as well as scaling tests based on, for example, appearance, basic tastes and/or texture, where the potential panellists are familiarised with two intensities and asked to scale a selection of unknowns. For non-foods, a similar approach is used for scaling handfeel or sound dependent on the products the panel will be working on. Sometimes product-specific acuity tests are also carried out using 10–15 discrimination tests such as the duo-trio or triangle. The panel leader is also trained as a panellist and provides training for the panellists on the product ingredients. Published lists of attributes are available and panellists are able to create their own as required. Product assessments are independently made in sensory booths and statistical analysis such as Analysis of Variance and multivariate methods are used to assess the panellists' output; however, consensus data can also be collected if desired.

The panellists are ready to begin work after around three months of training (Hootman, 1992), but this will depend on the modalities the panel need to be trained in and the complexity of the product. Rutledge and Hudson (1990) describe the process of training a panel for Spectrum analysis as taking a period of around five months (175 hours) and the collection of several hundred samples for the training process.

There are many similarities to the QDA method, but where the main difference arises is in the use of quantitative references. Both Spectrum and QDA use *qualitative* references where required, to ensure that the panellists are in agreement about which element of the product they are actually going to measure, but only Spectrum uses quantitative references. These quantitative references mean that all panellists will agree that a particular reference is, say 5.0, on the 15-point scale. This means that an attribute for a certain product would be scored the same for the same product by all panellists. Therefore, comparisons across many types of products and panels can be made. For example, you could produce a profile of one product and compare that product to a product profile you created some time back. It also means that the measurements for different attributes are also comparable: a '5' on one scale will have the same *intensity meaning* as a '5' for a completely unrelated attribute. Rutledge and Hudson (1990, p. 81) describe this: 'a panellist would ask "Is the egg flavour in this mayonnaise as strong in intensity as the 5.0 apple flavour in the reference apple sauce?"' However, Lawless and Heymann state that, as there are no published data to support the use of these absolute quantified references, they are 'somewhat sceptical of this equi-intensity claim'. They also mention, in regard to the calibration by quantitative references, that, 'We are not sure that this level of calibration can be achieved in reality'.

7.4.1.4   Quantitative descriptive analysis (Stone et al., 1974; Hootman, 1992)

The details for this method were published in 1974 although the method had been in use for some time before this. QDA was developed to help alleviate some of the issues the authors felt to exist with the first two previously published methods, such as consensus scoring, modalities assessed and choice of panellists. Generally, 10 to 12 panellists take part in QDA although this can range from eight to 15 depending on the experience level. Panellists are screened based on their product use, interest, communication skills, discrimination ability and consistency. Sensory screening is performed by the use of around 15–20 repeated discrimination tests (paired comparison or duo-trio) starting with easy differences and becoming more difficult. The panel leader acts as a moderator and trainer and does not take part in the sensory assessments.

The first step in the training is to develop the scorecard or attribute list for the product type. This is a common language that describes the panellists' perceptions of the products in the order that the attributes are perceived. The attribute list also includes the protocol for the assessment of the products so that all panellists are in agreement as to how the product should be assessed, as well as references, such as product ingredients, to help the panellists agree on the attribute being assessed. The number of sessions needed to create the attribute list will depend on the product complexity and the panellists' experience with the method.

The scale used is an interval scale, originally 6 in. in length (around 15 cm) with the anchor points placed 1.25 cm from each end where the word anchors such as weak–strong or not-very are placed. In the original paper, a midway anchor point was used and this was labelled 'moderate'. Panellists are trained to use the scale by being presented with the range of product intensities and they are encouraged to use the whole scale. During product testing the words used as the anchors for the scale demonstrate the scaling boundaries (Stone et al., 1974). Data are collected from the panellists' individual assessments in the booths which are repeated several times: the exact number is dependent on the project requirements, panellists' experience and the product type.

The authors state that they expect there to be significant differences between the panellists in their mean scores: panellists may use the part of the scale they wish to use as long as they are consistent within themselves. The variability between the panellists is not an issue as it is covered by the use of replications and in the statistical analysis: the ANOVA allows the panellist effect to be partitioned from the product effect, and the assessment of the panellist*product interaction effect allows us to document the agreement between the panellists about the rank order of the products for each attribute.

Stone (2015) states that a panel can be ready to start work on QDA profiling after around two weeks or nine panel sessions: three for screening, five for the language development and one session for a 'pretest'. If you are only going to use the panel once, this seems like a reasonable approach, however, if you plan to use your panel for a variety of projects and hope to employ them for several years, a more detailed training plan such as that described in Chapter 6 might be advisable.

A very good document that allows a comparison to be made between the Spectrum and QDA methods of profiling is one of the ASTM standards: E1490 – 11 *Two Sensory Descriptive Analysis Approaches for Skin Creams and Lotions*. This document is especially helpful if you are deciding between these two profiling methods.

There are many other methods that are used for product profiling. Some of the methods are 'static': they give a profile of the product at one point, and others are more temporal in nature. Dehlholm (2012) gives a very useful and nicely laid out time-sequence diagram for descriptive methodologies. Many of these methods are described in detail in the original paper by the author of the method, in sensory textbooks and sensory standards. A short introduction for each method and various references are given here.

7.4.1.5   Deviation from reference profile

The deviation from reference profile (or relative to reference) method can be helpful, especially in a quality control or matching type project (Larson-Powers and Pangborn 1978; ISO 13299, 2016). Samples are presented in pairs in comparison to a reference product following a similar procedure to a quantitative descriptive profile. Panel selection, screening and training are also similar to a quantitative descriptive profile. The attributes, generated by the panel or devised beforehand, are rated using structured or unstructured scales. There are two options for collecting the data for this method and in both methods replicated data are collected. In the first, the reference is marked on the scale (generally in the middle), and the sample is rated relative to the mark and to the reference, which is presented and identified as the reference at the same time as the sample. In the second, two scales are presented for each attribute, and the panellist is unaware which of the samples is the reference. They rate the sample and the reference sample consecutively and the 'comparison' to the reference is then calculated by the experimenter.

The reference is generally included as an additional sample as a 'blind reference' to check panel performance. The reference sample is selected based on its consistency and score: it should be in the middle of the sample set in terms of intensity. This can be difficult to achieve for all attributes, and a different reference could be used for each modality or for certain key attributes. Analysis of the data concentrates on the differences between the test samples and the reference(s). Certain aspects of the standard quantitative descriptive profile analysis methods can be used, depending on how the experimental data were collected.

7.4.1.6 Free choice profiling

Free choice profiling (FCP) was developed in the 1980s as a method for collecting profiling data from consumers, bypassing the need for training (Williams and Langron, 1984). The authors were also interested in comparing the consumers' use of language for the same element of the product, and if this is your method objective, this test can be really helpful in meeting your aim. Consumers are generally screened and selected based on their descriptive abilities and product usage. Trained panellists can also take part in this method. Each panellist (around 10 in total) creates their own list of attributes and then rates them on a scale of their choice (or a scale determined by the experimenter). Attributes generally are not defined or elaborated, as there is no need for consensus or agreement.

Sessions for the generation of the attributes can be held for each panellist: the only restriction being on the preparation and consistency of the products to be evaluated. There are generally three or four sessions: one to generate attributes and two or three to collect the replicated data. Obviously, this will depend on the number of samples and if it is possible to assess them all in one session. Samples are presented for rating one at a time in a similar way to a quantitative descriptive profile.

The analysis of the data is complex and time-consuming. Generalised Procrustes Analysis (GPA) is required to account for the different attributes and scale use. The output from the analysis is similar to a Principal Component Analysis (PCA) map. If you are interested in trying out FCP there is a useful description of the steps you need to take, as well as some very handy tips, available from the Sensory Dimensions website (Sensory Dimensions, 2017).

7.4.1.7 Flash profiling

The flash profile is a version of the free choice profile (FCP) but instead of using a scale, the products are ranked (Sieffermann, 2000). Many of the aspects of FCP are carried over into this method. Panellists are screened and selected based on their descriptive ability and their use of the product to be analysed. The first session includes some information about the method and then the panellist is given the sample set to assess. All samples are presented at once (one of the main differences to FCP) so that the panellists can directly compare and sort the samples. The panellist again generates their own attribute list, but attribute lists may be shared to allow panellists to develop more attributes. Samples are then ranked for each chosen attribute. This is generally replicated by including sample repeats or running another replicate session.

One of the main limitations of the method is in the need to present all samples at once. This can be more difficult for hot and cold products or for products that are fatiguing to assess. Different groups of products may need to be assessed when there are several samples to assess, as ranking more than around six samples at one time can be quite difficult and confusing, particularly for untrained panellists. Data analysis is again by GPA, and it can be quite difficult to interpret the results.

7.4.1.8   Polarised sensory positioning

Polarised sensory positioning (PSP) was developed for the rapid assessment of water by consumers (Teillet et al., 2010), and there are several adaptations of the method. The method involves the selection of 'poles' or references for the sample set to be compared to. Three poles are the recommended minimum, but more poles could be incorporated. The poles should represent the differences seen across the sample set (Teillet, 2015). The consumers are asked to score each sample in relation to the pole, on a scale from the 'same taste' to a 'totally different taste' or asked which pole a sample is the most similar to and which pole a sample is the most dissimilar to. The data can be analysed with multidimensional scale unfolding or Statis.

There is a good discussion about poles, which poles to use and whether the choice of pole has an impact on the resultant data, in the PSP chapter of *Rapid Sensory Profiling Techniques* (Teillet, 2015, p. 223).

7.4.1.9   Pivot profile

Pivot profile was first presented in 2007 as part of a PhD thesis and published in 2015 (Thuillier et al.). The method is similar in some ways to the free choice profile and relative-to-reference methods, as it combines elements of both. Panellists are generally trained, but can be untrained, but do need to be screened with a good descriptive ability for the sample set. Samples are presented with a reference sample, and the panellist is asked to list the attributes that are less than the reference (called the 'pivot' hence the name) or more than the reference. Panellists are asked to use words that can be analysed using text recognition and therefore they must be allowed to be prefixed with 'less' or 'more'. If a panellist used the term 'not greasy', for example, this would not be helpful as 'less not greasy' and 'more not greasy' do not make sense.

The initial list of words used are then grouped by the experimenter and checked by the panel. The number of times a descriptor is 'less than' or 'more than' the reference for each sample is counted, and the frequencies of 'less thans' are subtracted from the 'more thans'. The data are then converted to positive values and analysed using Correspondence Analysis to produce a product map.

The choice of pivot/reference product has some of the same difficulties as for the deviation from reference profile and has been the focus of a subsequent publication (Lelièvre-Desmasa et al., 2017).

7.4.1.10   Temporal dominance of sensations

Temporal dominance of sensations (TDS) is a temporal method for the creation of product profiles (Pineau et al., 2003). It is quite different to time intensity measurements (TI) because it records several attributes at one time and does not record intensity per se. The 'dominant' sensation is reported by the panellist over the period of the product assessment by selection of the relevant attribute from a list. This might change rapidly or may stay fairly static over the assessment period, depending on the product. The dominant sensation is not necessarily the most intense sensation but the one that catches the panellists' attention at the time. The output is a curve showing the dominance rate for each attribute from all panellists over time. Attributes can be selected from previous quantitative descriptive profiles or chosen from a qualitative description by the same panellists.

Panellists are often trained and generally there are more panellists required for TDS than for other descriptive methods. Training is often conducted on the attributes with reference standards to help identification and on the protocol for assessment. No training is required on the use of a scale, although the first TDS experiments included scales. Several replicates are performed to collect the data although this depends on the number of panellists in the study. For example, with a minimum of 30 panellists one replicate can be performed and with a minimum of 10 panellists, four replicates are required (Pineau and Schlich, 2015).

There has been a lot of interest in the use of TDS, and the technique has been investigated with various products and analysis methods (see for example, Ng et al., 2012; Albert et al., 2012; Galmarini et al., 2017). TDS has also been performed with naïve consumers (Meyners, 2016) with a short training session. The training involved information about the method and how to use the computer system as well as experience with the product and attribute assessment.

7.4.1.11   Progressive profiling and sequential profiling

In progressive profiling (Jack et al., 1994) the panellists assess the intensity of several attributes at certain time intervals rather than over a continuous period as for TI or TDS. The time intervals can be set depending on the product to be assessed. For example, in an assessment of chewing gum in-home, four attributes were measured at intervals of 45 seconds over a 10 minutes time period (Galmarini et al., 2016). The time period can be quite extensive depending on the product to be assessed. Panellists tend to be trained so that the protocol for assessment is the same for each panellist.

The method was adapted for an assessment of oral nutrition supplements to include multiple consumption points to determine the impact of 'build-up' attributes such as mouth drying and mouthcoating (Methven et al., 2010). This sequential profiling method might be useful for the assessment of home and personal care products such as hair waxes and conditioners to determine the effects of repeated use.

7.4.1.12   Ideal profile method

The ideal profile method was derived from consumer methods to develop the ideal product by asking consumers to rate their ideal or suggest improvements (Worch et al., 2013). It can be useful to understand consumers' opinions to optimise or develop new products. Consumers are asked to rate the intensity of each attribute and the *ideal* intensity of the attribute, as well as the usual acceptance type questions. In this way each line scale presented to the consumers is paired: the measurement of the attribute then followed by the measurement of the ideal level. The method has been trialled with the assessment of eight skin creams (Worch et al., 2014), and this paper is a useful starter if you are considering using this method.

7.4.2   Language development

7.4.2.1   Introduction

One of the first things to train panellists in can actually be one of the most difficult: the difference between objective and subjective judgements. It can take some time before the panellists get used to leaving their likes and dislikes to one side and learn to describe all the different elements of a product. For example, if the panellists were learning to describe cola drinks, the first time they assess the range of products, they might be discussing which one is 'their' brand and might only be able to describe the flavour as 'cola'. Only once they are introduced to the various flavour attributes of cola, for example, citrus, caramel, spice, and have had the chance to try the various references relating to these flavour attributes, can they be expected to give accurate objective descriptions. It is very difficult for consumers to describe something that they do not know the name of and because of this they also have the tendency to describe just the first level of the product: for example, cola and not the notes that make up the cola flavour. There are lots of aspects to a product's odour, flavour and texture that consumers may well be sensitive to but which they find very difficult to articulate.

During your training sessions, the best way to check if the learning is working is to check the data and outputs. But it can also be very useful to listen to the panellists discussing amongst themselves and their communication with you. If someone does not appear to be communicating well in the team, it might help to have a discussion with them one-to-one. This way you can determine if they are just a little nervous to voice their opinions in front of the group or if there is another reason. You might be able to boost their confidence by mentioning some of their data or a comment that they did make.

If the panel are going to be developing profiles and this involves creating lists of attributes (or lexicons) with definitions, you might like to demonstrate the importance of the language used for these elements to the new panellists. Language development can also be critical for quality control panels learning an off-note or taint language, a technical panel learning a lexicon for working with the Spectrum profiling method, or any panel learning to use an existing attribute list or lexicon. The attributes must be easily understandable by the panellists, the sensory team and whoever is going to be reading the report and acting on the data. Future panels may also need to use the lexicon and they might not always be based on your site. Therefore, the language used is critical. You will need to collect together the range of samples to help with the training and it can be useful to cover as much of the sensory space as possible at this stage (Griffin et al., 2017). The samples might be different production dates, different types of a similar product (for example, yoghurts: plain and fruited, set and runny), competitor products or different packaging, for example.

The attribute lists (or lexicons) used for profiling have several features. Firstly, there will be a list of **attributes**, generally in the **order** that they occur in the product or sometimes in the order that is the most sensible to assess. For example, sometimes the appearance attributes may be assessed at the end of the session for food products which are hot so that the aroma and flavour are assessed while the product is still warm. It does not matter how long the list of attributes is, only that it completely describes the products being assessed, and the panellists understand and can use each and every attribute to describe the product range. If the panellists have generated, defined and referenced the attribute list themselves, the more chance they will have of using it successfully (Lyon, 2002).

Secondly there will be a description for how the attribute will be assessed. This is generally known as the **protocol** and describes the amount of sample to be assessed and the way the sample should be assessed. This helps the panellists create good data as they are all assessing the samples in a similar way. For example, if the panellists were assessing washing powders and used different quantities of powder or different water temperatures, the data from each panellist would be very different.

Next would be the **definition** for each attribute. The definition describes the actual thing that is being measured (for example, the type of strawberry aroma) and sometimes includes the protocol for the specific attribute. The definition often includes a **reference**, particularly for aroma and flavour attributes. The **anchors** define the ends of the scale and for some methods will include quantitative references. Figure 7.5 gives a simple example of an attribute list for a food product demonstrating the six elements: the protocol, the attributes, the order, the definition, the references and the anchors.

The ® in the text indicates that the qualitative reference listed was actually assessed by the panellists and chosen from a range of other references to be indicative of the note picked up in the sample set.

When generating a list of attributes with your panel for a particular product there are several important factors to bear in mind. These are listed in Table 7.2 and described in more detail in the following paragraphs.

Firstly, the attribute list should **cover all aspects** of the product for the modality or modalities of interest. If key attributes are missing you may run the risk of 'attribute dumping': where the panellist, frustrated by the fact they have nowhere to rate a particular attribute, 'dump' their rating for that attribute within another attribute. This is particularly noticeable when the panel are naïve or mid-training. One way around this is to include the attribute 'other' and allow the panellists to rate and describe the 'missing' attribute(s). However, do not use this method as a replacement for good discussions or you might find some key attributes only measured by one or two panellists.

Flavour: Take a large sip, then hold and swirl around mouth for a couple of seconds and then swallow. Assess attributes.		
Attribute	**Definition**	**Anchors**
Almond	The nutty/almond flavour of marzipan ® Golden Marzipan	Not to very
Caramel	The creamy, sweet flavour of toffee notes as found in ® Cadbury's Caramel	Not to very
Cardboard	Flavour of a freshly opened cereal box (with the cereal removed) ® Empty Branflakes box	Not to very
Fresh mint	The flavour of the fresh green notes as found in ® Fresh Mint Leaves	Not to very
Tinned strawberry	The mushy over cooked strawberry flavour as found in ® tinned strawberries	Not to very

Figure 7.5 Example attribute list.

Table 7.2 Important factors to consider for attributes

Cover all aspects	Discriminate	Unipolar
Non-redundant (little or no overlap with other attributes)	Consumer or technical language	Simple and singular
Unambiguous	Precise	Reliable
Measured by one sense at a time	Suitable reference	Reference easy to create, use and store if possible
Agreed	Detected by the majority of the panel	Able to be measured via a scale
Relatable to consumer language	Understandable (by panellists and all users of the data)	Relate to real products

You might notice that some attributes are missed by the panel because all the samples demonstrate that attribute. For example, the term 'sweet' might not be used to describe a range of chocolates or a series of soft drinks as they are all similar in sweetness. You may need this attribute in the list if you are developing a description of the products as well as trying to discriminate between them.

The attributes should ideally **discriminate** between the samples being assessed. For example, if different jelly beans are being assessed for variation in sweetness, then it would be (very!) useful if the sweetness attribute showed the differences, if there are any, in sweetness. However, in some applications the need to show a difference is not always relevant. For example, if you were creating sensory profiles for a quality control project and the aim was to develop sensory specifications, the important aspect for the data to show is how sweet the jelly beans were so that level (or range) can be matched during subsequent production runs.

The scale used for the attribute should be **unipolar** as this allows for the centre of the scale to represent some aspect of the product. For example, if you were trying to use a scale which went from 'hard' on the left-hand side to 'soft' on the right-hand side, can you picture and define the midpoint? The attribute should also be **non-redundant** with all the other attributes of the same modality. This means that two attributes should not be measuring the same thing, i.e., little or no overlap with other attributes. For example, if a panel was developing a list of attributes for the assessment of tissue paper, the inclusion of the attributes rough and smooth would give the same (opposite) results. Another example would be for more poorly defined attributes such as rating red fruit aroma intensity and raspberry aroma intensity. As raspberry is a red fruit would that mean it would be rated under red fruit as well as under raspberry? The in-depth discussions with the panel can help alleviate these problems, i.e., asking them where they would rate a particular attribute and checking if everyone is doing the same thing (or not). You can use PCA of the data to determine redundancy, but at that stage, unless you have performed a 'practice profile' prior to the real rating experiment, it's often too late. Apply caution: if attributes are redundant in one study this does not mean they are always redundant in others. Also, some attributes may seem redundant but think twice before removing: dense and hard may seem redundant as they quite often go hand in hand – but butter is dense and not hard, and refrigerated Aero bars are hard but not dense.

Depending on the objective and the type of panellists, the **language** used to list an attribute may be either consumer or technically orientated. For example, if the panel is made up of a team of flavourists, the use of chemical terms such as 4-(4-hydroxyphenyl) butan-2-one when developing a raspberry flavour might be fine, however, the use of such terms with an external panel made up of local residents probably would not. It can also be useful if the attributes can be related to instrumental measures, particularly for quality control panels, as this can allow instrumental assessments to take the place of routine sensory assessments.

An attribute should also be **singular** rather than being a mix of several attributes. A classic example of this is 'softness'. Softness is actually a combination of at least four different measures: smoothness, compressibility, ability to spring back when compressed and lack of ridges when folded (Lawless and Heymann, 2010; Civille and Dus, 1990). Another example is freshness which has many different meanings dependent on the product category being assessed. Think about the aroma of a *fresh* cup of coffee compared to a *fresh* smelling shower gel: they are not the same. The main issue with measuring the compounded attribute is that it is very difficult for the product developers to act on the results. For example, for softness, if a project was related to making a tissue product softer, based on your panel's data, should the developers change the smoothness or the compressibility?

The attribute should also be **unambiguous, precise and reliable**. If an attribute is difficult to understand and panellists keep getting confused with the way to measure it or what the measurement means, it probably means that it needs reviewing and rewriting. The panellists will not be able to use the attribute reliably if it confuses them. Those people looking at and using the panellists' data may also be confused by the attribute and that does not help anyone.

An attribute should be **measured by one sense** at a time. For example, if your protocol starts with, 'Look at the sample, feel it and measure the...' then it will be combining two modalities: appearance and texture. An attribute can of course be measured within two or more modalities, but it must be measured twice (or more). For example, the panellists may well measure chlorine odour in water and also the intensity of the chlorine when the water is consumed, but they are measuring the attribute twice and not combining several measurements in one.

Some attributes require **qualitative references** so that the panellists know which element they are actually measuring. This is particularly the case for aroma and flavour, but there is less need for appearance and texture references as for these latter two modalities it is easier for the panellists to demonstrate and describe appearance and textural attributes. Although the idea is to find a reference that singularly (i.e., demonstrates one attribute in easy isolation) defines the thing that is being measured (for example, in the way that sucrose demonstrates sweetness) the panel may well need to assess a range of references so that they can see the whole picture and learn the boundaries. For example, as mentioned, sucrose demonstrates sweetness, but there are many types of sweetness and demonstrating the difference between natural and artificial sweetness may mean presenting the panel with many different examples. For more information about both qualitative and quantitative references see 6.3.7.

These references need to be **easy to create, use and store** where possible. If the reference chosen has a very short shelf life or a long cooking or preparation process, you may find that you spend a lot of time in recreating the reference. This can also lead to variability in the reference itself, so wherever possible try to use products that are very simple and ready prepared.

The attributes to be used need to be **agreed** by the majority of the panel: ideally the whole panel. A handy tip when there is a dispute about the presence or otherwise of an attribute, is to present two unknown samples and ask the panellists if they can identify which sample is which. For example, you might present samples X and Y. Sample X is sample 5 from the set of samples you are working on and is the one that some panellists feel has, for example, an off-note. Sample Y is another sample from the set. If the panellists are able to tell you that sample X is sample 5, then that can indicate that sample 5 does indeed have the disputed note.

Some attributes are not **detected by the majority** of the panel due to sensitivity or experience. These attributes may well be synonyms for other attributes and become incorporated with other terms. However, other attributes may well have been only generated by one panellist because they are the only person to pick up these notes. If only one or two panellists are rating the attribute, this does not give you the data necessary to make any conclusions. Explain this to the panellist so that they understand the issue: just disregarding their suggestions is not likely to make them feel like contributing in the future, particularly if it's 'always them' who generates these additional terms. The best option is to remove this attribute from the *measurement* list but do include it in your report if the panellist is always able to detect this note in that sample. If the attribute is related to a taint, you might be better off keeping it in the list of attributes to be measured: it could be that the panellist is particularly sensitive to that note and that might equate to 10% of the population. Another option is to include an 'other' attribute in all modalities for panellists to use to measure the additional attributes, but do not use this as an excuse for not having a thorough discussion.

Most attributes can be **measured by a scale**, but some are more difficult and the only way to determine sample differences is by ranking or a yes/no choice. Castura and Findlay (2006) created a useful classification of attributes in a pictorial form which can be downloaded from their website (www.compusense.com). Attributes that only seem to permit ranking are those that are only present in the sample set at two or three levels. The yes/no or binary attributes (called off/on in the Castura and Findlay poster) tend to be those related to a taint.

If the data are being used to compare to any consumer output, but in particular for preference mapping, it will be important that the terms can be **related to consumer language**. However, sometimes there appears to be little relation of the analytical sensory attribute to the consumers' liking or preference. As long as the attribute is **understandable** by all the panellists and by all users of the data, you will have a good chance of it being related to consumer language. Sometimes the attributes generated will not **relate to real products**. You might get attributes such as 'farmyard' (for example, for mature cheddar) or 'damp undergrowth' (for example, for potatoes).

You can try a visit to the local farmyard or a walk in nearby woods to try to reference these terms, but oftentimes they can be referenced by a closely related product. For mature cheddar, overly mature cheddar might give just the right note, if it is not hidden amongst other attributes, for example.

Some profiling methods have attributes and definitions already developed which can be adapted and developed for use with your product(s). An example of this is the Texture Profile (ISO 11036, 1994(E)) which is based on technical and rheological measurements. It includes the attribute titles (for example, hardness, cohesiveness, springiness), the definition (e.g., cohesiveness: mechanical textural attribute relating to the degree to which a substance can be deformed before it breaks) as well as describing the technique or protocol for the assessment (e.g., cohesiveness: place the sample between the molar teeth, compress it and evaluate the amount of deformation before rupture). You may also have attribute lists from previous projects, especially if you are recruiting panellists to join an existing panel. The critical thing when using a ready-built language is that the panellists understand what they need to do prior to the actual measurements. Therefore, the training will tend to focus on the learning of the protocol (see Section 7.2) and the use of the scale rather than the development of the language.

Some products also have a published language ready for use such as coffee (Chambers et al., 2016), herbs and spices (Lawless et al., 2012), car interiors (Verriele et al., 2012), beef flavour (Adhikari et al., 2011) and dry dog food (Di Donfrancesco et al., 2012), and there are even publications dedicated to one attribute (such as 'nutty': Miller et al., 2013). Some products have standards with a developed language such as the ASTM standard E2082-12: *Standard Guide for Descriptive Analysis of Shampoo Performance*. A quick search on the internet for your product and words such as 'lexicon', 'term derivation' or 'profile' may well give you some useful documents that can be used to develop an attribute list or help when the panellists get stuck.

There are a number of ways to demonstrate how important the language development phase is to the new panellists. One way is to use something similar to the repertory grid approach and another is building models from children's construction blocks. Both are described in Section 7.4.2.2 and 7.4.2.3, respectively, and make great starting points for the development of a profiling language. These sessions of product and language familiarity increase agreement among the panellists and give the biggest increase in panellists' discrimination abilities (Byrne et al., 2001).

For some handy tips on training a profile panel, please see 9.2.1 and for more advanced training for panellists developing product languages, please see Chapter 10.

7.4.2.2 Repertory grid approach to demonstrate the importance of language

The repertory grid approach uses triadic elicitation, which basically means that there are three products presented to help create descriptions by the comparison across and between the three samples. The repertory grid approach presents all three samples and asks 'in what way are two of the products similar, but in the same way different from the third?' It's in the comparison of the three products that people realise something about the other products that might have been missed if they had not had the third product in the mix. When described to the panellists, but only *after* the practical session, I call this 'what is not in the samples can be as important as what is'. If you tell the panellists at the beginning of the sessions that this is what they will learn by the end of the session, it does not work well. It can be much better to allow the panellists to learn for themselves. The practical session is described below using biscuits, but you may well have items that you can use from your product range. This adaptation of the repertory grid method involves presenting one sample at a time.

Sample 1: a square biscuit with three layers: biscuit, cream and biscuit, such as a Custard Cream.

Sample 2: a square (or rectangular) biscuit with three layers: biscuit, cream and biscuit, such as a Bourbon.

Sample 3: a round biscuit with particulates (for example, rolled oats), but no layers, with a chocolate coating such as a Chocolate Hob Nob.

Introduce as a training session about appearance but do not give any other clues about what you expect or the reason for the training. Give the panellists sample 1 (keep the other two samples hidden from view) and ask them to describe the appearance. Tell them they can do whatever they like with the biscuit, for example, pull apart the layers or break the biscuit pieces (but not eat it!). To make it more interactive, the panellists could work in pairs. Give the panellists five minutes and then write up the appearance attributes generated on the flip chart in a column for sample 1. Ask one panellist for one attribute, write it up, move to the panellist next to them and repeat until there are no more attributes. Query any attributes you do not understand to show the panellists you are very interested in what they have found about the biscuit.

Give the panellists sample 2 and repeat the process. When collecting in attributes for sample 2 notice how the panellists have shifted their descriptions from 'light cream colour', 'crumbly', 'snappy noise when broken' to 'darker than sample 1', 'not crumbly like sample 1', 'harder to break than sample 1'. Remind the panellists that you would like to list the attributes and not the intensities or comparisons, but do not worry if they find it hard to convert 'harder to break than sample 1' into 'hardness'. If they are a new panel, there is a lot more training to do before they can become quick enough to make this conversion. Once you have finished collecting the attributes for sample 2, ask the panellists what had changed between their descriptions for sample 1 and sample 2. See if they notice that they had begun to make comparisons. Tell them that making comparisons between samples is a very important part of their role and is of course impossible to do with the first sample and that is why, when they are working on creating a profile, they will see the samples a number of times in a different order each time to enable great comparisons to be made.

Give the panellists sample 3 and repeat the process. This third sample has no layers. If the panellists did not mention three layers in their previous descriptions, they may well do now. This third sample also has a chocolate coating. It's very unlikely that the panellists will have mentioned 'no chocolate coating' for samples 1 and 2, but they will probably request this to be added to the attribute list. As sample 3 also has particulates in the biscuit itself, the panellists may also request that samples 1 and 2 are also described as uniform (or a similar word) in the biscuit parts. This approach is a quick way for the panellists to learn for themselves how important it is to make sample comparisons and how, by comparing across samples, new attributes can be created for previously assessed samples: 'what is not in the samples can be as important as what is'.

7.4.2.3   Children's construction blocks language helper

For this section you will need two simple construction block sets such as those designed for ages 5–8 with around 200 pieces or less. These could be a small car, boat or space rocket. You will need to buy two identical kits of each (four kits in total, two of each model). You will also need a large screen to separate the two groups of panellists. If you do not have a screen you can get the two groups of panellists to group around a table each facing in opposite directions. Before you give the panellists the task, explain that they are not allowed to mention what the finished object is (e.g., boat, car, plane) and then ask them who would like to go first. Give the group who opted to go first the pile of bricks and other bits from the kit but not the instructions and do not let them see the box! Pour the pieces on to a tray and give it to them that way as it prevents small pieces being lost. Give the other group the finished model. Tell the group with the finished model (the 'model group') that they need to explain how to build the model to the other group who have all the requisite pieces in front of them (the 'building group'). Tell them they have 20 minutes to complete the task. You can always give them more time if they make no progress, but the demonstration can work well if they think there is a time restriction. Tell the building group with the pieces that they can ask questions of the model group but not what the model actually is.

Usually the group with the model panics and different panellists begin to shout out random instructions whilst constantly checking the time. The building group begin to ask questions and can get quite frustrated when they cannot work out what the model group means by 'bumpy bits up' or 'on the left' (the building group's left or the model group's left?) or 'small circle thing' (the steering wheel). Generally little progress is made and if there is progress the models do not resemble each other. Once the two groups are approaching being completely frustrated, call time and give them a short break.

Once the panellists return, ask them in their original groups to discuss what they could do better. Give them around 10–15 minutes for this. Then ask them what the other group could do better: give them around five minutes for this. Then get together as a big group and discuss all the ideas. For example, the model group might suggest that they set up a dictionary of terms and go through this with the other group. For example, 'bumpy bits up' means that the pieces that have round circles on (with the logo) to help the pieces fit together, should be laid on the table so that the circles are pointing at the ceiling. Or 'left' means where the windows are in the room and 'right' means where the door is. The building group might suggest that they could check and group all the different colours before the building commences and name certain odd-looking pieces so they are more recognisable. There might also be suggestions about how to give instructions and how to ask questions, or the assignment of 'builders' or 'instructors', or taking the time to check understanding before commencing to the next stage. All good suggestions to help the panellists learn how to develop a language that everyone will understand and be able to use to build the model.

Once the ideas have been discussed ask the panellists if they think they could do better if you asked them to do it all again. The answer will probably be yes, although I had one group of trainee panellists that said 'no' because there was a couple of people who hated children's construction kits! Split the panellists into the same groups as before, but this time give the builders the finished article and the model group the pieces. Use the second pair of construction kits for this part. For example, if you used the two cars for the first part, use the two rocket ships for the second part. They will probably notice quite early on that the kit is to make a different model and you may get some friendly complaints. But you should notice a marked improvement in communication, although not necessarily in model building, and particularly so if the second session takes place on the next day, perhaps because the panellists have the chance to consider what they might do and develop a better plan of attack.

7.4.3 Scale training

7.4.3.1 An introduction to scales for the panel leader

One of the main tasks in sensory science is to quantify the level of intensity of a particular attribute, say bitterness or greasiness, to understand the differences between samples. For example, we might be working on a project to improve a process, reduce costs or create a new product. The measurement of intensity is mainly achieved through the use of a scale. For example, you might like to ask one of the following questions:

- Which of the hand creams are greasy?
- Which is the greasiest and least greasy hand cream?
- How greasy are the hand creams?

The type of scale you choose to use will very much depend on the objective of your sensory study and also the people who are taking part, along with their level of training. There are many different types of scale to choose from and some of these are shown in Table 7.3. Some scales are only used in one type of method, whereas other methods use different types of scale to collect the data. BS ISO 4121, 2003 'Sensory Analysis — Guidelines for the Use of Quantitative Response Scales' gives some useful information about the choice of scales.

There are four main types of *data* gathered in sensory science: nominal, ordinal, interval and ratio. These are also shown in Table 7.3. Note that in the first example the *data* resulting from the scale are nominal, but the *scale* is a category scale. It's good to keep these two descriptions, data and scale, separate in your mind: the actual *appearance* of the *scale* used does not necessarily mean that the *data* coming from that scale will be the same. For example, you may use a category intensity scale with consumers and you may also use it with a highly trained panel: this does not mean that the consumers' use of the scale will be interval even if the trained panellists' data are found to be. And it's worth keeping in mind that not all scales fit into these four categories: there has been some criticism about this classification. Some scales are also classified as discrete (for example, category scales), and some scales are continuous (for example, line scales).

Table 7.3 Examples of scales

Data type	Example
Nominal	Nominal Please tell us when you think you would be most likely to consume this drink: ☐ With breakfast ☐ With a mid-morning snack ☐ With lunch ☐ With dinner ☐ At any time with or without any food
Ordinal	Please taste the samples from left to right. Sample codes: 591 | 624 | 730 | 819 Write the codes in increasing order of aroma intensity in the boxes below. Least | | | Most Sample code Comments:
Interval	Interval ├────────────────────────────┤ Not greasy Very greasy
Ratio	0 = no sweetness 10 = 'sweet' 20 = twice as sweet 100 = 10 times sweeter

When we use scales in sensory science we have a long list of requirements so that the scale will work and give us the data we need. Some of these are listed below:

1. Absolute
2. Low variance
3. No bias
4. Unaffected by context
5. No floor or ceiling effect
6. Shows differences between individuals
7. Does not show differences between individuals
8. Psychological distances between units on the scale are equal
9. Is relatable to a semantic explanation (e.g., moderate, strong)
10. Discriminates samples
11. Is easy to use
12. Is 'easy' to analyse data

And that is why there is a huge amount of research into sensory scales in an attempt to find the Holy Grail as shown in the list above.

If you asked an ordinary person to tell you if a sugar solution is weak or strong, or ask them to assign the sugar solution a number from 0 to 10, it might seem quite simple, but actually it is quite complex! There are many aspects to take into account from context and word meaning, to how people use numbers and scales differently. For example, think about the phrase, 'A small elephant was frightened by a large mouse'. If you had no prior knowledge (context) of mice or elephants you may well imagine the mouse to be larger than the elephant rather than a tiny mammal which was slightly bigger than the usual mouse! The adjectives (e.g., small and large) become *relative* and depend on the nouns they are used with: the *small* elephant is in fact much, much larger than the *large* mouse. And, depending on your previous experiences and your physiology, your use of the descriptors 'weak' and 'strong' may well be entirely different from mine and anyone else reading this book.

We tend to use category, interval and ratio scales for measurements in sensory science and these are described below in Table 7.4. Nominal scales are more often used

Table 7.4 Category, interval and ratio scales

Name:	Category scales		Interval scales	Ratio scale
Description	Semantic (e.g., none, very strong) or numeric scale used to rate intensity. The differences between each of the scale points are not necessarily the same.		Spacing of responses is equal, e.g., difference between 3 and 5 is the same as the difference between 7 and 9.	Numbers represent an actual quantity, allows the use of ratios: 'twice as much', 'ten times as much'. We are unable to make these statements with category or interval scales.
Data type	Ordinal but encouraged to be interval to allow parametric statistical analyses		Interval	Ratio
Example	Flavour Profile 1. not present 2. threshold 3. slight 4. moderate 5. strong	9-point hedonic scale 1. dislike extremely 2. dislike very much 3. dislike moderately 4. dislike slightly 5. neither like nor dislike 6. like slightly 7. like moderately 8. like very much 9. like extremely	Line scale ⊢————⊣ Not greasy Very greasy	Magnitude Estimation 0 = no greasiness 10 = 'greasy' 20 = twice as greasy 100 = 10 times greasier

in consumer sensory science and tend not to be numeric, such as gender. However, check-all-that-apply (CATA) scales are also nominal, and these scales are becoming more popular as an analytical sensory tool.

Psychophysics helps the sensory scientist to conduct perceptual measurements in a robust way but does not direct what we should do. For example, psychophysics may be interested in understanding differences in people's sensitivity to certain stimuli, but, although sensory scientists may also be interested in this area, the main focus in industry tends to be on sample or product differences and not the person per se. People working on genetic differences and their effect on taste would be interested in the differences between people – this distinction is worth keeping in mind as you continue through this section. Some element in the product (often called the stimulus) is perceived by the panellist (this is the 'psychophysical process': *psycho* – the mind or mental processes and *physical* – the physical thing being assessed (lotion, fruit pie, etc.)), but then the panellist considers (thinks about) this sensation prior to scaling the intensity. So it is really a two-step process. The panellists do not measure the amount of sugar in a solution: they measure the sensations they perceive. This is why the manner in which the perception is measured can change the measurement.

In the early days of psychophysics, researchers thought that people were not able to measure stimuli directly, so they developed (extremely time-intensive) methods of measuring differences *between* stimuli instead. After a few years of this, they decided to try the more direct method, basically asking people to measure things using scales with various categories that represented an increase in intensity. Some researchers also used scales where only the ends were labelled (unstructured line scales) to try to avoid the issue that the distances between the categories might not be psychologically equally spaced. Look at the scale in Figure 7.6. Do you think the difference between 'weak' and 'moderate' is the same as the difference between 'strong' and 'very strong'? The researchers' results compared well with the earlier work they had conducted using the indirect methods, but a new problem was introduced: people tended to avoid using the ends of the scales, just in case there were weaker or stronger samples somewhere in the mix. This was referred to as the 'floor and ceiling effects'. Psychophysicists have known for some time that ratings of perceived intensities can be affected by context and experience.

The researchers continued to experiment with these direct measurements of sensory perception, and in 1957 the use of magnitude estimation was suggested by Stevens and Galanter (1957) as the best way of measuring perceived intensities. Magnitude estimation is fairly simple. Let's look at the method to measure sweetness as an example. The assessor is presented with a 'standard sample' of sucrose in water (known as the 'modulus' – a known concentration) and is told to assign it a value of, say, 10. They are then presented with the first sample in the study and asked to assign a number to the level of sweetness proportional to the 'standard sample'. So, for example, if the

Figure 7.6 Category scale semantics.

sweetness is felt to be double that of the first sample, the assessor would assign a value of 20. If it was half as sweet, they would assign a value of 5. The next sample would be presented and the experiment continued. The scaling method appeared to give ratio data, did not require labels or anchors and had no ceiling effect. The method was best for determining the mathematical relationship (exponent) between magnitude and perception for the stimulus under test, but it was neither better than any other method at determining differences between samples nor comparing across modalities.

There were some reports of issues with the method: assessors needed more detailed instructions and some assessors used low numbers and some used high suggesting that bias had not been removed. Also comparing assessors' assigned numbers was meaningless. One major issue with the use of the method in sensory science was that the exponents varied for different stimuli ranges and with the modulus used, and this issue was made worse by the previously mentioned idiosyncratic use of numbers. The method also did not give information on absolute intensity as there was no way of judging whether any number assigned meant weak, moderate or strong.

Stevens and Marks (1980) devised the method of magnitude matching (known more recently as cross-modality matching) in an attempt to make absolute measurements to compare and measure sensations from individuals and groups: for example, PROP (6-n-propylthiouracil) tasters and nontasters; they wanted a 'uniform measuring instrument or yardstick'. Magnitude estimations cannot give us this information: your 20 might not be twice my sensation magnitude of 10, for example. And category scales were felt to be too relative and affected by the context of the experiment (remember the small elephant).

In an attempt to get around the scale-use issues mentioned above, several other researchers looked into various different approaches such as Borg (1982) who created a category-ratio scale for physical exertion. Unfortunately, the success of the scale was short-lived in terms of sensory science: although it compared well to magnitude matching for PROP taste, there were mixed results for different modalities. Green et al. (1993) were also interested in developing a scale to measure perception that gave ratio-level data with absolute intensities that would show differences between individuals but be useful across modalities. They felt that magnitude estimation did not give absolute ratings nor highlighted differences between individuals. They also felt that magnitude *matching* was not able to give absolute ratings and had the additional requirement that people needed to give a consistent response to the 'standard modality' as we discussed earlier. They were interested in Borg's scale and were curious to know if it worked in contexts other than perceived extreme exertion. In the process of their research they designed a new scale which they called the labelled magnitude scale (LMS). The LMS has six verbal descriptors (see Figure 7.7) with unequal quasilogarithmic (i.e., similar to a logarithmic scale) spacing with the upper point described as 'strongest imaginable'. The relative spacing of the verbal descriptors was derived using magnitude estimation. As mentioned before it produces ratio-level data and enables different modalities to be compared on the same scale. Notice that the scale has verbal descriptors and not numbers.

Bartoshuk (2000) reviewed the various methods that the previous authors had used to assess different people's sensory experiences, i.e., genetic variation in taste and in

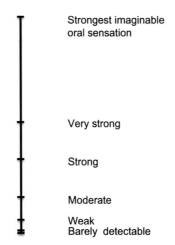

Figure 7.7 The labelled magnitude scale.

2004 published a further paper where the authors created a new scale based on Green's LMS (Bartoshuk et al., 2004).

The reasons behind the creation of the new scale are nicely demonstrated with the analogy used by the authors of an 'elastic ruler'. Bartoshuk suggested that the adverb/ adjective descriptors (as we mentioned earlier such as '**very** strong') vary depending on what is being measured – she referred to a 'very strong rose aroma' and a 'very strong headache' not being equivalent as an example, and hence the elastic ruler which might stretch and compress for each stimuli type.

Bartoshuk is especially interested in genetic variation in taste and so was interested in the fact that variations in taste ability also caused the elastic ruler to stretch and compress depending on whether people were supertasters, medium tasters or non-tasters. Using a scale with the adverb/adjective descriptors with these different groups of people would therefore not be useful but magnitude matching works due to the fact the comparisons are made to a stimuli different to taste (such as sound – supertasters would match tastes to louder sounds than nontasters, for example). Using magnitude matching and the LMS as a base they developed a scale to measure taste intensity on a scale of **all** sensory experiences, i.e., unrelated to taste, and called this the **generalised LMS (gLMS)**. At the top of the scale is the label 'strongest imaginable sensation of any kind' instead of 'strongest imaginable oral sensation' as in the LMS. The authors tested out this scale and found that the original LMS would underestimate the differences between the three taster groups: to supertasters the strongest burn of chilli was more intense than the strongest toothache! This also seemed to indicate that the labels are not actually measuring absolute intensities.

One of the main reasons for looking into these types of scales was to understand whether the points on a category scale were uniform; that the distance between, say 5 and 6, was the same as the distance between 7 and 8. One particular scale that was interesting to look at in this regard was the 9-point hedonic scale (see Figure 7.8) used in consumer research (or affective research). The scale is known to have issues with

Overall liking score	Verbal representation
9	Like extremely
8	Like very much
7	Like moderately
6	Like slightly
5	Neither like nor dislike
4	Dislike slightly
3	Dislike moderately
2	Dislike very much
1	Dislike extremely

Figure 7.8 The 9-point hedonic scale.

unequal scale intervals (the authors pointed this out when they derived it in the 50s), but it was also found to have issues with the fact that consumers avoid the use of the 'Extremely' points at both ends of the scale and therefore this makes the scale less useful when the objective is to compare extremely well-liked (or disliked!) products.

Schutz and Cardello (2001) and Cardello and Schutz (2004) devised the labelled affective magnitude (LAM) Scale for use with consumers. This was followed by several publications comparing the LAM scale with the 9-point hedonic scale and other hedonic type scales (for a good review read Lim, 2011); the main finding being that none of the scales were superior to the other, but the choice of which scale to use depends on the objective.

The LMS was found to be very useful in determining differences in genetic taste differences but has had a few hiccups in its use in sensory science applications. One issue has been around a clustering effect of the labels; participants tend to rate *on* the semantic label and around the label rather than elsewhere on the scale. Additional training and orientation sessions helped to get round this issue. One researcher (Hayes et al., 2013) attempted to replace the gLMS with a generalised visual analogue scale (gVAS) – simply the gLMS scale but with only the top and bottom descriptors (i.e., 'No sensation' at the bottom of the scale and 'Strongest imaginable' at the top – the same as any VAS scale). Using magnitude matching the authors found that the gVAS did not have the clustering issue associated with the gLMS, but of course the scale did not have one of the main requirements: the semantic labels (weak, strong...).

The LMS has also been shown to be affected by context effects (Lawless et al., 2000; Diamond and Lawless, 2001) – these authors found that a midrange intensity sample was scored higher when it followed a weak sample and lower when it followed a more intense sample, although this issue was helped with the use of reference standards. Delwiche et al. (2001) also found evidence of panellist bias when using the LMS and devised a correction factor for individuals by asking them to rate unrelated sensations such as loudness and weight, at the same time.

Bartoshuk et al. (2002) questioned some of the assumptions that the LMS was based on. She found that the maximum oral pain was not the same on average across all subjects. In a series of experiments, she found that maximum perceived intensities vary across modalities which contradict the assumptions of range theory – backing up her elastic ruler analogy, and her suggestion for the gLMS, as we mentioned earlier.

Schifferstein (2012) gives a very nice review of the LMS and is well worth a read. He gives some nice explanations behind the need for specific scales in sensory science and charts the history of the development of the scale as well. The author critically reviews the original research and the claims made by the researchers. He questions (1) whether the LMS gives ratio-level data; (2) the scale allows for comparison of individual and group differences and; (3) the claim that the LMS does not have ceiling effects.

Lyon (2002) state that there are three main methods for calibrating trainee sensory panellists on the use of the scale: the Spectrum method (we could also include the Flavour Profile and Texture Profile) with its absolute scales and extensive quantitative references; the use of quantified references for some attributes or some scale points (a kind of hybrid Spectrum) and 'auto-calibration with the samples under investigation'. A review of some recent publications relating to lexicon development suggests that there are a couple of other methods as well. Chambers et al. (2016), Ting et al. (2015) and Corollaro et al. (2013) have developed their own intensity references for their lexicons, mainly for the top and bottom of the scale. The scale used in Chambers et al. (2016) was a 15-point scale with 0.5 point increments, and the intensity references were determined by the panellists through reference assessment, discussion and agreement. Several references were assessed by the panellists, and the choice of these and their related intensity values were based on group consensus. Other authors use one of the samples as the intensity reference, having gathered data over six different replications for that reference, and present this sample blind on several occasions to check the panel's calibration (Griffin et al., 2017). As ISO 11132, 2012 states, 'In most cases, no true value is known and the overall bias for an assessor is taken to be the difference between that assessor's mean and the mean for the panel'.

What is the difference between the scales used in an absolute scaling-profiling method such as Spectrum and a method such as QDA or a generic profiling method? With the Spectrum method it seems quite clear. References within the scale for every attribute tell the panellist where to rate. For example, in the hardness scale mentioned earlier. But what about if you are using QDA or you simply cannot train your panel with quantified references because you do not have them? There are many reasons for not having access to quantified references, not least that the products listed for the Spectrum method are generally only available in the States, although of course you could purchase them and store them in your country much more easily today than say, 10 years ago. Another reason for not having references for the ends of scales, might be that you do not actually have the ability to create the references, as the end of the scale is actually what you are aiming at in your product development. For example, if you were working in hair care and needed a reference standard for the end of the scale for hair smoothness and your project was related to making the conditioner leave the hair more smooth, you might find the generation of the standard reference quite difficult. However, often the ends of the scales can be created by using the product in a different way. For example, in the smoothness example, you might be able to create a very smooth hair switch or mannequin head by over conditioning the hair: leaving the conditioner on the hair for far longer than usual or treating the hair with conditioner several times. But you might find it difficult to decide when is too much conditioning: have you gone over the top and no one's hair would ever be that conditioned or smooth? How do you decide which is the end of the scale? The Spectrum references' quantified published values were taken from the data of 10 different panels in industry and are reviewed and updated regularly (ISO 13299: *General Guidance for Establishing a Sensory Profile*, 2016) and in that respect the investment may be well spent. You could create your own quantitative references using this approach (conducting many replicates of the same products) for your scales, but you may only have your panel to take the values from. But this is certainly a solution for some time down the line once your panel have assessed many samples. Unfortunately though, the time machine is yet to be invented, so what do we do in the meantime?

Let's look at how panellists (and us!) might use a scale. Let's say we had a 5-point scale, and we were using it to measure the sweetness of a range of sugar solutions (let's keep it simple to start with). There are five solutions ranging from 0.05 to 0.4 g/100 mL. The choice of where in the scale to place the first sample will be based on the panellists' previous experiences: if you had never given them sweet items to rate before, they will be basing their judgements on sweet things they have tried and the placement of the samples might be somewhat random and unevenly spaced, especially if you presented the samples in a random order to each panellist and monadically, i.e., one at a time with no going back to previous samples. By unevenly spaced, I mean that if the solutions were created to be equidistant from each other, the scale use may well not show this. If we allowed the panellists to assess all the samples first, in a similar way to how we might conduct a profile prior to the rating, the panellists will probably distribute the samples across the scale so that the weakest is on the left and the strongest is on the right. And maybe now is a good time to mention that this is one of the main issues with a sensory scale: its lack of units. If we were measuring weights, the measurement units might be in grams, distance may be in miles, panellists' heights in metres and the amount of liquid presented in millilitres. But our profiling scales: what units are they measured in?

If we now gave the panellists a new set of sucrose samples (without telling them we were changing the concentration range) which went from 0.4 to 1.0 g/100 mL (a wider concentration range) and again we present the samples monadically without allowing the panellists to assess them first, how do you think the scale use would change? According to Torgerson (1958) the panellists will begin by placing the samples at the stronger concentration end of the scale. If you repeat the scaling exercise a few times, the panellists would gradually, with some variation, redistribute their scores across the whole scale range, with the 0.4 g/100 mL (which was at the top end of the first concentration trial) being rated near to the bottom of the scale and the 1.0 g/100 mL gets rated at the top end. In psychophysics this is called 'shift in position'. Without any instruction to do so the panellists will, over a number of assessments, build up their own impression of the product scale range and use it accordingly. However, this depends on the panellists being able to view the sample range prior to making their assessments. If the panellists are unable to see the product range prior to rating, as in single stimulus assessments (one product rated at a time and not assessed and/or discussed prior to being rated), you will certainly need to have the standard quantified references to mark the end of the scale. If you are in the multiple-stimuli assessments you can use the scale as described above: allowing the panellists to assess the samples and 'see the range of the ratings'. But then we have another issue. The ratings from different panellists may well be different: Mildred may use all of the scale to rate her samples, while Belinda may not. And that results in different mean scores over their replicates for different panellists for the same sample and the same attribute. Should we force Belinda to use more of the scale or to rate the samples exactly the same way as Mildred?

If we now gave the panellists a new set of sucrose solutions that covered the range of both experiments from 0.05 g/100 mL through 0.4 to 1.0 g/100 mL we would expect them, after several trials, to spread these concentrations over the same 5-point scale. In some ways this is a bit like having a ruler made of rubber, with no markings and no units, so that it can stretch and shrink to cover the range of concentrations required. In this way, the scale is determined by the range of concentrations present in the samples (Torgerson, 1958).

We know from Thurstone (1927) that panellists will give variable results themselves relating to the measurement of intensity: this is down to many well-documented reasons. It might be down to the fact that the product actually varies in intensity, or the differing amounts of product consumed (is it possible to always take exactly the same volume/amount in a 'sip' or a bite?) or a physiological effect such as sensory adaptation, palate cleanser/rinsing, product temperature, presentation order, carryover from previous products, or the strengths of the signals reaching the brain, or even the person's mood (van Hout, 2014; O'Mahony et al., 1994). As a result of this we can *expect* a panellist to be variable in the ratings that they give to a product. Luckily the use of replicates (repeat of session, in this case data collection, with the same panellists, samples, test method and conditions ISO 11132, 2012) in our profiling experiments and the use of Analysis of Variance helps us account for this variance amongst the panellists.

Product variability can be determined during the initial assessments of the samples. If the samples are simply called 1, 2, 3, etc., you can get some pretty quick feedback from the panellists during the second assessment of any sample if it does not quite match what they wrote for the description the first time around. You might also have previous data to show that a particular production line is more variable than the others, or you may expect variability from the inherent nature of the product if you were assessing fresh fruits or vegetables. Data about product variability can also be gathered during the rating stages by sampling different containers or batches and recording which panellists were served which part of each sample. For appearance assessments of variable products it can be useful to get the panellists to all rate the same sample of the product. For example, if you are rating the appearance of fried potatoes and there are many differences between the individual pieces, asking the panellists to all rate the same subsample can still give the data required for the appearance profile but will help minimise the variability in the panellists' ratings. But remember to note the variability in product appearance in your report. If there is variability in a product that is made up of lots of pieces, for example, bags of snack products such as crisps (potato chips) or nuts, asking the panel to evaluate several and then completing their final assessment can be one option (see for example, Griffin et al., 2017, p. 3), but this can also cause issues if one nut, say, was more salty or aged than all the other nuts. You may have to collect variable data and state in the report that the products were variable. Some products can be mixed prior to assessment. For example, if you are testing small pots of fruit yoghurt you could empty all pots and gently mix prior to portioning out for each panellist. Be careful that any mixing you do at this stage does not change the product texture (ASTM E1871, 2010).

Whichever type of scale you opt for to collect your data, it is worth thinking about the scale rules suggested by Lawless and Heymann (2010, p. 158–160):

1. Provide sufficient alternatives – a 5-point scale might not be enough to show the differences between the samples, especially for a highly trained panel.
2. The attribute must be understood – in consumer studies the attribute must be simple enough for all to understand and in descriptive panel work the attribute must be very well defined (see 7.4.2). There is no point collecting data if people are all measuring different things.
3. The anchor words should make sense – if the scale runs from low to high, what do the panellists rate if there is none of that attribute for a particular sample? They should also match the attribute title: if the attribute is 'clear' in appearance, having anchors 'clear' on the left-hand side of the scale and 'opaque' on the right-hand side would be very misleading!
4. To calibrate or not to calibrate – calibration of the panel is not always required – consider the project requirements. If you do need quantitative reference standards it might be worthwhile including them across the scale as well as at the ends. A good example is in the three standards for measuring pepper heat published by the ASTM (for example, ASTM, 2011: in this standard there is also a hidden control to check the panellists' performance). A very well written and interesting paper about context effects and the use of quantitative references is by Olabi and Lawless (2008). The research indicated that panellists will still change their intensity ratings when samples are presented in different contexts, even if quantitative references are used. Lawless and Heymann go on to say, 'People differ in their sensitivities to various tastes and odours and thus may honestly differ in their sensory responses'.
5. A warning: grading and scoring are not scaling – some scales appear to have a number relationship but do not. They are simply a grade given to a product and doing any statistical analysis or even taking simple means is meaningless.

7.4.3.2   Training panellists in the use of scales

When training panellists to use scales, rather than jumping straight into scaling, start with simple ranking tests with samples that are quite different. This way, by starting with easy tests you start to build confidence so the panellist feels they are competent and capable.

When you introduce the scale to your panellists you will need to explain to them how the scale is to be used. This can differ for different scaling methods, but the majority of scales start from the left-hand side at zero and finish at the right-hand side with the highest level. Generally, the scales will run from 'not' to 'very' or the word anchors can be adjusted to match the attribute title. Sometimes a scale can start at 'low' but this can cause issues when the attribute is not present for some samples. Do not be tempted to swap scale directions within the questionnaire/ballot to 'keep panellists on their toes' as this will just end up with messy data and confused panellists. If you give them time to learn how to use the scale to begin with, and plenty of practise, they will produce good data.

As a starter, you could repeat the scaling exercise that was used in the screening (see Figure 5.8) or you could create further shaded shapes, pictures or photos to scale. This can be particularly useful when training a quality control panel, as photos of products from the factory lines can be used to demonstrate the scale.

You could also give out some of your blank line scales to the panellists and ask them to mark on the line where they think various semantics might be positioned. For example, where they might rate 'slightly' or 'moderate' and then get the panellists in small groups to discuss their outputs.

When starting out training the panel on scales, start with one attribute. Sweetness is the easiest to set up for food panels as solutions of sucrose can be easily prepared and adjusted. If your method involves the use of an unstructured line scale, it can be easier to start with a 10-point category scale and gradually move to a line scale. For home and personal care products, simple initial tests could involve various grades of sandpaper, different types of fabrics, or your products where you have created known differences between samples. For example, hair tresses or fabrics which have been treated in different ways. Model odours and fragrances can also be useful at this stage. You could also train in scale use in comparison to a certain product or products (for example, see Figure 5.19).

Directly comparing ranking and scaling exercises can be very useful in demonstrating to the panellists how to use the scale for the profiling method you have chosen. You could use basic taste solutions for this and ask the panellists to rank the range of sweetness solutions (see Table 5.3: Suggested concentrations for basic taste ranking tests) and then, with the same set of samples but with different codes, ask them to rate the same solutions. The panellists will be able to mark their own results if they do the test on paper (see Figure 7.9). If you are working on using a rating system similar to that used in QDA, it will be the pattern of the panellists' ratings that will be important and the discussion will focus on the rank order and discrimination, not where on the scale the panellist has rated (ASTM, 1981). If you are working with quantitative and calibrated references, it will be the actual score that you will be monitoring and the discussion will focus on agreement around scale use.

The test can be repeated with the same set of samples but in a different order, different levels of sucrose and different ranges of sucrose so that you can demonstrate the type of scale you have chosen to use.

RANKING AND THEN RATING USING SCALES

Name _____   Date   _____

Ranking
Taste all of the coded solutions in front of you in the order presented and then rank them in increasing order of sweetness. Write the codes below from least sweet at the top to most sweet at the bottom.

Least Sweet   _____

                   _____

                   _____

                   _____

Most Sweet   _____

Rating 1
Taste each coded solution in the order presented and then rate the sweetness of each solution for intensity/strength of sweetness using the line scales below.

Code

908   Not   _____   Very

389   Not   _____   Very

013   Not   _____   Very

270   Not   _____   Very

422   Not   _____   Very

Figure 7.9 Example panellist questionnaire for ranking and then rating.

7.5 Training a time-intensity panel

The session plan below describes the steps required to train a panel for TI work. TI studies are usually conducted with trained panellists, however, Galmarini et al. (2016) found that the progressive profiling method used with consumers in home, although not as detailed as the output from TI, gave an efficient and cost-effective result. If your product is quite fatiguing or the measurements are to be done over a longer period, you may therefore wish to look at progressive profiling.

The panellists for TI are recruited and screened in the same way as descriptive profiling panellists (ASTM, 2013). In fact, it can be helpful if the panellists have previously created quantitative descriptive profiles as this gives them a good grounding in the measurement of sensory attributes. You will also need to have the attribute(s) that you plan to measure defined and documented before you start and these can be generated through descriptive profiling. However, it is not necessary to train the panellists in profiling prior to training them for TI as the attributes can be generated and defined by another panel prior to the TI analysis.

If the panel are completely new to sensory science, you may wish to do some initial training as described in Chapter 6.

The first part of training a panel in TI is to introduce them to the concept of measuring intensity over time but without showing them any TI curves in case this biases them. The second part of the training is to introduce them to the product type and the attribute to be measured. The samples are described by the panellists prior to any measurements and a comparison of the sample set is discussed. During this part of the training, the protocol for sample assessment is also discussed, agreed and documented. Samples can also be ranked to check for panel agreement. The third part of the training is in the use of the software used to collect the data. As the sensory panellists will probably create TI curves that are unique to themselves, it is very important to give them time to practice so that they can become consistent in their own assessments. Sometimes intensity references are used to calibrate the panellists to give similar values for the maximum intensity. If you would like to use reference standards, they will need to be assessed in the same way as the test samples and also prior to test sample assessments, so remember to plan for this in your experimental design.

For more general information about continuous time intensity please see Chaya (2017) for a good overview. For more information about training a panel in TI, see Peyvieux and Dijksterhuis (2001). The ASTM *Standard Guide for Time-Intensity Evaluation of Sensory Attributes* (ASTM, 2013) also has some useful information about TI data analysis and the assessment of panel performance.

7.5.1 Session plan for training a panel to create continuous time-intensity profiles for greasiness of skin creams during application

This is continuous TI profile measured in real-time to determine the greasiness perception of four creams through application to absorption. The objective is to train the panellists in TI and to compare the four samples' greasiness over a short time period.

Session 1: Introduction to time-intensity (2 hours)

Objectives: to introduce the concept of TI by demonstrating with various 'samples'.

- Start by playing a piece of classical music that begins very quietly and then gradually builds up in volume (intensity) and then gradually fades away. A good example is Claude Debussy's Prelude to the Afternoon of a Faun which has several increases and decreases in volume in the first two minutes.
- Play the first two minutes of Claude Debussy's Prelude to the Afternoon of a Faun (or your choice of piece and timing) to the panellists without any introduction. Just ask them to listen as they will need to describe it later.

- Then ask them to listen to the first two minutes again and to make notes about the volume/intensity while listening.
- Allow them to listen to the first two minutes again and add to their notes.
- Ask them to share and discuss their descriptions in pairs.
- As they are discussing, walk around the room and listen in to the descriptions.
- Choose two or three pairs that came up with some good TI descriptors (e.g., 'increases rapidly', 'increases slowly', 'plateaus') to share their descriptions with the rest of the group. Discuss what people found within the group.
- Ask the panellists to think about various products they use that may change over time. These products may be food or non-food depending on your objectives. Let them discuss in pairs or groups of three.
- Again, as they are discussing, walk around and listen in to the descriptions. Select a couple of groups that have come up with some good examples and descriptions.
- As they describe the products to the rest of the panellists ask them to describe in more detail: *what* actually changes, *why* they think it changes, *how* do they detect it changes and in *what* way does the product actually change.
- Have a general discussion about the various products and how the intensity changes over time.
- Give them a sheet of paper with the x and y axes marked (see Figure 7.10) and ask them if they could represent one of the products they chose or the music they listened to by way of a line on the graph. Some panellists may find this difficult. Do not share the panellists' outputs among the group.
- Introduce the product and the attribute (greasiness).
- Select two commercial products (samples A and B) that are very different in greasiness to demonstrate the attribute.
- Ask the panellists to apply one of the samples to the back of their left hand and the other to the back of their right hand and then to describe to their partner the difference in the feel of the products on the skin. Their partner will write down what greasiness means for them.
- Allow the panellists to swap: the other panellist tries the samples while their colleague writes down their description for them.
- While the panellists are doing these assessments, watch the application process carefully.

Figure 7.10 Blank time-intensity graph.

- Discuss the panellists' descriptions and the methods they used to apply the creams.
- Summarise the learnings from the session and explain the plan for the next session. Thank for the panellists for their work.
- Check back on the session objective – has it been achieved? Adjust next day's session plan accordingly.

Session 2: Development of protocol

Objective: to discuss and finalise the protocol for application

- Introduce the plan for the session and check everyone happy.
- Ask the panellists what they remember from the first session.
- Play them the piece of music from the first session.
- In the booths show the panellists the software for the collection of data.
- Ask them to use it to record what they remember about the volume/intensity of the piece of music. Tell them there is no wrong or right answer – it's just a chance to try out the software.
- Discuss how they found using the software. If they ask, explain why you cannot show them their curves – tell them that the output is far too complicated to show them.
- Give half the panellists sample A and half sample B and in their pairs ask them to discuss the protocol for application. Hopefully they will notice that it will be quite difficult to apply the cream and to rate the intensity of greasiness at the same time. See if they come up with a solution (the solution is that someone else applies the cream for them).
- Swap the groups so they have the other sample and ask them to design a protocol for the application of the cream by another person. This should also include an idea about the length of time that greasiness is perceived and therefore measured. The protocol also needs to include the washing and drying of the hands and deciding the length of time required before the applications can be made again on the same hand. This should fit in with the assessment of a product on the other hand.
- Listen in to the discussions and choose a couple that sound good to share with the rest of the group. Allow the others to add in their suggestions and write up the protocol on the board.
- Try the protocol in pairs with sample A and then sample B.
- Discuss the protocol – does it work for both commercial samples? Finalise the protocol.
- Thank the panellists for their contributions. Summarise the learnings from the session and explain the plan for the next session.
- Check back on the session objective – has it been achieved? Adjust next day's session plan accordingly.

Session 3: Trying out the protocol and software in the booths

- Introduce the plan for the session and check everyone happy.
- Ask the panellists what they remember from the first two sessions. See if there are any questions or queries about the protocol.
- Assess sample A again in the booths with the software to record the TI curves and check that the protocol works and the panellists are happy with the software.

- Allow the panellists to assess samples A and B again in their pairs and describe the differences in greasiness. Ask them how the two samples compare in greasiness overall.
- Conduct a practice profile session for the two commercial samples in the booths with the software to record the TI curves, with three replicates if possible.
- Summarise the learnings from the session and explain the plan for the next session.

Assess the panellists' consistency after the session and check what additional training might be required. Consistency can be measured by certain key curve measures such as maximum peak intensity, time taken to reach the maximum peak intensity and time taken to return to zero when greasiness is no longer perceived. Do not expect the panellists' individual curves to be similar: what you are aiming for is their individual measures to be consistent. At this stage of the training, the curves will not be consistent but this will give you a good starting point to measure the effectiveness of the later training sessions.

Session 4: Assessment of the sample set

- Introduce the plan for the session and check everyone happy.
- Ask the panellists what they remember from the first three sessions. Check if any queries.
- Tell them that the results from the practice session were very good and that you are very pleased with the progress so far.
- Assess the first sample in pairs so that a description of the sample 1 can be written in each panellists' book.
- Check the protocol is OK for the real samples. Edit if necessary.
- Repeat for samples 2, 3 and 4.
- Ask the panellists to make notes on how the samples compare as they assess each sample.
- Discuss how the samples compare.
- Assess the four samples in the booths using the software to record the TI curves.
- Check panellists happy with protocol.
- Replicate with all samples if possible in the time left.
- Summarise the learnings from the session and explain the plan for the next session.

Assess the panellists' performance after the session and check what additional training might be required. Do not expect the panellists' individual curves to be similar: what you are aiming for is their individual measures to be consistent. At this stage of the training, the curves should begin to be consistent for each sample. Check whether certain panellists need more guidance.

Session 5: Practice, practice, practice

- Introduce the plan for the session and check everyone happy.
- Ask the panellists what they remember from the first four sessions. Check if any queries.
- Assess the four samples in the booths with the software to record the TI curves.
- After each sample, check the protocol and discuss how the samples compare. Ask the panellists to individually rank the product in terms of greasiness intensity over time. Do the panellists agree?

- Assess the four samples again in the booths with the software to record the TI curves.
- Summarise the learnings from the session and explain the plan for the next session.

Assess the panellists' consistency after the session and decide if the data collection sessions planned can go ahead or if additional training session(s) are required.

Session 6 and 7: Data collection

- Introduce the plan for the sessions and check everyone happy.
- Check if any queries.
- Conduct four replicate evaluations of the samples.
- Thank the panellists for their hard work and ask them for some feedback on the training and sessions so that you can run better sessions next time.

7.6   Training for rapid methods

7.6.1   Introduction to rapid methods

The main point about rapid methods (for example, flash profiling, napping and ultra-flash profiling) is that they are intended to give faster results than conventional profiling methods, as in-depth panel training is not necessarily required and, as there is often no need for any group discussions, scheduling can also be done for each individual. Rapid methods are often proposed as a replacement for conventional profiling, and in some respects they are, but be aware that the output is generally in the form of a sensory space or map. If you need data to show in detail how sample X compares attribute-by-attribute across all modalities to sample Y, you might be better off with a conventional profile. If you are particularly interested in the difference in intensity of certain attributes you might like to select flash profiling or conventional profiling (not napping or ultra-flash profiling). If you simply wish to highlight the main differences and similarities between a set of products, or screen samples for more detailed analysis down the line, or if you do not have a descriptive profile panel trained and raring to go, rapid methods can be very useful.

There are many rapid methods in use and a great resource if you would like to find out more detail about each method is the book 'Rapid Sensory Profiling Techniques and Related Methods' edited by Julien Delarue, J. Ben Lawlor and Michael Rogeaux (Delarue et al., 2015). This book includes a huge amount of detail about each of the methods, along with some excellent industrial case studies, and also some additional interesting chapters about measuring emotions, understanding consumer behaviour and conceptual profiling.

You could say that the origin of rapid methods was Free Choice Profiling (FCP, Williams and Langron, 1984). This method involves consumers creating product profiles based on their own descriptive language (see Section 7.4.1 for more detail). However, one of the downsides to this method is that it involves consumers using a line scale and this can cause some issues. Flash profiling has some benefits in this aspect, as it is essentially FCP with ranking instead of the scale use. Two practical issues with both of these methods are creating the ballots/questionnaires and the product preparation required for each individual person: this can take quite a bit of time!

Some of the rapid methods, for example, napping and flash profiling, require all the samples to be presented at one time, which can cause issues with some products such as alcohol or some home and personal care products. This is an additional issue for flash profiling as all the products need to be assessed and ranked for each attribute, which can cause problems with some products (Petit and Vanzeveren, 2015). For example, if you are assessing hand creams it is quite easy to run out of skin, and assessment of many samples can actually be quite tiring and also a little confusing. Another issue with methods where all samples are presented at one time is with the serving of products that change rapidly; for example, frozen desserts, hot products, beers (for example, if the head is an important aspect in the assessment). For cold and hot products if you can keep your product cold (i.e., use specific containers) or keep the product hot (i.e., using an infrared lamp, Albert et al., 2011) you can still use these methods, but be careful that your product is not changing on standing and obviously you will take particular care about food hygiene issues. For products that change on standing, such as beer, you can prepare/pour each sample as the person requires it, but you will need to repeat this every time the person wishes to compare each sample to another.

If you already have a panel trained in descriptive analysis, the generation of a profile can actually be quite rapid, particularly if your panel has worked on that product before or they have a wide selection of attribute lists/lexicons to refer to. For example, it may take one to two sessions to assess all the samples and agree the attributes to be measured and then, however many sessions it might take to rate the samples, depending on how many samples there are and the project risk. If you are screening a range of samples to determine a smaller number for a consumer test, two replicates may well be enough and maybe even one replicate might suit if your panel are very experienced in that product and the project risk is very low. However, you may wish to use rapid methods if you do not have an experienced profiling panel, or your panel is busy with other work, or you wish to use consumers to assess the products.

Napping or projective mapping, as it is also referred, is one type of rapid method that can give descriptive information about your products. It can give a holistic view of your product (global napping), or the nappes can be created by modality or any other product usage category (partial napping). It generally takes just one session, but if you wish to create nappes for several modalities or you have several samples to assess, you may need more sessions. If you are conducting the test with consumers, you will need to run several sessions for the larger number of consumers required: the method might then not be particularly rapid (Lê et al., 2015).

Projective mapping was originally proposed by Risvik et al., in 1994. The main aim of the technique is to obtain a sample configuration in a two-dimensional space based on (dis)similarities among a sample set. Napping was later introduced by Pagès in 2005 with the application of a statistical analysis method, multiple factor analysis (MFA), to the projective mapping exercise. Napping is a type of projective mapping, which uses a certain shape and length of a two-dimensional space (40 cm × 60 cm) for its data collection and analyses the data only with MFA. The name comes from the French word for tablecloth, as the products are placed on the table (you do not need the cloth) in terms of how the panellist sees the similarities and differences. It has two main outputs. Firstly, you will get an idea of how the samples compare, which was

Table 7.5 Napping output

Sample	Panellist 1	Panellist 2	Panellist 3	Panellist 4...
1	X= Y=	X= Y=	X= Y=	X= Y=
2	X= Y=	X= Y=	X= Y=	X= Y=
3	X= Y=	X= Y=	X= Y=	X= Y=
4	X= Y=	X= Y=	X= Y=	X= Y=
5	X= Y=	X= Y=	X= Y=	X= Y=
...				

why the original projective mapping method was created, and secondly, if used with consumers, you will get an understanding of how the consumers view the products by their choice of attributes that discriminate (or do not) the sample set. This second application is one of the main reasons for the creation of the napping method in contrast to projective mapping (Lê et al., 2015).

The data from napping (i.e., the coordinates: see Table 7.5) need the additional descriptive information from the panellists/consumers about the products to help with the interpretation of the data and to produce an output that might be used instead of a conventional profile. This is when the method becomes called ultra-flash profiling as opposed to napping, as napping does not include descriptions.

Napping tends to be performed on 10 more or more products because that is where we see the most benefit from the method, but the number will depend on the product and the type of panellist. The method relies on the fact that the samples can be assessed throughout the testing period and therefore cooked hot products or products that change on standing are more difficult to assess using napping. Consumers can probably assess 10–15 products and a trained panel possibly more.

The training is minimal for a trained analytical panel and can be accomplished by giving the panellists a handout as shown in Figure 7.11 and having a short discussion. With consumers you may wish to do a short introduction to the method and tell them what aspects of the product you are interested in. For example, you might wish the consumers to place the products on the map according to their packaging only or you might be interested in just aroma, or just appearance (often referred to as partial napping). You could of course repeat the process for various modalities. You could also give the consumers or panellists an example of the output from this method as shown in Figure 7.12.

Once the panellists have finished, your job will be to measure the sheets of paper and put the information into a table form. The output will look something like that in Table 7.5. If you have computer software to run the test this will save you some time, but if you are running this test with consumers you may need to train them to use the software first.

The analysis of the data can be carried out using open source software which is another benefit of this method. Please contact the book author if you would like an instruction sheet on how to collect the data and do the analysis using SenoMineR and FactoMineR.

7.6.2   Session plan and information for ultra-flash profiling

The session plan for this section is for performing descriptive napping, or as it is some-times called ultra-flash profiling, on 15 soft drinks. The analysis can be performed with naïve consumers, semi-trained panellists or trained panellists depending on your objective for the test. For example, if you wish to find out which attributes or aspects of a product are important to consumers of your product, you will need to recruit consumers to conduct the test. These consumers will need to be users of the product but also good communicators so that you can use and understand their descriptions.

If you are using consumers you will need more people than if you are using trained panellists. If you wish to incorporate some element of liking into your analysis you will need to use even more consumers. Across the different publications there is a wide range of consumers used: from 8 to 100. The number of people you need will depend on your objective, project risks and what you intend to use the information for.

In some publications a second or third replicate is performed, however, many of these experiments were related to determining the robustness of the method and comparing it to conventional profiling. If you are interested in checking whether you would get the same result on another occasion, you might like to replicate your experiment; however, this does detract from the rapidness of the method. Another option would be to include a couple of samples twice to help check for consistency for each panellist. For a good discussion about replication see Hopfer and Heymann (2013) and Delarue et al. (2015).

Where panellists have not assessed the product type before, you might like to con-sider assessing a subset of samples as a group and discussing the descriptions (Liu et al., 2016). This will make life easier when analysing the various descriptors. For example, consider a situation where the consumers are assessing the appearance of hair by the use of mannequins. If some consumers use the term 'volume' to describe the differences in the sample set (for example, no volume, medium volume, lots of volume) and others use the term 'flat' (for example, very flat, not so flat, not at all flat), these descriptors may be quite easy to combine and compare, but in other cases it is not always so obvious that the words used mean the same thing. If a panellist does not use the term volume or flat in their descriptions can we assume, for exam-ple, that 'greasy looking' might be the same as 'flat?' Or did they not notice that the volume was different across the range of mannequins? Another solution if you think the descriptors may prove difficult to analyse, is to present the panellists with a list of words they might use to describe the products in a similar way to CATA methods.

You might also like to consider a short training exercise for the panellists similar to that suggested by Hopfer and Heymann (2013, p. 169). These authors wanted to make sure that everyone understood the concept of using a two-dimensional space to place their samples and so they carried out a task assessing various paper shapes. The shapes were made up of seven different colours, six different shapes and two different sizes and in each set of shapes there was an 'odd' sample that was a single colour or a single shape. With any of the training methods suggested, be careful that you do not change the original concept of the method and miss the objective of using the method (Kennedy and Heymann, 2008). This is particularly important if you are using nap-ping to understand which attributes consumers might find important in your product set, as any training you might give in this situation may bias the consumers. However,

Liu et al. (2016) found that training on the method or the products gave more robust results than classical napping.

Before the session

Assign the 15 samples a three-digit code and record the details on a spreadsheet.
Check the codes are not confusable (i.e., do not have 121 and 211).
Print out the labels.
Label each sample with its three-digit code.
Label glasses to match the samples.
Make sure the product brand is not visible to the people taking part.
Each person will need a sheet of paper at least 60 cm × 60 cm. We will be measuring these later so we need paper that is not too flimsy or easily destroyable.
Book a room where each person can have a table to themselves to lay out their paper-work and samples, as the sensory booths are too small to conduct this test easily. This room should meet as many of the requirements for sensory testing facilities as possible.

On the day

Give everyone the instructions (see Figure 7.11), a pencil, a decent pencil eraser and a sheet of paper at least 60 × 60 cm. The more samples you have the more paper you might need.
Get the samples ready.
Have some crackers and water available.

Instructions for today's assessment
- Please assess the samples according to your own criteria:
  - This can be based on both qualities (the description of the product) and also intensities (the level of the descriptors within the samples).
  - Please think about the appearance, aroma, flavour and mouthfeel of the product.
- You will be given a set of samples to assess for appearance, aroma, flavour and mouthfeel.
- Once you have assessed the first sample, place the glass on the large sheet of paper in front of you – *handy tip: as it's the first sample and you don't really know how the others will compare, just place it down anywhere – you can move it around any time.*
- You have some notepaper - note down the main characteristics of this sample (and its code!!) as this will be useful later…
- Remember to cleanse your palate – there is water and crackers are also available.
- Assess the next sample and write down some notes to describe it.
- Then place the second sample on the sheet so that if the samples seem identical to you they are near one another and those that are different are further away from each other.

- Repeat for all samples:
  - Place them on the paper so that if the samples seem very similar to you they are near one another and those that are different are further away from each other.
  - Remember to note down the main characteristics for each sample on your note paper.
  - You can move your samples around at any time to suit how you would like to group them.

- Once you've assessed all the samples check your positioning – are you happy? If not, shuffle the samples about.
- Now you are ready to write on the sheet you have placed the samples on.
- Pick up the first sample and write its code on the sheet where it had been located.
- Next to the code write the main characteristics of this sample.
- Put the samples in a row to one side as you finish copying down the code and the descriptions.
- If you have groups of pretty much identical samples, you can write down all the codes and draw a ring around them. Remember to write their main characteristics or why you grouped them together.

Figure 7.11 Panellists' handout describing ultra-flash profiling.

Example for sweets:

Figure 7.12 Example ultra-flash profiling for sweets (hypothetical).

Go through the instructions and example (see Figure 7.12) with the panellists and take questions.

If you are assessing a small number of samples first to discuss the attributes used, do that now. Then move on to the test itself.

While people are working, make notes of any issues – samples that are difficult to assess and people who found the task difficult.

You will also need a metre rule for measuring and a laptop to type in all the data.

7.6.3   Session plan and information for check-all-that-apply

The CATA method is also known as a rapid method to determine product profiles with consumers alongside hedonic data (Meyners and Castura, 2014; Ares and Jaeger, 2015). The method involves asking consumers to select words from a previously derived list that describe the product as opposed to completing open-ended comment-type questions. It is suggested that 100 or more consumers are required for the test (Meyners and Castura, 2014). The CATA list may include product attributes such as flavour or texture descriptors, product usage or emotional terms, for example. The lists can be generated from a descriptive profile, from a focus group or from the same consumers. They are also often created from previous consumer research. The lists need to be presented to the consumers in a sensible manner. For example, listing alternate flavour and texture attributes might not be a sensible approach if the flavour attributes are generally perceived before the texture attributes. The lists will also need to be balanced and randomised between consumers, although each consumer can have the same order for all products they assess. CATA data are generally analysed using contingency tables although other methods such as MFA and correspondence analysis can also be used.

This session plan for a CATA session is for a selection of commercially available granolas. To determine the list of CATA questions for the assessment, a group of 12 consumers were selected on the basis of their descriptive and communication abilities. These consumers did not take part in the main part of the CATA assessment. The session plan is for the simple training of these consumers to generate the CATA lists.

7.6.3.1 Session plan

Session 1
- Introduce the panel to the objective of the session: generation of the CATA list.
- Show a short presentation about the type of sensory attributes we would like the panellists to generate, differentiating between hedonic and descriptive terms.
- Ask them to assess sample 1 and make a list of words to describe the aroma, flavour and texture of the product.
- Ask each consumer to read out their list and mention that they are free to borrow words from other panellists if they would like to.
- Correct the usage of any hedonic terms.
- Ask them to assess sample 2 and describe it with a list of words.
- Ask each consumer to read out their list and mention that they are free to borrow words from other panellists if they would like to.
- Make a list for each modality on the flip chart/white board/laptop.
- Discuss how the samples compare.
- Repeat until all six samples have been assessed.
- As words are added to each modality, discuss which words are synonyms for others and which might be combined.
- Discuss the order of the terms in the list as this can help shorten the list if the attributes are sorted according to importance.
- Create the list of words for the CATA study.

Session 2
- Introduce the panel to the objective of the session.
- Present each of the samples to the consumers in a balanced order with the CATA list randomised between consumers.

- Gather the consumers together and ask for feedback. The feedback could also be gathered after the CATA session through the sensory software if you wish.
- Present the results to the consumers and discuss the CATA list.
- Refine the list for use with the consumers.

7.7 Validation of panel training

Once your panel is trained you will need to check that the training has been successful, or in other words validate your training. If you have trained the panel in discrimination tests, a simple way to validate the training is to present the panellists with various samples with known differences as detailed in Section 7.3. You may also have samples from previous tests (for example, consumer tests and descriptive profiles) that you can use for the discrimination tests. Arrange a selection of tests that will assess the panellists' capability in certain products and in different test types. If you have trained the panel in five different discrimination tests, develop two or three known sample comparisons and repeat some with different tests. For example, you might run three each of A-not-A, same-different, triangle tests, duo-trio tests and tetrads, with samples where the level of an ingredient has been reduced; with the reduction decreasing over the three tests and/or with diluted drinks gradually lowering the level of dilution.

If your training has involved quality control tests you could use the sample suggestions in Section 7.3 as well as previous samples that have caused issues or have been rejected, to check that the panellists give the answer expected. In fact, this type of test is important to include regularly for quality control tests as one of the errors associated with this type of testing is the habituation error (Kemp et al., 2009). For example, if your production is generally error-free the panellists may well expect each sample to pass and therefore may miss important differences. If you include an edited sample in a test every time tests are being run, you can confirm that your panel are working as expected and hence validate the results on test-by-test basis.

Validation of qualitative and quantitative profiling can follow the same type of approach but requires a bit more planning, as if you are able to repeat your validation study at certain intervals you will be able to validate your panel output year on year. ISO 11132, 2012(E) (*Guidelines for Monitoring the Performance of a Quantitative Sensory Panel*) suggests that validation can be carried out by using a recent data set (i.e., you do not need a validation sample set) if there are attributes which statistically significantly discriminate the samples. Alternatively, if you have a set of samples designed specifically for validation, the standard suggests that at least one pair of the samples should show differences for at least eight attributes. Either way, the product type should be similar to those the panel usually assess. If you decide to create samples for validation purposes, consider the sample set carefully so that the samples will be available in the future. For example, you could create samples specifically for validation studies with known formulations and differences. If you are able to create six or more samples for the validation you will also be able to use multivariate statistical methods to assess the panel's output.

You can also validate the results from one profile via the replicates collected and via panel performance measures (see Chapter 11). If the panel are about to complete their first ever quantitative profile, including several more replicates than would usually be planned, can be very helpful in assessing the panel performance. Or you can design their first ever profile to be on a set of samples that have known differences, for example, a product which differs in the level of an ingredient or has different processing or treatment. One or more samples repeated in the set can also be helpful to check that these two samples are seen as similar. You can also use discrimination tests to check that the panel can discriminate the samples by both methods.

Another approach to validating qualitative or quantitative profiles could be by simply repeating a previous study with your panel and determine if the results are similar. You will need to edit the sample set so that sample 1 becomes sample 5 and sample 5 becomes sample 3, for example, or the panellists will quite quickly notice that the sample set is the same. Repeating a sample to increase the sample number by one (or two) can also be helpful in disguising the fact that the profile is being repeated.

The validation of rapid methods can be performed by simply repeating the test and seeing how consistent the results are. For example, if you were repeating a napping study you could compare the RV coefficients for three or four test replicates as an indicator of repeatability (see, for example, Louw et al., 2013). Repeating the same samples in each experiment can also help to validate rapid methods on a test-by-test basis.

You might be involved in a sensory ring trial or proficiency testing (see Chapter 11 for more information), and this can be used as your validation if the product type is close enough to the products you assess regularly. Of course, if you have several panels within your company you can set up your own ring testing methodology that can be used to validate the language generated as well as the data produced by each panel (for example, Drake et al., 2002; McEwan et al., 2002, 2003). If you have access to instrumental data for your samples, this can also be used to validate your results. For a detailed example see Martens et al. (2013).

Product assessment/orientation for sensory panels

8.1 Introduction

There are many different types of sensory panel. Some panels are recruited from internal staff (for example, the finance manager or a team administrator) and these tend to work on sensory methods that are quick to complete, such as discrimination or quality control type tests. Other types of panels might be recruited from external applicants because the methods the company needs to use take several hours, such as quantitative descriptive profiling or time-intensity methods. For more information about various panel types see Section 2.1 and their recruitment processes, Section 2.6.

Both types of panel, internal or external, might work on one or two product types, or in some companies they might work on many hundreds, depending on what the company makes and where. Sometimes a company may well have several panels working on different products or types of products. This is common for sensory agencies.

This chapter assumes that the panel has been through the training process (see Chapters 6 and 7) and has completed several discrimination or quality control type tests, or if they are training to be a quantitative profile panel, that they have completed at least one profile and perhaps also a validation profile (for more information see Chapter 7).

When a profile panel is about to start creating a profile for a new set of products, this is often referred to as the 'training' session, however, this terminology can be misleading for other people in the company who might hear the word 'training' and assume that the panel is inexperienced. A better word might be orientation.

The information in this chapter can also be used when conducting orientation or product training for an experienced panel. For more information about the different steps involved in creating product profiles please refer to Chapter 7. The chapter also gives information on how to introduce an established panel to using a new method. BS EN ISO 13299 (2016) 'General Guidance for Establishing a Sensory Profile' is a very informative document when developing qualitative or quantitative sensory profiles. The document also gives a very useful overview of the various descriptive profiling methods.

8.2 Session plans for a discrimination or quality control type panel

8.2.1 Session plan for a discrimination panel

This session plan is for a panel who generally conduct discrimination tests but are being introduced to a new method for odour masking: the ABX task (Greenaway, 2017). The ABX task has three test samples: all are labelled with a three-digit code but two are presented together first. One of two samples is the control product (A)

ABX Task

Name: **Date:**

Please taste the samples in the order presented on the tray, from left to right.

In between assessing each product please take a drink of water.

Once you have assessed both products coded as shown below, please request your third sample.

Your task is to identify if this third product more closely resembles the first product you assessed (517) or the second product you assessed (823).

Please indicate your response by circling your answer below:

              517   **823**

Please write any comments below:

Figure 8.1 Panellists' worksheet for the ABX task.

and the other is the modified product (B). After the panellist has had time to compare the samples they request to see the final three-digit coded sample, sample X, which is either the control or the modified product. The participant has to decide which sample X most closely represents: A or B. An example of the panellist's worksheet is shown in Figure 8.1. In this example of training an established panel in a new method, there is one panel session which includes both the training and the collection of data.

Session plan

- Introduction to the new method and a demonstration of the method in the discussion room using Farnsworth Munsell colour chips or printed squares of coloured paper (15 minutes).
- Dummy test in the booths with known samples that are fairly easy to match (see details below). Panellists get immediate feedback on their result and time to reassess the samples once they know the answer (20 minutes).
- Questions and answers back in the discussion room (10 minutes). Samples available to reassess if required.
- Dummy test 2 if necessary dependent on the results of dummy test 1 (20 minutes).
- Break (10 minutes).
- Completion of the first three discrimination tests (50 to 60 minutes) with a 10-minute break between each test. No feedback on results is given. A selection of puzzles (crosswords, word searches, quizzes) is available in the discussion room for the panellists to complete if they wish while they wait.
- Finish up (10 minutes): ask if any questions. Thank the panellists for their hard work. Explain the plan for the next session.
- Check back on the session objective – has it been achieved? Adjust next day's session plan accordingly.

Panellist	Set 1 (dummy)	Set 2 (odour masker X)	Set 3 (odour masker Y)	Set 4 (odour masker Z)
Panellist 1	AB-A	BA-B	BA-A	AB-B
Panellist 2	BA-B	BA-A	AB-B	AB-A
Panellist 3	BA-A	AB-B	AB-A	BA-B
Panellist 4	AB-A	BA-B	BA-A	AB-B
Panellist 5	BA-B	AB-A	AB-B	BA-A
Panellist 6	AB-B	AB-A	BA-B	BA-A
Panellist 7	BA-A	AB-B	AB-A	BA-B
Panellist …	AB-B	BA-A	BA-B	AB-A

Figure 8.2 Example presentation design for ABX training.

Training samples: Sample A (standard with added malodour) and Sample B (standard with no added malodour).

Test samples: Sample A (standard) and Sample B (three trial products with odour maskers X, Y and Z).

There are four possible sample presentation orders where the latter sample is the 'X' in the ABX nomenclature:

- AB-A
- BA-B
- BA-A
- AB-B

Each presentation order should be used an equal number of times and rotated across the design to reduce presentation order bias (Rousseau et al., 1998). The presentation design used for the training is shown in Figure 8.2. If a second dummy trial is required, reverse Set 1.

8.2.2 Session plan for training a quality control panel on a new product line

This is a three-session plan for the training of a group of panellists who work in quality control in a factory that produces fragrances. The panel are regularly assessing and checking samples from various product lines. The purpose of the session is to introduce the assessment of a new product and to create a sensory specification for that product. A sensory specification is similar in format to other product specifications: the only difference being that the panellists describe what the product should look like, smell like and taste like, for example. Sensory specifications are a vital part of ensuring product quality and if developed in line with consumer measures, they can be even more valuable. The results from tests involving sensory specifications are assessed by the sensory scientist and do not involve the panellists making any decisions about the quality of the product (Muñoz, 1992). The method therefore gives very actionable results which can also be correlated to instrumental measures.

The first of the 30-minute sessions involves conducting difference from control (DFC) tests (Whelan, 2017a) for each of the three batches from first production of the fragrance to understand production variability. As there is not enough time for a third DFC within the session, the panellists are called back later in the day to take part in the final DFC. The DFC data are assessed prior to the third session, and the number of samples needed to create the sensory specification for the second production run decided. The second session includes a reminder about sensory specifications and a qualitative assessment of a product. The third session creates the first draft of the sensory specification for use during the second production run. Once the draft specification is ready to go, the panellists will also assess the fragrance ingredients to check. The whole process will be repeated and the final specification drawn up.

Session plan

Session 1

- Welcome and thank you to the panellists for their attendance.
- Introduction to the session and agreement for the plan (30 minutes).
- Panellists are given a reminder about the DFC test and given a printed instruction sheet to read through and ask any questions (see Figure 8.3).
- Panellists conduct three DFC tests from the first three batches of the first production run. Within each DFC test the control is repeated as a hidden control to validate the data.
- The first DFC compares samples across the three batches to determine the variability for production run 1.
- The second DFC compares products within the third batch to determine variability within a batch.
- The third DFC compares products within the second batch to determine variability within a batch (conducted later in the day).

Session 2

- Welcome and thank you to the panellists for their attendance.
- Introduction to the session and agreement for the plan.

Part 1: Reminder about sensory specifications (15 minutes).

- Presentation about sensory specifications.
- Read through two or three specifications.
- Discussion about the important elements of a sensory specification.
- Write list up on the flip chart for everyone to refer to when making assessments

Panellist ID:_____   Date:_____   Time:_____

Instructions:

   a) You are provided with five samples, a control sample labelled 'C' and four test samples labelled with three-digit codes.

   b) Evaluate the control sample first, wait for 30 seconds and then evaluate the first test sample.

   c) Determine if the test sample is different from the control and record the magnitude of that difference on the scale below by adding a cross to the appropriate box. If you do not perceive a difference, please add a cross to the box marked 'No Difference'.

   d) If you determine there is a difference between the control and test samples, write in the comments section, in what respect they are different.

   e) Repeat for the remaining three test samples

Sample code	0 No difference	1 Very slight difference	2 Slight difference	3 Moderate difference	4 Large difference	5 Very large difference
771						
209						
384						
526						

Remember that a duplicate control may be the test sample some of the time.

Comments:

_____

_____

Figure 8.3 Example questionnaire for the difference from control test.
Adapted from Whelan, V.J., 2017. Ranking test. In: Rogers, L. (Ed.), Discrimination Testing in Sensory Science. A Practical Handbook.

Part 2: Qualitative assessment of product (15 minutes).

- Each panellist is given a sample of the product from the middle of the batch and asked to describe the aroma of the product.
- Ask two or three panellists to read out their descriptions.
- Collect attributes on the board by asking each panellist for one descriptor.
- If time, begin to discuss the attributes that could go into the specification.

Session 3

- Welcome and thank you to the panellists for their attendance.
- Introduction to the session and agreement for the plan.

Part 1: Qualitative assessment of product (10 minutes).

- Each panellist is given a further sample of the product based on the DFC results and asked to describe the aroma of the product.
- Collect attributes on the board by asking each panellist for one descriptor.

Part 2: Draft sensory specification (20 minutes).

- Discussion and consensus about the sensory specification.

8.3   Session plans for a qualitative descriptive profile panel

Qualitative profiles can be very helpful for many different types of sensory projects, particularly new product development (NPD). For example, if the objective of the NPD project is to make a product with a natural peach flavour, asking the panel to describe the new product (or a series of prototypes, e.g., from a flavour house) and determining if it is described as natural peach, could be a very useful first step, saving time on a more detailed quantitative profile. If you are planning a qualitative assessment of products in relation to a reference product, you might be interested to read more about the pivot profile (see Section 7.4.1.5).

The example below is a plan for the qualitative assessment of a range of twenty malodours. The panel have been trained to assess various malodours but have not yet worked on the malodour used in this study. The panel have been in regular attendance completing malodour assessments for the past year. The objective of the study is to describe each of the malodours to determine which will go forward to quantitative profiling.

8.3.1   Session plan for a qualitative profile

Objective: to remind panellists about the key pointers for writing a good qualitative description. To assess and compare the twenty malodours.

There are twenty malodours to assess, but as only five can be assessed on the first day, we need to include some malodours that repeat every day (for example, malodour E) and some that repeat over sessions to check panellist consistency.

Malodour A	*Malodour B*	Malodour C	Malodour D	**Malodour E**
Sample 21 (session 4)	Sample 7 and 12 (session 2)	Sample 19 (session 3)	Sample 15 (session 3)	Sample 5, 11, 14 and 24 (sessions 1, 2, 3 and 4)
Malodour F	Malodour G	Malodour H	Malodour I	Malodour J
Sample 1 and 3 (session 1)	Sample 8 (session 2)	Sample 13 (session 3)	Sample 9 (session 2)	Sample 17 (session 3)
Malodour K	Malodour L	Malodour M	Malodour N	**Malodour O**
Sample 15 and 18 (session 3)	Sample 23 (session 4)	Sample 10 (session 2)	Sample 22 (session 4)	Sample 2 and 16 (session 1) and (session 3)
Malodour P	Malodour Q	Malodour R	**Malodour S**	*Malodour T*
Sample 20 (session 4)	Sample 24 (session 4)	Sample 6 (session 2)	Sample 4 and 9 (session 1) (session 2)	Sample 21 and 23 (session 4)

Bold, sample is repeated across sessions: E, O, S; *Italics*, sample is repeated within a session: F, B, K and T.

Session plan

Session 1

- Project leader from previous project coming in to say thank you for their work on Project Y (5 minutes)
- Describe the plan for the next five sessions and get the panellists' agreement (5 minutes)
- Give out three qualitative descriptions (OK, good and excellent) to each group of three panellists and ask the panellists to critique them (15 minutes)
- Group discussion about what makes a good qualitative description (15 minutes). Write up a list of the panellists' ideas on the flip chart/screen. Make sure the list includes: detailed, understandable descriptors; references; comparisons across samples; headings where needed, etc.
- Note the sample table and that several samples are assessed twice. This is so that the panellists' descriptions can be checked/validated day-to-day and to also allow comparison across days and samples.
- Assessment of **samples 1 to 5** (60 minutes): assess sample 1 in the malodour booths and then return to the discussion room. Ask three to four panellists to read out their descriptions. Ask for clarification of any words used, e.g., where have you smelled that odour before? Check that everyone is happy to assess sample 2. Return to the discussion room and ask three to four panellists to read out their descriptions.
- Ask the panellists how samples 1 and 2 compare. Ask the panellists to check through their descriptions to see if they meet their earlier requirements they listed.
- Break (5 to 10 minutes).
- Assess sample 3. Return to the discussion room and ask three to four panellists to read out their descriptions. Ask the panellists how samples 1, 2 and 3 compare.

- Continue until all samples have been completed. Ask the panellists how samples 1 to 5 compare. You could ask them to group the samples by similarity using Lauren's buckets or any other method of sample comparison (see Section 6.3.8 for more information).
- Finish up (15 minutes): ask the panellists to check through their descriptions to see if they meet their earlier requirements they listed. Ask them to think about references for homework for any of the more difficult to describe aspects of the malodours. Thank them for their hard work and explain what they will be doing in session 2.
- Check through the panellists' notebooks and comparison sheets. Have they grouped samples 1 and 3 as expected?

Session 2

- Recap from yesterday. Check that all panellists were present and go over the description criteria if not. You could also go over the description criteria if you think that the descriptions you read yesterday were not meeting your expectations. Ask two or three panellists to read out some descriptions to remind everyone of the type of descriptors being used (10 minutes).
- Assessments of **samples 6 to 12** (90 minutes). Assess sample 6 in the booths and then return to the discussion room. Ask three to four panellists to read out their descriptions. Ask for clarification of any words used? Ask the panellists how sample 6 compared to session 1's samples. Check that everyone is happy to assess sample 7. Assess sample 7 in the booths
- Return to the discussion room and ask three to four panellists to read out their descriptions. Ask the panellists how samples 6 and 7 compared. Ask them to create Lauren's buckets for today's samples. (It's best not to ask panellists to do buckets covering several days unless they are very experienced or have assessed the samples several times.)
- Repeat the assessments, reading and comparisons, giving time for the panellists to think about their sample groupings for the day, for the next samples. Plan in a 10-minute break midway.
- Finish up (15 minutes): ask the panellists to check through their descriptions to see if they meet their earlier requirements they listed. Ask them to think about references for homework for any of the more difficult to describe aspects of the malodours. Thank them for their hard work and explain what they will be doing in session 3.
- Check through the panellists' notebooks and comparison sheets. Have they grouped samples 7 and 12? Does the description from session 1 for sample 5 match that of sample 11?

Session 3

- Recap from yesterday. Ask two or three panellists to read out some descriptions to remind everyone of the type of descriptors being used (10 minutes).
- Assessments of **samples 13 to 19** (90 minutes). Assess sample 13 in the booths and then return to the discussion room. Ask three to four panellists to read out their descriptions. Ask for clarification of any words used. Ask the panellists how sample 13 compared to session 1 and 2's samples. Check that everyone is happy to assess sample 14. Assess sample 14 in the booths

- Return to the discussion room and ask three to four panellists to read out their descriptions. Ask the panellists how samples 13 and 14 compared. Ask them to create Lauren's buckets for today's samples.
- Repeat the assessments, reading and comparisons, giving time to think about their sample groupings for the day, for the next samples. Plan in a 10-minute break midway.
- Finish up (15 minutes): ask the panellists to check through their descriptions to see if they meet their earlier requirements they listed. Ask them to think about references for homework for any of the more difficult to describe aspects of the malodours. Thank them for their hard work and explain what they will be doing in session 4.
- Check through the panellists' notebooks and comparison sheets. Have they grouped samples 15 and 18? Does the description from session 1 and 2 for samples 5 and 11 match that of sample 14?

Session 4

- Recap from yesterday. Ask two or three panellists to read out some descriptions to remind everyone of the type of descriptors being used (10 minutes).
- Assessments of **samples 20 to 24** (60 minutes). Assess sample 20 in the booths and then return to the discussion room. Ask three to four panellists to read out their descriptions. Ask for clarification of any words used, e.g., where have you smelled that odour before? Ask the panellists how sample 20 compared to session 1, 2 and 3's samples. Check that everyone is happy to assess sample 21. Assess sample 21 in the booths
- Return to the discussion room and ask three to four panellists to read out their descriptions. Ask the panellists how samples 20 and 21 compared. Ask them to create Lauren's buckets for today's samples.
- Repeat the assessments, reading and comparisons for the next samples, giving time to think about their sample groupings for the day. Plan in a 10-minute break midway.
- Finish up (30 minutes): Thank them for their hard work and arrange for refreshments and cakes/biscuits.
- Check through the panellists' notebooks and comparison sheets. Have they grouped samples 21 and 23? Does the description from session 1, 2 and 3 for samples 5, 11 and 14 match that of sample 24?
- Write your report including the summarised descriptions of each of the malodours and your recommendation as to which samples should go through to quantitative profiling.

8.4   Session plans for a quantitative descriptive profile panel

Two examples are given in the section: one for a food product (gravies) and the second for a home and personal care product (creams).

Prior to developing the session plan you will need to make sure that you are clear on the project objective and action standards. Check that a quantitative profile is required and go through the steps as outlined in Section 4.2.10. For example, ask yourself if a qualitative profile might be good enough. Think about the experimental design and consult a statistician about how you might analyse the data. Do this *before* collecting the data as you might find that you are unable to analyse the data how you imagined because of the way you designed the experiment.

You will also need to have asked all the relevant questions about sample preparation and presentation, plan how the test is to be conducted (for example, how many sessions will be needed, when the panel sessions will be, if replicates are required, how much sample is needed, how the data will be collected, etc.) and when the report is required. You will then be able to work on a plan that will hopefully meet all the requirements of the project. If you do not think there is enough time in the plan to collect the data you feel the client needs, it will be worthwhile looking at different sensory test options and discussing these with the client. You could show the client some expected outcomes and see if they are happy to change to a different method or if they are able to extend the deadline to enable you to complete the original plan.

If your panel has never worked on a descriptive profile before, you might like to follow the outline plan as shown in Figure 8.4. If the panel has created several profiles before on various product types and is now required to create a profile on a product they have never worked on before, you could follow the outline plan as shown in Figure 8.5. If the panel has created several profiles before on different product types and is now required to create a profile on a product they have worked on before, you could follow the outline plan as shown in Figure 8.6.

Decide on your profile approach: are you using a pre-developed language or are you developing the language with the panel? Are you using a Quantitative Descriptive

Initial training	Introduction to ways of working
	Introduction to sensory science and test methods
	Introduction to scaling
Method training and validation	Training how to assess products
	Language development training
	Create language for product
	Conduct validation profile
Conduct profile	Assess range of products and develop language
	Collect practice profile data
	Check panel performance
	Feedback and remedial actions
	Collect final data set

Figure 8.4 Steps to create a descriptive profile – naïve panel.

Analysis (QDA) type of scaling approach or quantitative references? Do the panel require any training prior to the study starting? Check the facilities, resources and equipment you might need for the study. Once you have all that information ready, you are good to go. A more detailed overview for the development of a QDA-type profile with trained panellists is given in Figure 8.7.

The example session plan below is for a quantitative descriptive profile panel who have previously created several profiles for yoghurts and seasonal vegetable products

Conduct orientation	Assess a range of products
	Develop the protocol for assessment
	Develop the list of words/attributes to be used to describe the products
	Present qualitative references
	Finalise attribute list (to include: references, definitions, protocols for assessment, anchors and order of assessment)
Conduct profile	Collect practice profile data
	Check panel performance
	Feedback and remedial actions
	Collect final data set

Figure 8.5 Steps to create a descriptive profile – trained panel, new product category, QDA-type approach.

Conduct orientation	Assess the range of products
	Check existing language. Edit if necessary.
	Finalise attribute list (to include: references, definitions, protocols for assessment, anchors and order of assessment)
Conduct profile	Collect practice profile data
	Check panel performance
	Feedback and remedial actions
	Collect final data set

Figure 8.6 Steps to create a descriptive profile – trained panel, same product category, new sample set.

Step	Detail
1	• Panellists assess the first sample in the set for the modalities of interest. • Samples may be competitor samples, different versions of the product (e.g., low fat, increased foaming), different points across shelf life, etc. • Attributes (words to describe their peceptions) are collected from every panellist. • A protocol (agreed way of assessing samples) is drafted.
2	• Panellists assess the second sample in the set for the modalities of interest. • Attributes (words to describe their perceptions) are collected from every panellist. • Attributes begin to be discussed and defined, starting with easier modalities first. • The protocol is agreed (but can be edited if new samples require a change).
3	• Panellists assess each of the samples in the set for the modalities of interest. • The protocol is agreed and finalised. • Qualitative references are assessed to help define the attributes so that the panellists agree and understand the element of the product being measured. It's best to leave the qualitative references until all the samples have been assessed so as not to introduce bias. • Panellists discuss, compare and describe the sample set.
4	• All samples are assessed at least twice while the attribute list (or language) is developed and finalised.
5	• A practice profile session is designed. • It includes a subset of the samples to check panel performance, replications, agreement and interactions and scale use. • Data are gathered for the practice profile subset.
6	• Analysis of the practice profile data. • Feedback session held with the panellists to correct any issues.
7	• Replicated assessments of the sample set. • Generally three replicates are collected, but if the panel are new to descriptive profiling you might like to conduct more.
8	• Panel performance checks. • Data analysis. • Report writing and presentation.

Figure 8.7 Detailed overview for the generation of a QDA-type profile with an experienced panel.

but have never worked on the product to be profiled (gravy). If your panel has not conducted profiling sessions before, you might find it useful to read the session plan in Section 9.2.1. You will notice in the training/orientation sessions that the products are labelled 1, 2, 3 and so on instead of labelling the products with three-digit codes. I tend to use this approach with all my profiles for several reasons. Firstly, it allows easy reference back to samples, for example, you might hear a panellist say, whilst looking at their notebook, 'I think this sample is very similar to sample 3 we saw last week'. Secondly, it allows the panel to use the sample 'name' (1, 2, 3...) to build up a picture for themselves, and to also share with other panellists and the panel leader, of how the samples compare by using comparison tables or sketches (see Section 6.3.8 for more detail). You can also use the sample 'name' to check for attribute understanding or detection by presenting an unnamed sample and asking, 'Is this the sample with attribute X or not?' or 'Which sample is this?' If all the panel can recognise the sample as sample 4 (and it was sample 4!) then you can be pretty sure that the sample is a good demonstration of that attribute. Of course, the samples are labelled with 3-digit codes during the rating sessions.

You might also notice that if you add up all the times set aside for each activity they do not add up to the total session time (three hours). This is because it's always a good idea to give yourself a bit of leeway when developing the plan as things can often take a little bit longer or even be quicker than you imagined. If you do find yourself finishing early then take a quick look at the plan for the next session – can you bring anything forward? If not, then have a series of tasks ready for the panellists to do. For example, discuss the assessment protocol in pairs and check that the protocol is the same, or go through the attribute list and consider the order in which the attributes should be assessed. In these example sessions I have written lots more information that you will need to write for your sessions because I also wanted to share some handy hints and explain my reasons for doing certain things.

8.4.1   Session plan for a quantitative profile example 1: food product

Product orientation sessions profile panel

Session 1

Objectives: Assess three of the samples, generation of attribute list, draft protocol (3 hours).
You will need:
   Serving cups, serviettes, pens, notebooks, etc.
   Water and cups

Samples

Sample number	Description
1	Gravy competitor 1
2	Gravy competitor 2
3	Current gravy product
4	New gravy product prototype 1
5	Gravy competitor 3
6	New gravy product prototype 2

Introduction (10 minutes): describe the plan for the next few sessions, for example, which are attribute generation sessions and when the profiling/rating will take place. Explain the plan for today and get agreement.

- Give the panellists **sample 1** and ask them to write down attributes for appearance, aroma, flavour, texture and aftertaste.
- Start with appearance (only) and write the attributes on the board. Ask them to tell you the attributes in a rough order – so the ones they would need to measure first would be good to be written down first – this saves time later. Remember to write up all attributes even if the word sounds quite similar to another you have already captured – the discussions about whether people are using the words in the same way can be had later.

- One of the best ways to do this is to start with a random panellist and ask them to give you *one* of their attributes (remember to mention about the order of assessment again before you start). Write it on the board. Then ask the person sitting next to them for one attribute and keep going around the table until you get a couple of 'I do not have any more' statements. Then ask the whole group: 'Any more to add?' Remember to ask the panellists, 'what do you mean by that?' or 'can you define that' or 'can you explain that to me?' if you are unsure what they mean or think the others might be unsure, or simply to check understanding.
- For the next modality, choose a different panellist and go round the table in the reverse direction.
- Tally any attributes that people mention several times if you would like to check how many people used each attribute, but this is not critical as it will come out in the discussion later.
- Once you have a list of attributes for that sample, start drafting the protocol for the assessment. It's a good idea to get the protocol decided as soon as possible to prevent people doing the assessments differently and hence having different attributes or sample comparisons because they were eating more/less or assessing in a different way.
- Write the draft protocol on the board for each modality and ask the panellists to check through it as they assess the next sample.
- Check the palate cleanser requirements as well.
- Move onto **sample 2** – and ask them to write down attributes for appearance, aroma, flavour, texture and aftertaste again. Remind them to think about the protocols and palate cleansers for assessments.
- Going back to the list of attributes you already have, go around the table again collecting new attributes to add to each modality. (Tally any that people have already mentioned if you would like to keep a track of the number of panellists using each attribute.) It can be useful to write sample 1 or sample 2 or sample 3 next to the attributes to help you keep a check on which attributes are different for which samples.
- Go through the draft protocol and check what needs to be changed, edit accordingly.
- Activity: Ask the panellists in pairs to discuss and describe the main differences and similarities between samples 1 and 2 for all modalities (10 minutes).
- Ask each pair to tell you how they think the two samples compare – take notes.
- *Getting the panellists to work together in this way gives everyone a chance to air their views, gives a break from listing attributes and gives you a quick view on how the samples compare.*
- Give them **sample 3** and ask them to write down attributes for appearance, aroma, flavour, texture and aftertaste.
- Remind them to think about the protocols for assessments.
- Then start adding their attributes to the ones you wrote down for sample 1 and 2 – tally any attributes that people mentioned for both samples if you would like to. It can be useful to write sample 1 or sample 2 or sample 3 next to the attributes to help you keep a check on which attributes are different for which samples.
- Activity: Ask them in pairs to discuss and describe the main differences and similarities between sample 1, 2 and 3 for all modalities (15 minutes).

- Ask each pair to tell you how they think the three samples compare – take notes (it can be useful to do this in a table format to help you with any report writing).
- Summarise the session's work and thank the panellists.
- Explain the plan for the next session.

Session 2

Objectives: Assess last three samples, add to attribute list, finalise protocol and begin attribute definitions (3 hours).
Introduction: Explain the plan for the session to the panellists and check they agree. Ask if any questions or queries from session 1.

- Give them **sample 4** and ask them to write down attributes for appearance, aroma, flavour, texture and aftertaste.
- Remind them to think about the protocols for assessments.
- Then start adding their attributes to the ones you wrote down for the first three samples – tally any attributes that people mentioned for both samples if you would like to. It can be useful to write the sample number next to the attributes to help you keep a check on which attributes are different for which samples.
- Check the protocol. Cannot finalise until we have seen the last sample but just check everyone is still happy with the assessment instructions so far.
- Leave the list of attributes up and start **defining the attributes** on the spreadsheet/whiteboard.
- Write up on the sheets/whiteboard: attribute title, definition, protocol, reference, anchors or you can use a computer and projector in the room and fill in the information directly (if your typing and editing is up to speed). See Figure 8.8 for an example.
- It can be easier to pick the easy attributes to title, define, reference, protocol and anchor first and work through to the harder ones. Appearance is generally an easier modality and it's easier to start with flavour and do aroma once flavour is finished.
- The protocol for appearance may well be 'Look at the gravy' but they may want to 'swirl the cup' or 'tip cup to look at greasy layer' or perhaps some stirring to assess thickness and particles.
- The anchors will quite often be 'not to very' and it's useful to remind the panellists to have the attribute title in a suitable format (e.g., 'not opaque' works but 'not opacity' does not). For appearance the panellists might like to use 'light to dark' for colour intensities, for example.
- When discussing the attribute definitions make sure that you have all the elements covered (see Section 7.4.2) and that the attribute meets all the requirements (see Table 10.1).
- You might find that one or two attributes were only generated by one or two panellists. These attributes may well be synonyms for other attributes and become incorporated with other terms. However, other attributes may well have been only generated by one panellist because they are the only person to pick up these notes. If only one panellist is rating the attribute, this does not give you the data necessary to make any conclusions. Explain this to the panellist so that they understand

_____ Attribute List

Attribute	Definition	Reference (where needed)	Anchors
Aroma or appearance protocol:			
Aroma or appearance protocol:			
Flavour protocol:			
Texture protocol:			
Aftertaste protocol:			

Figure 8.8 Blank attribute list for completion.

the issue: just disregarding their suggestions is not likely to make them feel like contributing in the future, particularly if it's 'always them' who generates these additional terms. The best option is to remove this attribute from the *measurement* list but do include it in your report if the panellist is always able to detect this note in that sample. If the attribute is related to a taint, you might be better off keeping it in the list of attributes to be measured: it could be that the panellist is particularly sensitive to that note and that might equate to 10% of the population. Another option is to include an 'other' attribute in all modalities for panellists to use to measure the additional attributes, but do not use this as an excuse for not having a thorough discussion. A list of attributes for the panellists to select but not rate is another option.

- Once you have done around five attributes or spent around 20 minutes on the definitions, move on to assessing sample 5.
- Give them **sample 5** and ask them to write down attributes for appearance, aroma, flavour, texture and aftertaste.
- Remind them to think about the protocols for assessments.
- Then start adding their attributes to the ones you already have.
- Go back to your definitions so far and go through and quickly check what you already have – all OK with them?
- Start defining the next attribute and carry on until you have done around five attributes or spent around 20 minutes on the definitions, move on to assessing sample 6.
- Give them **sample 6** and ask them to write down attributes for appearance, aroma, flavour, texture and aftertaste.
- Remind them to think about the protocols for assessments.
- Finalise the protocol.
- Then start adding their attributes to the ones you already have.
- Activity: Ask the panellists in pairs to discuss and describe the main differences and similarities between samples 4, 5 and 6 for all modalities (15 minutes).
- Continue with attribute definitions.
- Summarise the session's achievements and explain the plan for the next session.

Session 3

Objectives: Continue to develop the lexicon for gravy
The samples will be assessed in a different order: 6, 4, 2, 3, 5, 1

- Introduce the session and explain the plan. Check everyone agrees. Give out the typed-up protocol.
- Assess sample 6 and ask the panellists to check through their descriptions for this sample they wrote previously. As they do this, ask them to check the protocol.
- Ask the panellists how they got on with sample 6 – was their description the same or different? This is useful for checking for sample variability.
- Go back to your definitions so far and go through and quickly check what you already have – all OK with them?
- Start defining the next attribute and carry on until you have done around five attributes or spent around 20 minutes on the definitions, move on to assessing the next sample.
- Ask the panellists how they got on with sample 4 – was their description the same or different?
- Start defining the next attribute and carry on until you have done around five attributes or spent around 20 minutes on the definitions, move on to assessing the next sample. If you have moved on to flavour remember to ask the panellists to start suggesting qualitative references.
- Ask the panellists how they got on with sample 2 – was their description the same or different?

- Start defining the next attribute and carry on until you have done around five attributes or spent around 20 minutes on the definitions, move on to assessing the next sample.
- Ask the panellists how they got on with sample 3 – was their description the same or different?
- Start defining the next attribute and carry on until you have done around five attributes or spent around 20 minutes on the definitions, move on to assessing the next sample. Make a list of qualitative references to try in the subsequent sessions.
- Ask the panellists how they got on with sample 5 – was their description the same or different?
- Start defining the next attribute and carry on until you have done around five attributes or spent around 20 minutes on the definitions, move on to assessing the next sample.
- Ask the panellists how they got on with sample 1 – was their description the same or different?
- Activity: ask the panellists to individually group (using Lauren's buckets for example) all six samples.
- Ask them to share and describe their reasons for the groupings.
- Thanks for the panellists for their hard work and say you will type up the attribute list so far to go over in the next session.

Session 4

Objectives: try various references, finish the attribute list
Samples will be assessed in the following order: 3, 1, 6, 2, 5, 4

- Give out the draft attribute list which includes the attribute titles, draft definitions and anchors and the finalised protocol.
- Assess sample 3 and ask the panellists to check through the attribute list and make any edits or note down any questions or queries they have.
- Begin by going through the attributes that you have not spent as much time discussing. Finish off each definition, anchors and references.
- Assess the qualitative references as required. One way to do this is to focus on a particular attribute and ask which references the panellists would like to assess. The panellists assess each reference writing a description of the reference. Once everyone is ready, ask for a show of hands for panellists who think that the reference is suitable for that attribute. If there is little agreement, discuss the descriptions and why the reference was not suitable. Remember that sometimes you may need to try several references until the panellists agree that the note they are measuring is present in the reference as well as the sample(s). See Section 6.3.7 for more information about references.
- Another option is to work through assessing each of the references on the list and to ask the panellists which reference most closely matches the sensation they described. If an attribute is left without a reference, ask the panellists for further suggestions.

- In between the assessments of the references, work through the samples in the order shown above.
- Continue to work through the attribute definitions, anchors and attribute order.
- Finalise the attribute list/lexicon for the six gravies.
- Ask the panellists to check their sample comparison notes and agree, through discussion, a summary of how the samples compare.
- Thank the panellists for their hard work and explain the plan for tomorrow.

Session 5

Objectives: To practise ranking and then rating gravies and finalise the attribute list/ lexicon

Ranking
- Give the panellists the three samples that are the most different based on the sample comparison discussion in the previous session.
- Pick a selection of attributes (3 or 4) that you think they have defined very well and highlight them on the board. Pick another 3 or 4 that you think probably require more work and highlight these too. Or maybe you have some key attributes that you wish to focus on based on the project objectives. Mix these attributes up (well defined, not so well defined, well defined, etc.) and do not tell the panellists why you have selected them. Place the attributes that can be assessed together next to each other on the board. For example, if you have two appearance attributes, the samples can be ranked for the first and then the second attribute without a break in between. If you have two texture and then two flavour attributes, the panellists might manage to rank all four attributes one after another. However, if there is a long time in between the appearance and the flavour/texture attributes, you may have to prepare fresh samples as the gravy will become cooler.
- Ask the panellists to rank the samples in front of them for each attribute – moving the cups around.
- Once the panellists have completed the ranking, compare the orders – does everyone have the sample order (e.g., 6, 4, 1, for example) or are there lots of different orders?
- If there are any attributes where they do not agree check:
  - The protocol – were they doing the same thing?
  - The definition – does it need editing if people do not read it the same way?
  - The anchors – have they got them the wrong way round?
- Repeat for each attribute, discussing and making changes to the attribute list as required.

Converting ranks to ratings:

- Show the panellists how ranking only tells us the sample order and not the intensity of the difference, by picking one of the attributes and showing them using the line scale – see below. Rank order: 3, 2, 1

Scale use:

- Give out the same three gravies from the ranking. Give them a sheet of blank scales (see Figure 8.9) and ask them to write in the attributes they ranked. It can also be useful to get them to write out the definitions on the sheet as well as this really makes them focus on what is written.
- Now they need to rate the same attributes.
- Go through the rating results. One way to gather this information is to get the panellists to split their scale into quarters (top quarter, bottom quarter, middle left quarter and middle right quarter: draw the scale on the board) and just check they have them in the same area (see Figure 8.10). Remember that it is the rank order that is important and not the actual place that the panellists are scoring in, however, if the panellists can be persuaded to use more of the scale they will find it easier to discriminate. Do not create the impression that all panellists should score all samples the same (for more information please see Section 7.4.3).
- You can assess panel agreement for the rank order and the scale use by counting how many panellists are using each quadrant for each attribute and each sample.

Figure 8.9 Blank line scale sheet showing three of the attributes.

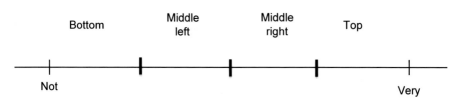

Figure 8.10 Scale quarters.

- Look at attributes where there was very little agreement – what is going wrong?
  - The protocol – were they doing the same thing?
  - The definition – does it need editing if people do not read it the same way?
  - The anchors – have they got them the wrong way round?
  - Ask if we need to change any protocols and definitions – if so then do so…
- Go through the attribute list a final time, checking that each attribute is sensibly defined and the anchors work. Ask the panellists which sample is the least and which is the most for each attribute as you go through to check agreement and understanding. Edit and finalise the attribute list.
- Thank the panellists for their hard work and explain that tomorrow will be the practice profile.

Session 6

Objective: conduct practice profile, feedback, finalise attribute list.

- Conduct the practice profile. This can be done with all the samples or a selection of samples as required.
- Give the panellists their results (for more information see Section 12.4.1).
- Make changes to the attribute list as required.

Sessions 7–9: Gather the data. Three or four replicates depending on how long each assessment takes and the general agreement or disagreement from the practice profile.

8.4.2 Session plan for a quantitative profile example 2: home and personal care product

The objective of this profile was to compare the characteristics of six creams from the dispensing, appearance of cream, through rub in and for 10 minutes after use. The creams were various products on the market for use with dry skin. The panellists had worked on cosmetic creams prior to this profile. Each session was 1 hour. A circular template was used for application of the creams to the arm.

Outline Plan for Creams Project

Sessions	Outline plan
1	Introduction to project, reminder about profiles and protocol development. Assess two samples in booth room and start protocol development. Assess next sample.
2–9	Continuing and finalising protocol discussion including an assessment of the correct weight of sample to be assessed. Assessment of samples while agreeing attribute list.
10	Practice profile on 3 samples replicate 1
11	Practice profile on 3 samples replicate 2
12	Feedback and edits for attribute list.
13 to 18	Collecting profile data, 3 replicates.

Detailed Plan

Session	Detailed plan
1	Handwashing/drying, template drawing, introduction to project, profiles and protocol development. Project introduction to include asking for informed consent forms and if everyone is happy to start, tell the panellists that the project is a profile of six cream samples. Do quick reminder about profiles and the plan moving forward. Assess two samples in booth room (samples 1 and 2). Ask panellists to read out descriptions of each sample and ask them how they think the two samples compare. Start protocol development (can use one of the two samples in the discussion room to aid protocol discussion). Assess sample 3. Ask panellists how three samples compare.
2	Handwashing/drying, continuing protocol discussion from yesterday. Assessment of remaining three samples (4, 5 and 6). Reading out descriptions. Panellists working in pairs to discuss how the six samples compare.
3	Handwashing/drying. Assessment of sample 4 in three weights: 0.3, 0.2, 0.1 g. Discussing protocol and decision on weight for profile. Collating attributes and writing up list of attributes on board.
4	Handwashing/drying, discussing first draft of protocol. Assessment of one sample to discuss protocol for each of the stages in discussion room. Assessment of a sample (see presentation design at end of session plan) in booths, with panellists reading through their descriptions they have already written and underlining attributes. Collating attributes and writing up list of attributes on board.
5	Handwashing/drying, assessments of samples and attribute discussions. Attribute and definitions discussions. Ask for comments on profile plan for the assessment of three samples per session.
6	Handwashing/drying, protocol checks. Assessment of samples. Attribute and definitions discussions.
7	Handwashing/drying. Assessment of samples. Protocol finalising and attribute discussions. Sample comparison summary.
8	Handwashing/drying. Assessment of samples. Protocol finalising and attribute discussions.
9	Handwashing/drying, protocol check, attribute discussions. Assessment of samples. Protocol finalising and attribute discussions – final attribute list.
10	Practice profile on three samples: two that are quite different and one that is similar to one of the other two.
11	Practice profile on three samples: replicate 2. Assessment of panel performance.
12	Feedback session and edits of the attribute list where required.
13	
14	
15	Replicates
16	
17	
18	

Laboratory Plan

Sample code	Description
1	Competitor Cream 1
2	Competitor Cream 2
3	Competitor Cream 3
4	Competitor Cream 4
5	Competitor Cream 5
6	Competitor Cream 6

1. Weigh 0.3 g (±0.05 g) of a sample into a little black weighing boat just prior to assessment.
2. Record all weights in the sensory weights book.
3. If samples are left for more than 15 minutes, discard.

Amount needed for all assessments: number of training assessments, number of replicates, plus discussions about protocol.

Assume 0.3 g for each assessment=$0.3 \times 13 = 3.9$ g per panellist=**39 g in total** (assuming 10 panellists).

If we need to use two packets of a product, open one and use to about halfway through training and then open new packet, ensuring there is enough left in the packet to complete the training and the profile. This way, if there is a difference between packets, panellists will have noticed this during training and we can adjust the attributes accordingly and make notes for the project leader.

Profiling Plan

- Samples assessed in blocks of 3 over 6 days.
- 3 replicates in total.
- One of the two designs from Table 8.1 randomised for the next replicate: see Table 8.2.

8.5 Session plans to determine shelf life

This plan is for a small-scale consumer study to determine the cut-off point for shelf life for an ambient product. It is well known in sensory science that the use of quantitative descriptive profiling with a trained sensory panel, combined with consumer liking or acceptance methods, provides the project team with valuable consumer insight into product behaviour over time. For example, once the key drivers of consumer liking over shelf life are known, a descriptive trained panel can then be used to predict the end of shelf life for both new and any adapted products. To assign shelf life to a new or adapted product, a review of data from previous experiments or new shelf life studies could be conducted to determine how stable the product is over shelf life. Appropriate storage conditions and the estimated storage period can then be recommended.

Sensory quality is often not given as much attention as it deserves, mainly because microbiological safety and nutrition specifications take a rightly deserved first and

Table 8.1 Sample presentation order for assessment of creams

Session	Samples				Date	Samples
1	1	2	3		13	Replicate 1: 3 samples
2	4	5	6		14	Replicate 1: 3 samples
3	Weight comparisons sample 4				15	Replicate 2: 3 samples
4	2	5	3	6	16	Replicate 2: 3 samples
5	6	3	1	5	17	Replicate 3: 3 samples
6	2	6	1	3	18	Replicate 3: 3 samples
7	1	4	6	2		
8	3	4	2	5		
9	5	1	6	2		
10	Practice profile					
11	Practice profile					
12	Feedback					

Table 8.2 Presentation design for first two replicates

Panellist	Presentation design for replicate 1			Presentation design for replicate 2		
P1	2	3	5	6	1	4
P2	3	1	4	2	5	6
P3	1	6	5	3	4	2
P4	1	2	4	3	6	5
P5	5	4	6	1	2	3
P6	5	1	3	6	4	2
P7	4	5	2	1	6	3
P8	6	2	1	5	3	4
P9	2	6	3	4	5	1
P10	4	3	6	5	2	1

second place. However, the sensory aspects of shelf life are incredibly important for product success in the market place. Sensory quality comes in a very close third place due to the fact that consumers will not repurchase a product if they do not like it the first time they try it. Imagine an ambient product that has a shelf life of 6 months. A consumer buys it at 5 months and does not even look at the shelf life printed on the pack. They try the product and find that it does not taste like the pack description seems to suggest and so they decide to never purchase it again.

Shelf life testing can be quite resource intensive and using a risk-based approach can be helpful in planning each study so that you make the best use of resource and facilities by only conducting comprehensive shelf life testing where required. For example, for a new product in a new range, the risk assigned may be 'high', as there is no existing data to base the shelf life determination on. But for the change to a new powdered ingredient supplier from an existing supplier, the risk of the shelf life being affected is probably very low and therefore the minimum amount of testing could be conducted.

Action standards can be very useful when determining shelf life. For example, an action standard may state: 'the product should fail when it no longer represents the product concept'. This can be very useful for short, medium and long-life products, as when they no longer match the concept, they are essentially outside a shelf life that is deemed acceptable by consumers. For more information about action standards please see Section 4.2.10.

For this study, sensory profiling has been planned for each relevant time point and if the samples are found to be different, a small-scale consumer study is planned to determine if the changes are consumer perceivable and if the change results in a drop in consumer liking. The approach is based on survival analysis (Hough et al., 2006b). If the small-scale consumer testing indicates that there is a difference in acceptance or preference between the 'fresh' product and the stored product, an external consumer study will be initiated.

8.5.1   Plan for small-scale consumer study

Consumers are recruited from company employees in the head office who are part of the testing database and who stated that they purchased soft drinks regularly. These employees do not work on the product type being tested but in a different part of the organisation. Each consumer is given the control sample which has been stored at 4°C and the ambiently stored product that was found to be different by the trained panel. Half the consumers are asked to assess the control sample first, and the other half are asked to assess the ambiently stored sample first. Samples are presented in bottles labelled with a three-digit code. The consumers are asked a series of questions for each sample. The ASTM *Standard Guide for Two-Sample Acceptance and Preference Testing with Consumers* (ASTM, 2014) is a very useful resource for planning consumer studies.

An information sheet for the consumers is shown in Figure 8.11 and the questionnaire in Figure 8.12.

Thank you for taking part in this study. We would like you to complete an assessment of two soft drinks which should take around 15 minutes. The information provided by you will be treated in confidence.

Please read the instructions before you begin the test. If you have any questions please ask: we will be happy to help you.

---

1. Each sample is labelled with a three-digit code. Please make sure you are assessing the right sample by matching the online questionnaire with the code.

2. We will give you two samples. Please assess them in the same order as the online questionnaire.

3. Please drink some water before you begin and after the first sample.

4. Please click on the link below to take you to the relevant questionnaire.

Figure 8.11 Instruction email for consumers.

Please assess the first coded sample

Q1. What is your opinion of this product **overall?**

Please use the scale from 1 to 9, where 1 means that you "dislike the sample extremely" and 9 means that you "like the sample extremely", and answer by tapping the number that best corresponds to your answer.

Dislike
extremely   Like
                                                                                        extremely

1	2	3	4	5	6	7	8	9

Q2. What is your opinion of the **overall appearance** of this product?

Dislike
extremely   Like
                                                                                        extremely

1	2	3	4	5	6	7	8	9

Q3. What is your opinion of the **overall flavour** of this product?

Dislike
extremely   Like
                                                                                        extremely

1	2	3	4	5	6	7	8	9

Q4. And is the **strength of flavour**…

Not nearly strong enough		Just right		Much too strong
1	2	3	4	5

Q5. What, if anything, do you particularly like about this product? Please type in your answer.

Q6. And what, if anything, do you particularly dislike about this product? Please type in your answer.

Please have a drink of water and then assess the second coded sample.

(Once the consumers have tried both samples they are presented with the following final question.)

Q13. Which of the two products do you prefer?

Figure 8.12 Consumer acceptance and preference questionnaire.

Refresher training for sensory panels

Your panel will need refresher training from time to time and not just when they have not been working on a particular product or method (or at all). The refresher training is useful to help remind all the panellists of the ways of working, the best ways to take part in a test, a selection of references or even to share learning among the panellists. For panels with complex assessment protocols, running a refresher session to remind them how to assess the product can be vital to prevent them slipping back into their usual mode of product use. If you have several panels, a refresher training session which mixes and matches different panel types can help spread handy tips and reference ideas among all panellists. If you work with professional panels that are made up of hairdressers, flavourists or chefs, you might ask them to help you design the refresher training. For example, they might be able to suggest which references or products they would like to assess to remind them of particular protocols or discuss the scaling of a particular set of samples in detail.

The examples included in this chapter include a quality control panel refresher training on off-notes and refresher training for a quantitative profiling panel.

9.1 Refresher training for a quality control training session on off-notes

This is a one-session plan for the training of a group of panellists who work in Quality Control and are regularly assessing and checking samples from the line. The purpose of the session is a reminder about the possible off-notes present in the products and to check that each panellist is able to correctly detect and also correctly identify the off-notes. To check for identification of the off-notes in the product, you could also set up a series of discrimination tests such as an A-not-A (de Bouillé, 2017). The samples for these tests could be previously rejected products or created by spiking or mistreating samples. These sessions are organised every week and panellists must attend at least one of the sessions every month. Data about attendance and number of correct assignments are recorded for each panellist. Different off-notes are selected for each of these validation sessions using a random number generator, or are chosen due to production issues or recent panel performance issues. The sessions take around 30 minutes. You might also wish to remind the panellists about the best way to assess odours (see Section 7.2). The session plan is based on BS ISO 5496:2006.

When running training sessions like this, it can be helpful, if you can, to assign images to each of the off-notes or to discuss with the panellists where they have detected the note(s) before, as this can help the panellists when it comes to recalling the information. Also consider the similarity of the attributes or off-notes in each session: when notes are more similar panellists will need to describe the attributes in

Sensory Panel Management. https://doi.org/10.1016/B978-0-08-101001-3.00009-4

more detail to be able to identify them (Civille and Lawless, 1986). For example, consider a line-up of men: if everyone is a different height it can be very easy to differentiate between them – 'It was the tall one'. But if they are all the same height, you may need to describe the people in more detail: one has brown hair, one has blonde hair, that chap has a beard, etc. And if there were two brothers in the line-up who looked very similar to each other, you might need to go down another 'level' and describe the colour of their eyes or the size of their nose.

9.1.1 Session plan for a quality control panel training on off-notes

- Welcome and thank you to the panellists for their attendance.
- Award giving for the panellists who have attended the most tests this quarter.
- Introduction to the session and agreement for the plan.

Part 1: Off-note refresher identification

- Assessment of six named off-notes (see sample details below). Panellists are all presented with six off-notes in jars which are labelled by their names and not by 3-digit codes at this stage.

Off-note name	Typical off-note description	Coded sample
Earthy	Earthy, mouldy, musty, undergrowth and potato skins	672
Chlorine	Chlorine, cleaners, sanitisers and swimming pool	194
Dimethyl disulphide	Sewage and rotten vegetables	385
Chlorophenol	Mouthwash, antiseptic, hospitals and plasters	406
Acetaldehyde	Apples, paint, solvent and fruity (when at low levels)	817/270
Flavour contamination	This off-note is caused by carry-over from previous samples on the production line. There are several options for description depending on the choice of flavour contamination in the training session	933
Blank	No off-note – just standard product used for Part 2 of the orientation training.	528

- Panellists are asked to assess each jar using the standard aroma assessment methodology and write down their descriptions for the named off-note using the lined paper provided (20 minutes).
- Each panellist chooses one off-note to describe to the other panellists and the different descriptors used are discussed, with the discussion lead by the trainer (20 minutes). The discussion gives time for the panellists to think about their descriptions, steal other panellists' descriptions to help them identify the off-note, and also have a break from smelling.

Part 2: Validation of training

- Each panellist is given eight jars with three-digit codes as shown in the last column in the sample list. One of the six off-notes is repeated (code 270) and a blank sample is also included.
- Each panellist is given the samples in a different order.
- The panellist is given a worksheet as shown in Figure 9.1 and asked to assign an identified off-note to each code.
- The panellists are allowed to leave once they have assessed all the samples; however, everyone is reminded to take their time and that there are no prizes for quickly returned incorrect sheets.
- Alternatively, if time, the panellists can stay and discuss their results. This is beneficial, as it gives the panellists the chance to reassess those off-notes that they might not have identified correctly and discuss where they went wrong with their colleagues.

Off-note training worksheet 1

Name: Date:

- Please assess the eight samples in the order the samples are presented to you.
- Please use the standard aroma assessment methodology and take a short break between each sample.
- You may write a description in the second column if you wish, but you must also write the off-note name.
- You may use your notes from the first part of the session.
- Once you have completed your assessments, please check you have written your name and date at the top of the page and give the sheet in to the session leader.
- Thank you for participating today. Remember to pick up your attendance voucher as you leave.

Code	Off-note identified
672	
817	
528	
385	
406	
270	
194	
933	

Figure 9.1 Panellist worksheet for off-note identification.

9.2   Refresher training for a profiling panel

These sessions are for a profile panel that have not been working for the last three months and prior to that they had been working on temporal methods. They have not worked on this product type before. This is the first profile they have conducted for six months and therefore they require a reminder about the process and handy tips to enable them to create a profile for this new product type. This is a three-session training profile with known samples. The known samples will help you direct the protocols for the assessment of shelf life and help identify the key attributes and references that may be required for the next profile. The next profile is related to determining the shelf life of a newly developed bread product.

9.2.1   Session plans for refresher training for a quantitative profiling panel

Session 1. Reminder about the profiling process through a presentation and three activities (3 hours).

Objective: To ensure that everyone understands the profiling process and has a chance to generate appearance, texture, aroma and flavour attributes for bread.
You will need:
  Presentation and equipment or just a print out of the presentation
  Plates, serviettes, pens, notebooks, etc.
  Water and cups

Samples

Sample number	Description
1	cheap white bread mid shelf life
2	expensive white bread mid shelf life
3	a product that is a mixture of brown and white bread
4	cheap fresh white bread
5	cheap white bread one day from end of shelf life or on last day of shelf life
6	expensive fresh white bread
7	expensive white bread one day from end of shelf life or on last day of shelf life

Introduction (10 minutes): ask people what they have been doing for the last few weeks since you last saw them. Anyone had a nice holiday they might share details on? Give them an update on any company or sensory team information that might be relevant to them (e.g., there was a computer upgrade). Allow the panellists to chat amongst themselves.
Presentation (20 minutes): include a reminder about the senses and give examples related to bread assessments. Include a reminder about the profiling steps and a couple

of slides describing 'the attributes of a good attribute'. Also allow time for the panellists to ask you any questions.

Activity 1 (40 minutes): Appearance

- Give the panellists two labelled products – Sample 1 (which is the cheap white bread mid shelf life) and Sample 2 (which is the expensive white bread mid shelf life) and ask the panellists to write down some appearance attributes individually in their notebooks.
- Then ask them to chat to their neighbour and share their descriptions. Ask them to count how many words they generated on their own and how many they have once they discussed their findings with a neighbour. Hopefully they have more words now to describe the bread.
- Then give them product 3 (a product that is a mixture of brown and white bread) and ask them to write ONLY how 3 compares to 1 and 2. How does this change the way they generate the attributes? You may find that they add attributes to their descriptions for samples 1 and 2 because sample 3 is quite different in appearance (for example it has a different type of crust and brown specks).
- Then go round the table and ask each panellist to read out their comparisons for sample 3. Listen to their descriptions and allow everyone to 'steal' ideas from other panellists to add to their own lists.
- Remind the panellists that using the other panellists' ideas is good – several heads are better than one!
- And how comparing samples can really help generate new words.
- And how saying that a product is 'NOT' something is really useful.
- And when you focus on a modality it can really generate some good descriptors! (They should have generated words for bread colour, internal texture, hole-iness, presence of flour, crust colour, crust depth, slice depth, thickness, density…).

(5 to 10 minutes break within the section above or afterwards – when the panellists come back ask them to sit next to someone else.)

Activity 2 (40 minutes): Texture

- Give them product 4 (cheap fresh white bread) and ask them to describe the texture of the bread:
  a. when looking at it (ask them to take their time and really examine how the appearance gives them clues to the texture when eating)
  b. and when eating it.
- Watch them while they examine the slices and while they eat, and make some notes for yourself to help develop the protocol. What to do they do? How much do they put in their mouth, what do they do with the crust, how long do they chew, and do they bite off a piece of bread first or tear a portion to eat?
- Then listen to their descriptions and allow everyone to 'steal' ideas
- Ask them HOW they examined/ate the bread – just listen to their descriptions about what they did.
- Do they think they will need to use a palate cleanser in between the samples at all?

- Give them product 5 (cheap white bread one day from end of shelf life or on last day of shelf life – you might be able to buy a bread approaching end of shelf life – if not then leave some out in the kitchen for a short while and it will start to stale).
- They need to do (a) and (b) again for sample 5.
- Watch them and see if they are trying each other's methods of examining/eating (they should do) and this will help for session 2.
- Then ask them to write a comparison of the texture of samples 4 and 5. Listen to two or three panellists' descriptions.
- Write up a draft protocol for the assessment of the appearance and texture of bread.

Activity 3 (50 minutes): appearance and texture

- Product 6 (expensive fresh white bread) and 7 (expensive white bread one day from end of shelf life or on last day of shelf life).
- Ask them to assess the appearance and texture of each product one at a time individually, writing descriptions in their notebook.
- Then ask them to talk about the appearance and texture attributes they generated, with their new partner – just 5 minutes.
- Go round the table collecting a word from each pair to describe the appearance. Then repeat with texture.
- Write the words on the board/laptop, but as you do ask them, 'what do you mean by that?' or 'can you define that' or 'can you explain that to me?' – just to start the ball rolling on attribute definitions.
- Discuss the possible protocols for the assessment of the appearance and texture. Remind the panellists that it's very important that they all follow the same protocol, for example, how much bread they will eat and how they will eat it.

Finish up (10 minutes): checking learning

- Ask the panellists questions about what they learnt today. For example: what senses can be used to describe the profile of bread? What are the steps in developing a profile?
- Check back on the session objective – has it been achieved?

After session: check tomorrow's plan and edit if necessary.

Session 2: Taking notes, making sample comparisons and keeping a 'picture' of the samples to allow rating. Continue to develop the lexicon for bread – starting with aroma and flavour so references can be introduced.

Objectives:
- Get panellists to write excellent notes and sample comparisons
- Remind panellists about references and how to include suggestions for references while making notes

Activity 1: Keeping notes (30 minutes)

Ask the panellists to get into pairs and discuss good ways to make notes in their books. Go through with the panellists about their suggestions. They might have some good suggestions and you can also use those below.

- It's a good idea to write the date and project/product name on each page.
- Try to use intensity descriptors such as none, slight, moderate, strong, very strong before attributes so you can more easily remember and compare sample to sample. This is difficult with the first sample but you can edit your notes when you assess the sample the next time.
- When you have assessed all the samples in a set, do a sample comparison. Write down a list of the attributes and next to each write, for example, 'sample X was the most and sample Y the least and the others fairly similar'.
- You can do this in a table format using your attributes or later in the project when the attributes are closer to being finalised by the panel:

Attribute	Least				Most
Crust colour	2 (lightest)	6, 7	3	4, 5	1 (darkest)
Slice colour	2, 6, 7		1, 4, 5		3
Hole-iness	6, 7	2		3, 5	1, 4

- Explain to the panellists that they could also do this by drawing scales in their book instead of the table format.
- Another way to do this is to use 'Lauren's bucket method' – draw some buckets on a page in your book and write in the samples that are the most similar, so that those that are different end up in different buckets. You can do this as an overall comparison, by modality and even by attribute or key attributes. This can really help you keep track of the samples. See below for an example (Figure 9.2)
- Ask the panellists what ways might be useful.

Figure 9.2 Lauren's buckets for the bread samples.

Activity 2: References (10 minutes)

Give out the handout about qualitative references as shown in Figure 9.3 and discuss.

Activity 3: aroma, flavour and aftertaste (part 1) (40 minutes)

- Give the panellists two different white breads, samples 1 and 2, one at a time and ask them to assess the samples for aroma, flavour and aftertaste. Remind the panellists that when they are thinking about aroma, flavour and aftertaste they need to also write down their ideas for references. So if a bread smells 'malty', for example, maybe we need some malted milk biscuits or some beer to try, to make sure we are all measuring the same thing.

Qualitative references:

- Qualitative references are used for aroma, flavour (and aftertaste) attributes
- They can also be used for texture
- They help us to identify and agree exactly what we are measuring for each attribute
- They are not related to the intensity of what we are measuring

Figure 9.3 Handout for panellists about references.

- Watch the panellists while they examine the slices and while they eat and make some notes for yourself – are they working to the protocol developed yesterday. What to do they do? How much do they put in their mouth, what do they do with the crust, how long do they chew and do they bite off a piece of bread first or tear a portion to eat?
- Between each sample write the list of attributes on the board/laptop, adding and discussing and asking questions (What do you mean by that? Do we need to add in some protocol? Should we do aroma first and then appearance? Which attribute should come first?).
- After sample 2 ask a couple of panellists how they think the two breads compare in flavour.
- Discuss the protocol for the assessment of the aroma and flavour. Edit the existing protocol for appearance and texture and discuss the order the modalities could be assessed in.
- Break (from 5 to 10 minutes).

Activity 4: aroma, flavour and aftertaste (part 2) (40 minutes)
- Give the panellists sample 3 and ask them to write notes about the aroma, flavour and aftertaste.
- Go round the room asking each panellist for one attribute adding attributes to the list where necessary.
- Give the panellists sample 4 and ask them to write notes about the aroma, flavour and aftertaste.
- Go round the room adding attributes to the list where necessary.
- Ask the panellists to write a comparison of the samples so far. Could be useful to split the panellists into groups of three to discuss the similarities and differences here – especially if they are getting tired… They could try out the table and bucket methods above and see which method they find the most useful.

Activity 5: aroma, flavour and aftertaste (part 3) (40 minutes)
- Give the panellists sample 5, 6 and 7 one at a time and ask them to write notes about the aroma, flavour and aftertaste.
- Go round the room asking each panellist for one attribute, adding attributes to the list where necessary. Keep adding to the attribute definitions, protocols, anchors, etc. as each bread is assessed. Develop the aroma, flavour and aftertaste attribute list on the flip chart with gaps for the references to try in the next session.

After session: Type up attribute list so far, so it can be used in the next session.

Session 3. Starting to develop the lexicon (3 hours):
Objective: To ensure that everyone understands the lexicon development process and to start developing the lexicon for bread

- Give each panellist a print out of the information below which describes the various parts of the lexicon (or attribute list):
  - The **attribute** is the word that summarises the description of the appearance, aroma, flavour, texture or aftertaste. For example, white colour, caramel aroma, moistness of internal crumb, butter flavour…
  - The **reference** helps us identify and agree exactly what we are measuring for each attribute. For example what *type* of malty flavour.
  - The **definition** helps everyone understand what the attribute means – not just us but the person who will be reading the report! For example: caramel aroma: the sweet and toffee-like aroma associated with lightly cooked granulated sugar (reference).
  - The **protocol** describes the actions prior to the assessment of the intensity of the attribute. For example, for the assessment of vegetables: Aroma: assessed from the bowl – cup the bowl in the hands and bring to the nose and take small bunny sniffs. Assess the intensity of the aroma (no cutting – assess the product whole).
  - We also need **anchors** for the ends of the scales such as 'not' and 'very' or 'light' to 'dark' so that we know which direction the scale goes and therefore how to rate the samples.
  - Finally we need to check that the **order** we have listed the attributes will work. Because we wrote up the attributes in the order they appeared, it should all work OK but sometimes we need to shift words around for practical reasons.
- Go through the words in bold and make sure everyone understands by asking the panellists questions.
- Write up an example on the board or give out an example on paper – choose an example from a previous project that the panel have worked on or take one from the literature. See Figure 9.4 for an example.

Attribute	Definition	Anchors
Flavour: cut-off bottom of stalk with a knife to leave florets intact and held together. Place piece of stalk in the mouth and assess the attributes below		
Strength of flavour	The intensity of the overall flavour	Not-very
Pea flavour	Intensity of cooked frozen pea flavour. Reference 3: frozen Birds Eye Peas	Not-very

Figure 9.4 Excerpt from an attribute list.

- Give the panellists three of the samples that showed the most differences, one at a time, and ask them to write notes about the appearance, aroma, flavour, texture and aftertaste.
- In groups get the panellists to discuss the main similarities and differences between the samples.
- Work on the attribute list/lexicon, finishing the definitions, anchors and attribute order.
- At the end of the session thank the panellists for their hard work and explain the objectives of the three sessions. Tell them that the next session will be the start of a new profile with new samples, but they can take their learnings, protocols and attributes forward for use with the new samples

Advanced training for sensory panels

10

10.1 Helping panellists to generate great attributes

If the sensory panel is going to be developing quantitative descriptive profiles and this involves creating lists of attributes (or lexicons) with definitions, you might like to demonstrate the importance of the language used for these elements to the new panellists. Language development can also be critical for quality control panels learning an off-note or taint language, a technical panel learning a lexicon for working with the Spectrum profiling method or any panel learning to use an existing attribute list or lexicon. The attributes must be easily understandable by the panellists, the sensory team and whoever is going to be reading the report and acting on the data. Future panels may also need to use the lexicon and they might not always be based on your site. Therefore the language used is critical. You will need to collect together a range of samples to help with the training and it can be useful to cover as much of the sensory space as possible at this stage (Griffin et al., 2017). The samples might be different production dates, products at different stages of shelf life, different types of a similar product (for example, different types of tea or coffee), competitor products or products stored in different packaging.

When generating a list of attributes with your panel for a particular product, there are several important factors to bear in mind. These are listed in Table 10.1 and described in more detail in Section 7.4.2.1.

There are a number of ways to demonstrate how important the language development phase is to the new panellists. One way is to use something similar to the repertory grid approach and another is building models from children's construction blocks. Both are described in Sections 7.4.2.2 and 7.4.2.3, respectively, and make great starting points for the development of a profiling language. If your panel has been working on descriptive profiles for some time, you may wish to give them additional training to improve their abilities. A session plan to describe this training is shown below.

10.1.1 Session plan for advanced attribute training for descriptive analysis panels

- Start with one modality. **Tip**: flavour is a good modality to start with. Aroma can be difficult as it can be quite fleeting and therefore is not so easy for trainees as it is harder to explain to other people what you are experiencing. Appearance and texture are easier as you can point and say, 'Look it's thicker'. You can repeat the study for aroma after you have assessed a range of samples for flavour. The plan can also be used to help panellists describe textural differences in more detail.

Sensory Panel Management. https://doi.org/10.1016/B978-0-08-101001-3.00010-0

Table 10.1 Important factors to consider for attributes

Cover all aspects	Discriminate	Uni-polar
Non-redundant (little or no overlap with other attributes)	Consumer or technical language	Simple and singular
Unambiguous	Precise	Reliable
Measured by one sense at a time	Suitable reference	Reference easy to create, use and store if possible
Agreed	Able to be measured via a scale	Relatable to consumer language
Understandable (by panellists and all users of the data)	Relate to real products	Detected by the majority of the panel

- Present the panellists with two quite different samples. For example, two juices that are the same flavour (e.g., orange) but quite different in that flavour (e.g., one is fresh orange and one is artificial orange).
- Ask the panellists to describe the flavour of sample 1 and then the flavour of sample 2.
- Ask them in pairs to discuss the differences and similarities in flavour between the two samples.
- Collate the descriptions on the board, asking questions as you go through. For example, 'what made you think that?', 'which sample was the most?', etc.
- Then ask the panel which elements (things) about the flavour can they describe – this is like digging down to the next level of the description. It should help to refer back to their discussions about how the samples compared.
- They will hopefully generate the following list with gradual prompting and may even tell you more aspects as their training progresses:
  - Type of flavour (e.g., orange)
  - Details about flavour (e.g., confectionery orange like in ice lollies)
  - Intensity of flavour (e.g., sample 1 is more orange than sample 2). It's useful to introduce the panellists to a simple intensity scale such as none, slight, moderate, strong and very strong, that they can use to indicate the level of each attribute. Although the intensities would not necessarily mean the same to everyone, they will help each panellist makes notes in their books and allow them to compare across samples.
  - Flavour timing (e.g., the orange in sample 1 hits you after the sweetness but the acidity in sample 2 seems to suppress the sweetness and you get orange first).
  - Flavour links (e.g., it seems like a more natural orange because there is **also** a zesty element or it seems like a fresh orange because it is **not** artificially sweet).
  - Flavour lingering (the orange flavour in sample 1 stays in the mouth the whole time, whereas sample 2, although still quite strong in orange, disappears more quickly).
  - The intensity words mentioned earlier can also help with understanding the differences in lengths of certain attributes which can be helpful for some products, e.g., orange flavour lingering. Figure 10.1 shows how the use of a simple tabular format of intensity descriptors allows us to understand that sample 2 is more intense in orange flavour to start with, but quickly decreases in intensity, while for sample 1 the orange flavour lasts longer.

Sample	Time								
	1	2	3	4	5	6	7	8	9
Sample 1	Strong	Strong	Strong	Strong	Medium	Slight	Slight	Slight	None
Sample 2	Very strong	Very strong	Very strong	Slight	Slight	None			

Figure 10.1 The use of intensity descriptors over time (see text for full details).

10.2 Training panellists in gas chromatography-olfactometry

Gas chromatography-olfactometry (GC-O) is gas chromatography (GC) with a human acting as a detector via a specially designed odour port. Other detectors such as flame-ionisation or specific sulphur detectors may also be present to record the eluting peaks, but the human can detect, describe and also measure the intensity of the odour perceived. GC-O is used for several different applications, but its main use is in understanding the contribution of various volatiles to the aroma of fragranced products or flavour of foodstuffs. GC-O can also be performed with animals to assess the impact of certain volatiles for inclusion in pet food formulations. This session plan is a training programme for a group of internal panellists who were recruited to be trained to take part in regular GC-O sessions. The GC-O work is for identification purposes. Emails were sent to all staff to determine who would be interested in taking part in the role. The email included details about the role, for example, the training process and time commitments, as well as information about GC-O.

The session plan is based on publications such as Bianchi et al. (2009), Delahunty et al. (2006), Vene et al. (2013), as well as Lawless and Heymann (2010). The session plan refers to the sensory elements of the training and does not include the analytical elements of setting up the instrumentation. The first five sessions involve group work and the subsequent sessions can be run for individuals or for small groups.

10.2.1 Session plan for training panellists in gas chromatography-olfactometry

Session 1: (1 hour) Group work

- Welcome and introduction to the training programme.
- Introduction to sensory science and the senses with particular attention to odour.
- Short presentation on GC-O including how GC works by showing a video (for example, the Royal Society of Chemistry https://www.youtube.com/watch?v=08YWhLTjlfo&t=83s).

Do	Don't
Listen	Use powerful fragrances, soaps, deodorants, etc.
Maintain good hygiene	Smoke before a session
Rest your senses between samples	Eat or drink within 30 minutes of a test
Take sensory testing seriously	Eat or drink any strong flavours within an hour of
Do not rush – take enough time when carrying out	a test (e.g.,mint, chilli)
tests	Do not participate in a test when you cannot smell
Switch off your phone	Do not participate in a test if you are unwell
Respect and follow test protocols, procedures	Do not participate if you have too much prior
and instructions	knowledge
Ask questions if you are unsure	
Tell us if you have any issues	

Figure 10.2 Panellist dos and don'ts (gas chromatography-olfactometry panel).

- Introduction to ways of working and panellists' dos and don'ts (see Figure 10.2).
- Visit to the analytical laboratory to see GC in action.
- Thank the panellists for their time, answer any questions and explain what will happen in the next session.

Session 2: (1 hour) Group work

- Welcome and introduction. Ask panellists what they remember from the first session and remind them of important things they may have forgotten.
- Explain that they will be assessing a series of aroma volatiles from vials and that they will be asked to describe the aroma individually and will later work in groups. Explain that they will need to commit the odour to memory so they can recall it quickly.
- Explain how to assess the aroma vials by giving the panellists the handout as shown in Figure 10.3.
- Give each panellist the questionnaire as shown in Figure 10.4 and the eight odour vials. The eight vials are made up of odours that are present in the product range for assessment. Choose the odours for this part carefully. Start with odours that are easy to describe and discriminate. You may need to repeat sessions 1 to 5 with further groups of odours if you are working with complex products. This is not always necessary if the panellists are able to meet and discuss their aromagrams. Explain to the panellists that there will be more volatiles present in the samples they assess and probably many unknowns, but this set of samples will get them started and will give them the chance to practice.
- Allow the panellists time to write their descriptions of each odour.
- Once everyone is finished, ask the panellists to get into pairs and discuss the descriptions. Tell them that they can 'steal' the other person's ideas if they wish.

- Remove the cap from the first bottle and gently sniff in the space above the bottle. If you can detect and describe the odour, replace the cap and write down your description in the box next to the code of the sample you assessed.
- If you cannot detect any odour, bring the bottle a little closer to your nose and again, sniff gently. If you can detect and describe the odour, replace the cap and write down your description in the box next to the code of the sample you assessed.
- If you cannot detect any odour, bring the bottle under your nose and sniff gently. If you can detect and desribe the odour, replace the cap and write down your description. If you cannot detect any odour, replace the cap and move onto the next bottle.
- DO NOT sniff too hard if you cannot detect an odour, as this may affect your ability to detect the odours in the later bottles.
- Remember to replace the cap on each bottle before moving to the next.

Figure 10.3 How to assess odours.

Odour descriptions worksheet 1

Name: Date:

- Please assess the eight samples in the order the samples are presented to you.
- Please use the standard aroma assessment methodology and take a short break between each sample.
- Please write a description in the second column.
- Once you have completed your assessments, please check you have written your name and date at the top of the page.

Odour description worksheet 1

Code	Odour description
139	
728	
540	
601	
773	
416	
205	
394	

Figure 10.4 Panellist worksheet for odour descriptions.

- Discuss the descriptions with the group and then tell them the name of the odour. It can be helpful for the panellists to have a name that is easy to remember and a summary descriptor. For example, if the odours included (2E)-3-phenylprop-2-enal, it would be best to refer to it as cinnamaldehyde and use the descriptor 'cinnamon' as that would be easier to remember. If the odour does not have a common name, ask the panellists to create one that will help them remember it.
- Allow the panellists to try each of the odours again with the names and descriptions.
- Thank the panellists for their time, answer any questions and explain what will happen in the next session.

Session 3: (1 hour). Group work

- Welcome and introduction. Ask panellists what they remember from the previous sessions and remind them of important things they may have forgotten.
- Tell them the plan for the session.
- Remind them how to assess the odours.
- Give the panellists their odour descriptions worksheet from session 2 and the eight odours to reassess. They can reassess the odours in their pairs or individually whichever they prefer. Give them time to remember each odour.
- Then give them the eight odours again, with two odours repeated, with new 3-digit codes and the worksheet as shown in Figure 10.5. The panellists' job is to recognise each of the odours from session 2, not just describe them.
- Once the panellists have completed the worksheet, allow them to mark their own work and discuss with a neighbour. Give them the eight odours labelled with the name (e.g., cinnamaldehyde) so that it is easy for them to cross-check their learning.
- Thank the panellists for their time, answer any questions and explain what will happen in the next session.

Session 4: (1 hour). Group work

- Welcome and introduction. Ask panellists what they remember from the previous sessions and remind them of important things they may have forgotten.
- Tell them the plan for the session.
- Give the panellists their odour descriptions worksheet from the previous session and the eight odours to reassess. They can reassess the odours in their pairs or individually whichever they prefer. Give them time to remember each odour.
- Then give them the eight odours again, with two odours repeated, with new 3-digit codes and the worksheet as shown in Figure 10.6. In a similar way to session 3, the panellists' job is to recognise each of the odours but this time they are only allowed to sniff the odours once. This is to get the panellists used to the speed of the assessments when working with the GC.
- Once the panellists have completed the worksheet, allow them to mark their own work and discuss with a neighbour. Give them the eight odours labelled with the name (e.g., cinnamaldehyde) so that it is easy for them to cross-check their learning but, again, they should try to only sniff each odour once.
- Thank the panellists for their time, answer any questions and explain what will happen in the next session.

Odour recognition worksheet 2

Name: Date:

- Please assess the ten samples in the order the samples are presented to you.
- Please use the standard aroma assessment methodology and take a short break between each sample.
- You may write a description in the second column if you wish, but you must also write the odour name.
- You may use your notes from the earlier session.
- Once you have completed your assessments, please check you have written your name and date at the top of the page.

Code	Odour identified
672	
817	
528	
749	
951	
385	
406	
270	
194	
933	

Figure 10.5 Panellist worksheet for odour identification.

Session 5: (1 hour). Group work

- Welcome and introduction. Ask panellists what they remember from the previous sessions and remind them of important things they may have forgotten.
- Tell them the plan for the session.
- Give the panellists their odour descriptions worksheet from session 4 and the eight odours to reassess. They can reassess the odours in their pairs or individually whichever they prefer. Give them time to remember each odour.
- Then give them the eight odours again, with two odours repeated, with *new* 3-digit codes and the worksheet as shown in Figure 10.7. In a similar way to session 4, the panellists' job is to recognise each of the odours but this time they are not allowed to check their notes. This is to check that the odour names are easily recalled by the panellists.

Odour recognition worksheet 3

Name: Date:

- Please assess the ten samples in the order the samples are presented to you.
- You must only sniff the vial ONCE!
- Write down the name of the odour in the column next to the code.
- You may use your notes from the earlier session.
- Once you have completed your assessments, please check you have written your name and date at the top of the page.

Code	Odour identified
903	
175	
288	
391	
652	
866	
445	
607	
594	
710	

Figure 10.6 Panellist worksheet for odour identification.

- Once the panellists have completed the worksheet, allow them to mark their own work and discuss with a neighbour. Give them the eight odours labelled with the name (e.g., cinnamaldehyde) so that it is easy for them to cross-check their learning.
- Thank the panellists for their time, answer any questions and explain what will happen in the next session.

Session 6: (1 hour) Individual work or in small groups

- Starting to work on the GC-O!
- Ask the panellist to sniff the effluent from the GC column without injecting any volatiles.
- Explain how to breathe (normally) and how the information about the descriptor will be captured.

Odour recognition worksheet 4

Name: Date:

- Please assess the ten samples in the order the samples are presented to you.
- You must only sniff the vial ONCE!
- Write down the name of the odour in the column next to the code.
- You may not use your notes from the earlier session.
- Once you have completed your assessments, please check you have written your name and date at the top of the page.

Code	Odour identified
182	
305	
749	
611	
837	
296	
884	
573	
934	
710	

Figure 10.7 Panellist worksheet for odour identification.

- Remind them that there may well be additional unknowns interspersed with the odours they have already tried. Tell them at this stage to just ignore them.
- When the panellist is ready, inject the mixture of odours that have been previously assessed and ask them to detect the odours as they elute from the column. They will find this difficult the first time around, as it will be a rather alien situation and they will need to get used to breathing and thinking about the volatiles at the same time.
- Give the panellist a short break and then repeat.
- Thank the panellist(s) for their time, answer any questions and explain what will happen in the next session.

Before the next session assess all the panellists' aromagrams and determine if anyone would benefit from some additional training with the odour vials (recognition errors) or further work on the GC-O (technique and speed issues). For some odours, the next odour may well be chasing it from the column and the time between the odours is minimal. This can mean that the recognition for these peaks can be harder for the panellist. If you can slow down the GC programme to allow the peaks to elute more slowly, this can help the panellist have time to recognise the first odour before the second odour appears. Other odours may cause issues as the odour may disappear before the panellists have had the chance to recognise it. The best way to deal with these peaks is to repeat the assessment with the panellist only assessing the effluent just prior to the peak eluting. This way they will be poised and ready to detect, recognise and output the information.

Session 7: (1 hour) Individual work or in small groups

- Ask the panellist if they have any questions from last time.
- Show them the aromagrams from all the other panellists and discuss their results.
- Remind them that there may well be additional unknowns interspersed with the odours they have already tried, but this time they might like to try and describe them.
- When the panellist is happy, inject the mixture of odours that have been previously assessed and ask them to detect the odours as they elute from the column.
- Give the panellist a short break and then repeat.
- Thank the panellist(s) for their time, answer any questions and explain what will happen in the next session.

Before the next session assess all the panellists' aromagrams and determine if anyone would benefit from some additional training with the odour vials (recognition errors) or further work on the GC-O (technique and speed issues). By this stage all panellists should be able to recognise the known odours easily, but some may still have issues with the speed of the assessment.

Session 8: (1 hour) Individual work or in small groups

- Ask the panellist if they have any questions from last time.
- Show them the aromagrams from all the other panellists and discuss their results.
- Discuss the results for the 'unknowns' as well as the 'knowns'.
- When the panellist is happy, inject the mixture of odours that have been previously assessed and ask them to detect the odours as they elute from the column.
- Give the panellist a short break and then repeat.
- Thank the panellist(s) for their time, answer any questions and explain what will happen in the next session.

Before the next session, decide which panellists will be taken on as GC-O panellists. Contact each person personally and explain the next stages of the assessments. Move on to the assessment of real extracts and run some 'sharing' sessions with all panellists so that they can look through their aromagrams and discuss and share the descriptions. This will improve consistency between the panellists and make your life a little easier in the interpretation of the data.

Ensure that the panellists get to take part in GC-O assessments at least two times per month, or their abilities to recognise and recognise quickly will soon disappear. Even if there are no 'real' samples to evaluate, get the panellists to assess model mixtures to keep them well practised.

Part three

Performance of sensory panels

Panel performance measures: setting up systems and using data and feedback to monitor performance and act on any issues

11

Carol Raithatha
Carol Raithatha Limited, Norwich, United Kingdom

11.1 Preface

Although sensory panels are made up of people, they have something in common with analytical instruments in that a sensory panel carrying out objective evaluations should be expected to produce valid sensory data at a predefined level of reliability. This means that the output of the panel is measured against performance targets and action taken in cases where the targets are not met.

This chapter introduces panel performance measures and focuses on panels carrying out objective sensory evaluation rather than consumer sensory panels. The main themes covered include how to check your data, monitor performance and address any issues. The chapter has four parts:

- An introduction which explores why measuring performance is necessary and how performance is defined;
- A section on panel performance measures for different types of sensory test methods;
- A section discussing how performance can be monitored, maintained and improved in a range of contexts and over time;
- A look at newer developments and tools for measuring panel performance.

The objective of this chapter is to provide the reader with background knowledge, some pragmatic advice, and examples and further references to aid in the initial design of their own panel performance management system.

11.2 Introduction

11.2.1 The core panel performance measures

In carrying out sensory analysis with screened, selected and trained panellists, there is an expectation that the output will meet certain requirements in terms of measuring what it is intended to and be reproducible. Without setting up some sort of panel measurement and monitoring systems that involve inspecting data, there is no way of ensuring this. This is true for all types of sensory testing including systems for quality assurance or control, discrimination testing and sensory profiling. Although panel performance measurement is often disregarded due to a focus on producing results quickly and cheaply, it is a vital element in the overall expert sensory evaluation offer.

For most methods, there are some common elements that should be monitored, although how these are defined and measured varies depending on the methodology itself and the context of the testing. These core performance measures include

- Discrimination or accuracy – Are the panellists and panel finding the elements within, or differences between, samples that they should?
- Repeatability or reproducibility – If the test was to be rerun with the same or similar type panel and the same samples would the results be the same or similar?
- Agreement – Are individual panellists, having received the same training and with the same experience, giving broadly similar responses when presented with the same samples: Could they be interchangeable?

In addition to the above, there are also important behavioural and attitudinal measures that should be considered such as how well panellists work together as a team, and the motivation and effort level of individual panellists.

11.2.2 Different contexts require different measurement systems

Although all objective sensory panels ideally require some sort of panel measurement system, this will look quite different depending on the methodologies carried out, the product/sample category, the company resources and expectations, the size of the panel and a range of other possible factors.

Some of the key factors to consider when designing a panel performance measurement system are introduced in Table 11.1. These include objectives of testing, context of testing, method type, panel size and type, resources available and sample type. In real-life scenarios, each panel performance system design is unique and takes into account many factors specific to the situation. A few typical examples of design options linked to key factors are highlighted in Figure 11.1.

Because of the large influence of contextual and other factors on effective panel monitoring design, any system should be reviewed and updated to be appropriate to the situation on a regular basis.

11.2.3 Panel performance monitoring as a key tool in panel management

Although often an afterthought, panel performance measurement should be an important part of planning an experiment and general panel management. In an online survey (Rogers and Raithatha, 2012) approximately two thirds of respondents (who all carried out or commissioned sensory evaluation and used panel performance measures) reported evaluating panel performance measures on every project. But respondent comments in the survey indicated that there was a need for more time efficient and clear solutions, and that existing software could be improved.

Table 11.1 Key factors to consider when designing a panel performance measurement system

Factor	Variations that will affect design
Objectives of testing	The overall objectives of the sensory test will help to decide the level of rigour of the panel performance system. Sensory testing carried out to deliver very specific quantitative information or to inform high value commercial decisions will have much more rigorous panel performance measures.
Context of testing	Panel performance systems for sensory testing carried out in the field or during a production process are likely to be simpler than that designed, for example, for a sensory panel supporting research and development and carried out in a controlled setting.
Method type	Every sensory method delivers different outputs and the performance measurement method is designed to ensure accuracy and reliability of these. The more complicated and multidimensional the output of the sensory method the more options available for evaluating performance of these outputs. Therefore, monitoring systems for more comprehensive methods, such as descriptive analysis, are likely to be more detailed.
Panel type	Linked to method type, panels can be composed of panellists with different levels of screening, sensory training and product expertise. For example, panellists taking part in ad hoc discrimination testing may have little sensory training and only be required to show a basic level of acuity and repeatability, while those employed in some types of descriptive panels may have years of training and be calibrated for testing on many scales for which their performance will need to be monitored and measured.
Panel size	The performance measurement of larger panels may allow for more statistical evaluation (such as multivariate statistics) when checking for consistency between panellists. Performance monitoring for smaller panels may need to be more in depth because the impact on decision-making for each panellist is high. In addition, the resource required for in-depth monitoring of each panellist may be more possible with small panels.
Time available for testing	The time available for performance measurement tests, as well as the gap allowed between a sensory test and taking action based on the output, is related to the possible complexity of the panel measurement system.
Sample amount and type	Smaller amounts of available sample will require simpler and more efficient panel measurement systems, often that are part of operational sensory testing. In addition, sample type is closely linked to design of a sensory test and therefore the panel performance measures of that test.
General resources available	In general, the more resources available, the more comprehensive the panel performance measurement system can be. This in turn increases the overall validity and reliability of the data/outputs.

• Objectives of testing

• A dairy quality control system for milk may require a minimum detection level of 'out of specification' samples as validated in regular off-line testing, whereas that for a milk research and development profiling panel may be focused on general discrimination abilities and overall repeatability and agreement.

• Panellist time available

• Dedicated professional panellists (those employed to carry out sensory testing) are likely to have more time for ad hoc assessments. For example, it is very difficult to use a non-dedicated panel to carry out descriptive analysis, unless there is prior agreement for the extended amount of time needed.

• Product type

• Using spiked samples and other accuracy/discrimination tests is easier in some categories compared to others. For example, it is easy to add sucrose to a soft drink but not so easy to create a sweeter apple.

• Resources available

• A small start-up bakery is using a panel of employees on an ad hoc basis for simple descriptive and comparative tests to support new product development. Panel performance in this case might be limited to repeating initial screening tests from time to time.

Figure 11.1 Real-life scenarios of a panel performance options specific to the situation.

Figure 11.2 Panel performance as steps within the project process.

Panel performance is a key element of a sensory service. Performance should be maintained at a minimum and it is worth considering whether performance should in fact improve over time, and if so, in which ways. Once targets are accepted and agreed with management, these should be dovetailed with objectives and targets for the panel as a whole; individual panellists; and also, the panel leader. Many organisations include an element of actual sensory assessment performance within a larger system of regular panellist reviews and feedback sessions.

Panel performance measurement should be closely linked with experimental design; part of the experimental design process should include setting expected panel performance targets. Using the analogy of a sensory panel taking the place of an instrument, understanding the accuracy and reliability of the instrument to be used in an experiment will inform the experimental design in terms of number of replications, expected predictive power of the outputs, etc. Panel performance measurement should be considered as a step (or multiple steps) in the process for any individual project as outlined in Figure 11.2.

There can be various ways of monitoring panel performance. Three of the main structural areas for panel performance systems are

• Time period: Performance at one point in time versus monitoring trends over time
• Intra- or extra project work: As part of normal panel work versus specific performance test sessions
• Target: For individual panellists versus the whole panel

Table 11.2 Structural options for monitoring panel performance

	Performance at one point in time	Monitoring trends over time
Performance measured within normal panel work	Performance criteria that individual panellists and panel need to meet to use data from any test.	Targets for performance over time for individual panellists and the panel based on operational data from tests within the time period.
Performance measured in specific panel performance tests	Performance criteria to check panel and panellists at regular intervals compared to objective/ expected result.	Targets for panellists and panel performance over time based on specific performance tests and closeness to objective/expected result.

Table 11.2 outlines how the application of these factors might look in practice. Ideally all areas of the table will be covered with a panel performance monitoring system, although in some cases the focus will be on one or two quadrants over the others.

11.2.4 Panellist screening and its relationship to performance monitoring

All panellists who take part in objective panels should be screened during the recruitment phase to ensure they have the sensory acuities and other competences and lack of restrictions as needed for their role. This screening should complement planned forms of performance assessment and monitoring for the working panel. For example, ISO 6658 2005 *Sensory analysis methodology – general guidance* states that performance of selected assessors should be monitored regularly to ensure that the initial criteria for their selection continue to be met. Two of the main areas to measure ongoing in terms of performance are core acuities (such as taste and odour perception) and test performance.

Test performance is a more direct route to understanding future performance, but ongoing monitoring of core sensory acuities may show longer terms trends or explain areas of specific poor performance. This could allow for any problems detected to be investigated and rectified in good time. It is known that some senses (such as odour perception) may deteriorate with age. Although it can be difficult to know how to manage a situation where the core acuity of a panellist has fallen below target, it is best to be aware of the problem.

Existing panellists may be asked to carry out screening tests as part of regular training days or complete these tests when they are run for potential new candidates. This gives a common starting point from which to be able to compare areas of strengths and weaknesses between panellists. It also allows for all the panellists within a panel to have had common experiences, especially where there is an element of familiarisation within the screening. This helps the whole panel be 'on the same page'.

11.3 Performance measures for the major panel types and methodologies

Performance measurement is more defined for some sensory methodologies compared to others. Although there is general guidance relating to panel performance measurement in sensory text books and industry standards, it will often be advisable to consult the standards and guidance attached to each methodology to decide the best performance measurement tests and data to collect in each case. In almost every situation it will necessary to make some adaptations linked to the specifics of a panel and its operation. This section presents key considerations; and examples of how data can be collected and analysed, and what sort of actions might be taken; for three of the main methodology areas/contexts of sensory evaluation.

11.3.1 Panel performance measures for profiling methods

Panel performance measures and approaches for profiling methods and panels have been well researched and documented in academic papers, text books and standards. For example, ISO 11132:2012 *Sensory analysis – Methodology – Guidelines for monitoring the performance of a quantitative sensory panel* gives guidelines for monitoring and assessing the overall performance of a quantitative descriptive panel and the performance of each member. New approaches and guidance in this area are often published, for example, ASTM WK8435 *New Guide for Measuring and Tracking Sensory Descriptive Panel and Assessor Performance* is currently under development. Another example is the chapter on panel performance, monitoring and proficiency (co-authored by Carol Raithatha and Lauren Rogers) in the textbook *Descriptive Analysis in Sensory Evaluation* (editors Hort et al.) which is due to be published in January 2018.

Although each organisation carries out sensory profiling in its own specific way, the majority of approaches fall into two main areas – those establishing vocabulary and scales relative to the samples being evaluated within a given project, and those using predefined vocabulary and absolute scales. The overall approach to panel performance for these two types of profiling is similar, but there may be some key differences linked to the methods themselves. For example, there will be more weight attached to consistency in scaling in the absolute scale methods than in the relative scales scenario.

In addition, approaches to panel performance for profiling panels will vary depending on resources, use of the panel outputs and all the various factors mentioned earlier. In practice, performance monitoring systems can range from very simple calculations attached to a testing session, to complex and multivariate outputs and databases linked to unique assessments, and can include evaluations carried out over time, and may be complemented with specific proficiency tests. Practices within these systems can vary too: For example, some sensory laboratories require a practice profile for panel performance checks to take place between vocabulary development and agreement and the actual data collection, but this is not a universal practice.

According to ISO 11132:2012, performance in the context of a descriptive panel and individual panellists comprises of the ability to detect, identify and measure an attribute, use attributes in a similar way to other panels and within panellists, discriminate between stimuli, use a scale properly, repeat their own results and reproduce results from other panels and other panellists.

In most cases, performance measures for profiling panels use indices or applied statistics to measure each of the three core areas of panel performance as introduced above; discrimination, repeatability and agreement. How each of the areas is monitored depends on the type of sensory profiling being carried out, the number of repetitions, the number of attributes in the sensory profile and the nature of the samples being tested (e.g., how different are they to each other, are any differences known prior to testing and how homogenous replicate samples are expected to be). Most of the basic indices used are linked to analysis of variance (ANOVA) outputs for each attribute in the profiling vocabulary or lexicon. Measures and indices are usually specified for the panel as a whole and for each individual panellist.

Table 11.3 gives some examples of how each of the core areas may be evaluated for profiling tests. It should be noted that there is not only one set of statistical indices or measures to use in panel performance. The choice of which to use ongoing will in part depend on how data analysis is carried out and what information is readily available and if the focus is on absolute or relative performance. Indices can be designed to

Table 11.3 Examples of how analysis of variance (ANOVA) statistical output can be used to measure profiling panel performance in three core areas

Panel performance area	Example of typical measure from ANOVA for each attribute	Example of several options for using measure
Discrimination	Sample or product F-value	Significant F-values as a percentage of all attributes or percentage of attributes where differences are expected.
Repeatability	Root mean square error (RMSE)	RMSE below a percentage of measurement scale or below a percentage of overall mean for attribute.
Agreement	Panellist by product or sample interaction F-value	Lack of significant panellist by product interaction or interaction F-value below a specific level.

measure absolute targets or with respect to relative measures. For example, repeatability targets for an individual panellist might be set at a specific maximum scale range for all assessments of a sample and attribute combination or could be for a maximum percentage more than the total panel range for that attribute.

In addition to statistical methods to evaluate discrimination, repeatability and consistency; most panel performance approaches will also consider whether the information generated from the panel is overall delivering what it is expected to and has some

	Maximum of Attribute 2	Minimum of Attribute 2	**Range of Attribute 2**
Product A			
Panellist 1	62	58	4
Panellist 2	64	58	6
Panellist 3	75	63	12
Panellist 4	56	51	5
Panellist 5	76	60	16
Panellist 6	51	40	11
Panellist 7	75	65	10
Panellist 8	55	29	26
Panellist 9	77	68	9
Panellist 10	37	17	20
Product A total	77	17	**60**
Product B			
Panellist 1	82	80	2
Panellist 2	78	55	23
Panellist 3	88	52	36
Panellist 4	49	30	19
Panellist 5	49	43	6
Panellist 6	70	62	8
Panellist 7	78	27	51
Panellist 8	45	33	12
Panellist 9	65	46	19
Panellist 10	22	17	5
Product B total	88	17	**71**
Product C			
Panellist 1	99	69	30
Panellist 2	81	65	16
Panellist 3	90	67	23
Panellist 4	55	52	3
Panellist 5	69	62	7
Panellist 6	85	72	13
Panellist 7	81	60	21
Panellist 8	81	50	31
Panellist 9	60	49	11
Panellist 10	31	7	24
Product C total	99	7	**92**

Figure 11.3 Pivot table showing the range of each panellist over replicates for each sample for Attribute 2.

validity based on the measurement context. For example, in a sensory profile designed to benchmark a new drink category; are differences being found for a number of attributes and does a multivariate analysis and segmentation of the data show different profile styles and groupings of samples that make inherent sense.

Steps in an example of a basic approach to determine and monitor panel performance for profiling data with at least two replications might be:

- Examine data for outliers and inconsistencies and identify any panellists with potential repeatability problems;
- Examine product by attribute (interaction) graphs to look for larger and expected product trends and to spot any large problems with panellist agreement;
- Carry out two-way (products by panellists) ANOVA for each attribute and inspect output measures to determine discrimination, repeatability and consistency of the panel and relevance of overall findings;
- Repeat ANOVA by individual panellists to clarify source of discrimination problems found for the panel as a whole;
- Look at correlations of individual panellist means to panel means for products by attribute to clarify source of agreement problems;

Figures 11.3 through 11.7 and the points below show examples of typical analyses and outputs (using pivot tables and specialised sensory software) related to some of the steps in the above approach, featuring the evaluation of one attribute (Attribute 2) from a profile of five attributes, three samples/products, carried out with 10 panellists, and with 3 replications.

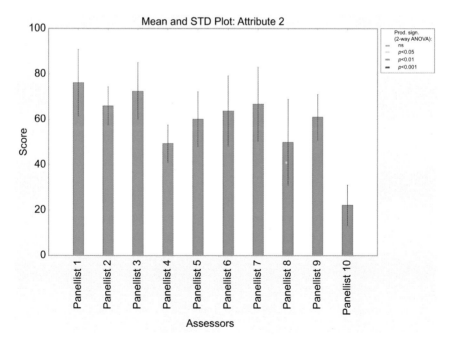

Figure 11.4 Mean and standard deviation plot (across all samples) for Attribute 2 from PanelCheck.

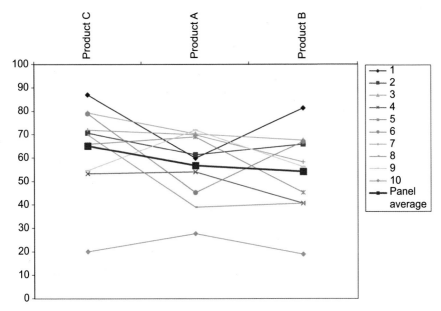

Figure 11.5 Interaction plot for Attribute 2 from SenPAQ analysis.

Attribute 2

Source	DF	type II SS	Mean sq	F-value	Pr > F
Assessor	9	19257	2139.6	6.98	0.0003
Sample	2	1966	983.2	3.21	0.0644
Assessor*sample	18	5521	306.7	2.95	0.0009
Error	60	6234	103.9		
Total	89	32978			

	Product A	Product B	Product C	LSD	Prob	Scale type	Low scores, %	Interaction F-value	Interaction p-value	RMSE	Pre scaling
Attribute 1	53.9	59.2	65.5	8.8	0.0420	0–100	0.0	4.3	<.0001	7.8	No
Attribute 2	56.9	54.2	65.2	9.5	0.0644	0–100	0.0	3.0	0.0009	10.2	No
Attribute 3	37.4	55.3	70.5	10.3	<.0001	0–100	0.0	4.6	<.0001	8.9	No
Attribute 4	25.9	28.9	24.1	3.7	0.0394	0–100	0.0	1.3	0.2045	5.9	No
Attribute 5	49.4	30.6	28.2	4.8	<.0001	0–100	1.1	4.0	<.0001	4.4	No

Figure 11.6 Analysis of variance for Attribute 2 and summary means table from SenPAQ.

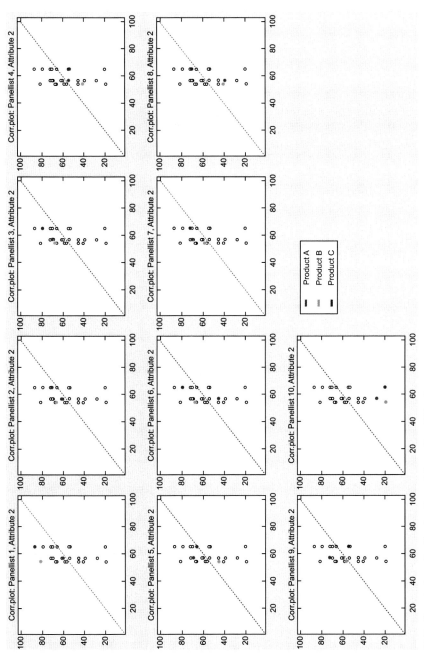

Figure 11.7 Correlation plot for Attribute 2 from PanelCheck.

The data have been converted to a 100-unit scale. The data set has been chosen to demonstrate realistic panel issues.

- Figure 11.3 presents the range of scoring (from a pivot table) over replicates for each panellist and for each sample for Attribute 2. This indicates some repeatability problems with quite a wide range (over 20 units) over replications for quite a few panellists and a particularly large range for Panellist 7 when scoring Product B (51 units).
- Figure 11.4 shows the PanelCheck (www.panelcheck.com) output of the mean and standard deviation for Attribute 2 for each panellist across the samples and replications, and that there was no significant difference found between the samples for Attribute 2 (indicated by the figure border colour), and that Panellist 10 tends to score lower than all other panellists.
- Figure 11.5 shows the SenPAQ (from Qi Statistics www.qistatistics.co.uk/software/senpaq) Interaction plot for Attribute 2. The plot shows that the scoring range over panellists for each product accounts for a large percentage of the total scale and that there is not consensus in sample ranking across panellists, indicating potential problems with understanding of this attribute. The plot again illustrates the outlying position of Panellist 10.
- Figure 11.6 gives the ANOVA for Attribute 2 and summary means data for all attributes. Attribute 2 is the only one of the five where no statistically significant difference has been found at the 95% confidence level, although it is showing borderline significance. Attribute 2 also has the highest root mean square error (RMSE) of the attributes showing some potential problems with repeatability. Both the Assessor and Assessor by Sample factors were found to be significant, indicating problems with panel consistency and scaling agreement.
- Individual ANOVA carried out for each panellist for Attribute 2 also highlighted poor repeatability for Panellist 7 compared to the other panellists.
- Figure 11.7 is a correlation plot which reinforces that Panellist 10 is a low-scoring outlier for Attribute 2. The plots shows the mean score for each sample for each panellist plotted against the panel mean score for that sample, with each panellist's own scores highlighted in their individual plot.

The evaluations above indicate that poor panel performance may have been a contributor to no significant difference being found between the samples for Attribute 2. There may be some problems with the definition and/or evaluation procedures and/or scaling references for Attribute 2 and these should be explored in further discussions with the panel and training. The potential that an element of the performance issues could be due to inherent sample variability should also be explored. The performance of those panellists showing poor repeatability should be monitored.

Action standards can be applied using statistical outputs such as those above from panel performance measurement. For example, a threshold level can be set for number of attributes with in target repeatability, discrimination and agreement for data from the profiling test to be judged to be within acceptable quality standard for further use. Case study 11.1 presents how hypothetical panel performance targets are set and used for a breakfast cereal descriptive panel.

Case study 11.1: Core panel performance targets for a multipurpose descriptive panel

Company A (a breakfast cereal producer) uses a trained descriptive panel in a multipurpose mode: To evaluate competitors, new product concepts and prototypes, shelf-life study samples, etc. Profiling is carried out in a relative way; the vocabulary and scales are developed and relative to each sample set evaluated. Evaluations need to be turned around in a very tight timescale, but because sensory outputs feed into important business decisions, the sensory panel leader is still concerned with ensuring a minimum level of data quality. He therefore has insisted on performing duplicate evaluations in every test and also has decided to apply an overall panel level hurdle for performance, below which, the data from the testing session should not be used. After inspecting the last 6 months of profiling data outputs he decides on the following panel level targets:

- Discrimination: There must be significant discrimination between samples for at least 75% of attributes in the profile
- Repeatability: The RMSE must be less than 15% of the scale for each attribute
- Consistency: No more than 25% of attributes should show significant sample by assessor interaction

When the above targets are not met, the data are evaluated in more depth to determine where problems may lie; within sample variability or preparation, the panel performance as a whole, individual panellist performance, assessment of very similar samples, etc. In some *urgent* situations, data from specific poor-performing panellists are removed and statistics recalculated and evaluated. But in most cases corrective actions are taken with the panel or individual assessors (retraining, redefinition of attributes, or recalibration on scaling, etc.) and the test rerun and a new set of data collected.

11.3.2   Panel performance measures for quality methods

Quality control or assurance panels may use a range of methods including variants of discrimination testing and/or profiling. But these types of panels often use adapted methods that are based on a more global evaluation of a product compared to a reference or specification and decision-making using a small number of panellists. In this context, management of response bias and product knowledge is as big a factor in panel performance as well as discrimination ability and reliability.

For example, in the standard *Sensory analysis – General guidance for the control, by sensory analysis, of the quality of a product during its processing* (AFNOR, 2013), it is acknowledged that possibly only one assessor will be making judgements on products taken in real time from a manufacturing line. The standard also recommends a panel of three for data collection for raw material acceptance or positive release quality assurance contexts. This creates quite a unique scenario when it comes to panel and panellist performance measures.

It is important to keep in mind that the core objective of sensory quality assessments is usually to screen out batches of product that will be rejected by consumers. A secondary objective is identifying the problem and correcting it. This means that detection and description of 'negative' attributes is key to success of many sensory quality methods, whereas 'negative' attributes in methods such as sensory profiling are normally considered equally as important to 'neutral' and 'positive' attributes. Screening, training and panel performance monitoring of sensory quality assessors therefore usually entails ensuring detection and recognition of typical taints, defects and off-notes for the sector of relevance.

Arguably the most basic sensory quality method is the In/Out or Pass/Fail method, in which trained and experienced assessors evaluate a product and decide if it falls within specification or is the same as a reference or not. A middle category may also be commonly used which indicates a borderline sample. According to Everitt (2010a), the red, amber and green 'RAG' system is one of the most popular sensory quality methods and comprises of three grades into which finished product can be allocated: green = target quality, amber = borderline quality and red = unacceptable quality.

Some quality methods are more complex and designed to work on two levels – objective evaluations followed by decision-making using the data obtained:

1. The panellist level: Where sensory evaluations are made and decisions are taken by panellists about differences between samples or levels of key attributes within samples are determined.
2. The supervisor level: Where outputs of panellist evaluations are used to make final decisions about whether to release product or not. The decision criteria may be explicit and predefined based on sensory outputs as compared to specifications or be more flexible and based on a range of factors affecting the business and the product.

In the case of two level systems, panel performance monitoring may be designed around screening and regular repeat testing as per screening to ensure discrimination and scaling abilities of each assessor for relevant attributes and defects. It should be noted that this sort of two-level system requires more time and resource than a one-level approach and therefore will not be practical in many cases.

Limiting bias is a key consideration within sensory evaluation of in production (one-level) quality scenarios, as it is often not possible to blind code samples, and panellists are likely to be highly involved in the manufacturing process and subject to pressures of maintaining production volumes. Performance measurement therefore needs to be designed to take this unique situation into account.

According to Rogers (2010), the use of panel monitoring techniques for In/Out type methods is critical and can be easily implemented by the use of rejected products which are usually kept from previous batches. This can work well in an off-line positive release panel but is not practical to implement in the online scenario, implying that separate/off-line performance assessment sessions are likely to be required for in production panellist. But, performance versus other online panellists and with respect to actual product quality can also be monitored on an ongoing basis, i.e., do panellists find (or miss) approximately the same number of quality issues as other panellists within a set period and are the 'hits' later confirmed as linked to problems with materials or processes.

It is possible to set targets per session for off-line quality evaluations in terms of the accuracy of hits of spiked or previously rejected sample. Meeting these targets will ensure that the panel and individual assessors are sensitive enough and using the correct decision criteria when categorising a sample as In or Out. Case study 11.2 is hypothetical scenario of a cereal manufacturer using an RAG method.

Case study 11.2: Performance monitoring of an off-line quality assurance panel

Company A produces a number of batches of a branded dry cereal product per week. A positive release system is operated on finished batches in which each week a sample of each batch of product produced is evaluated by a small panel of assessors. All panellists are screened to be able to detect and recognise common taints and defects in the cereal. In addition, all the panellists have received training to be familiar with the product and its specification and have been shown to be able to discriminate between 'in specification' and 'out of specification' product and have an understanding of the range of variations that is allowable and to describe defects effectively.

For each ongoing evaluation session, samples from each batch are compared to a gold standard control using a three-point scale: (1) – meets the specification and is similar to the control in terms of sensory quality (green), (2) – borderline in terms of sensory quality in one or more modalities compared to the standard (amber) and (3) – lower sensory quality compared to the control and out of specification (red). Within each evaluation session, known/previously agreed out of specification/rejected product samples and/or gold standard samples are included blind.

The panel is monitored in terms of performance, within each session, and on an ongoing basis, with the target being set for maximum misses and false hits. When there is a problem this is investigated and the samples may be re-evaluated. Individual assessors are also monitored in a similar fashion over time. Any issues as identified by a missed performance target are investigated and a training plan for each assessor and the panel is reviewed and implemented regularly.

In addition, these quality panels may have more in-depth evaluation/training sessions dedicated to finding areas of detection and discrimination weakness and improving performance. For example, a session may consist of introduction or re-introduction of common defects followed by evaluations of series of triangle tests with samples with and without common defects or taints, or with slightly varying concentrations of key ingredients, followed by feedback and a discussion on best evaluation approaches for detecting differences and off-notes or defects. This is similar to the training procedure described in Chapter 9 using the A-not-A procedure.

Key elements of the performance monitoring system in this case are the training with respect to the method and product; choosing gold standard control products and having clear and defined specifications; collecting rejected product for use in later performance evaluations; the in-built monitoring system and the implementation of a training plan.

11.3.3   Panel performance measures for discrimination testing

There are few references with respect to panel performance for discrimination test-ing. This may be because discrimination testing is sometimes carried out with con-sumers and performance of consumers is not generally monitored or considered relevant. The ISO standard for the triangle test (ISO 4120:2004), which is arguably the most well known and commonly used discrimination test, states that experience and familiarity with the product to be tested can improve the performance of an assessor. The standard also suggests that monitoring of assessors over time may be useful for increased sensitivity but does not give further information as to how to monitor these assessors.

Findlay and Findlay (2017) suggest that a screening phase added to a discrimina-tion test can help to increase overall performance on the test. They give an example of recruiting a large pool of employees or consumers, and in the first instance asking them to carry out basic discrimination testing on a model product or solution. This phase serves to screen out those participants with the highest acuity and to allow for a 'warm-up' in the discrimination method. Then advanced discrimination testing on the product in question is carried out using the smaller, but more discriminating pool of 'qualified assessors'. The authors argue that this approach is effectively improving the sensitivity of the test in real time, and with modern automated sensory tools, it need not be costly or time-consuming.

The ability to discriminate is arguably the most important panel performance consideration in discrimination testing: It is vital to be confident that the panel detects differences between samples when they are present. A logical way of mon-itoring panel performance in this context would be to monitor that a panel picks up an expected difference every time it is exposed to one or at least a target propor-tion of testing events. This would also be true for individual panellists making up the panel. Ongoing monitoring can therefore involve setting up tests with expected differences (spiked samples or samples with known sensory differences) for both the panel as a whole and individual panellists. As some methods of performance monitoring of individual panellists may require multiple evaluations of the same samples, performance monitoring for discrimination testing could be quite rigorous and time-consuming and therefore difficult to implement in the context of a normal busy sensory department.

A possibly less time-consuming method of monitoring individual panellists in the case of employee or repeat discrimination panels is to monitor percentage of 'correct' assessments within a time period and compare this between panellists and possibly against an action standard. Findlay and Findlay (2017) again argue that assessing panel performance in this way has become much easier and more efficient with the advent of modern database-driven tools.

Some applications of sensory evaluation, such as taint detection, rely heavily on dis-crimination testing type techniques so it can be informative to look at these specialist areas with respect to how are panellists selected and monitored. For example, accord-ing to ASTM E1810 – 12 *Standard Practice for Evaluating Effects of Contaminants on Odour and Taste of Exposed Fish*, panellists should be selected for their experience and ability to detect and quantify the off-notes from a suspected contaminant source; they should be trained in the evaluation procedure; and their performance should be validated before testing. In this situation, because of the specificity of the sensory evaluation being carried out, one could argue that panel performance should be validated before or during every testing occasion and/or for detection and/or recognition of the compounds most commonly causing problems.

From another perspective, discrimination testing is often used as a selection method and ongoing evaluation tool for panellists carrying out other types of sensory testing such as sensory profiling or quality evaluations. For example, screening protocols for quality panellists will often include several discrimination tests of typical 'In' and 'Out' of specification samples. Discrimination testing can also be an important part of training for many sensory panels. ISO 8586:2012 *Sensory analysis – General guidelines for the selection, training and monitoring of selected assessors and expert sensory assessors* suggests using triangle tests in training for detection of a stimulus; and paired comparison, triangle and duo-trio tests to demonstrate differences in special tastes and odours at high and low concentrations and to train panellists to recognise these stimuli.

11.4   Monitoring, maintaining and improving performance

Panel performance measurement is not a one-time activity, it is something that needs to be continuously addressed and monitored.

There are four elements of monitoring and maintaining performance to consider as outlined in Figure 11.8:

- Inspecting performance statistics over time
- Monitoring motivation levels and teamwork
- Comparisons with other panels and proficiency testing
- Communication and two-way feedback systems

This section looks at each of these elements in turn.

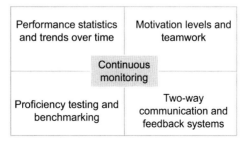

Figure 11.8 Areas to consider when monitoring panel performance.

11.4.1   Inspecting performance statistics over time

Setting up a database system to be able to inspect individual and panel trends over time can provide a lot of useful information to help with panel monitoring and training. When designing such systems careful consideration needs to be given to the key performance criteria to measure, time intervals and options for summarising and/or visualising these. The design of the system should relate to frequency of panel project work, panel training intervals and panellist movement.

Using a database where queries can be made to evaluate panel and panellist performance based on date, product type or other relevant factors can be a helpful aid in longer term panel monitoring for all types of sensory methodologies. For example, it may be possible to determine how often an individual panellist contributes to problems with the usefulness of overall panel data and why, therefore allowing the panel leader to plan for retraining and concretely improve the efficiency and validity and reliability of the panel.

A common approach is to take statistical outputs or measures from ongoing project work and plot individual panellist or overall panel statistics over time or testing session instance showing movement within specified control levels. The control levels or targets can be absolute and determined in advance based on required repeatability, accuracy and/or consistency or may be relative and calculated based on data collected within a time period. Plotting the data in this way helps to clearly see trends and the effect of events on panel performance.

Figure 11.9 shows a control chart that could be set up to monitor panel level discrimination targets as outlined for the ad hoc breakfast cereal descriptive panel as described in the case study earlier. In this hypothetical case, the panel starts with good discrimination

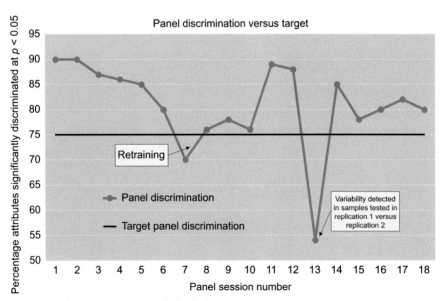

Figure 11.9 An example of a panel discrimination control chart.

ability, but this gradually deteriorates over time as some assessors leave, structural changes happen within the business and the panel loses motivation and the panellists forget some of their training; in particular, the process of choosing attributes that discriminate between a sample set, how to gain consensus on attribute definitions and how to use scales consistently. The panel leader decides a retraining and panel motivation session is necessary and after a half day intense but fun training event, performance improves. Performance following the training is quite variable but above target. On one testing session discrimination performance shows a marked drop below the target, but when samples tested in that session are inspected, it is obvious that, due to storage problems, some of the samples given to the panel in replication 1 are quite different than those for replication 2, and this is believed to be the main cause of the poor discrimination performance.

A range of tables, coding and graphing techniques can be used for tracking performance data over time. For example, Szczepanski et al. (2016) presented a two-tier system for longitudinal tracking of sensory panel performance. The first part consisted of a table that presents and summarises individual panellist and study/test performance via colour codes and summary statistics. A second table shows the progression of each panellist over time allowing diagnosis of whether performance is improving or deteriorating.

11.4.2   Monitoring motivation levels and teamwork

Understanding how and why assessors are enjoying their work or not, and whether they are performing well as a team, can be very important in the global view of panel performance monitoring. This is because any problems with team cohesion and individual and panel motivation are likely to affect data quality. It is important to remember that a sensory panel consists of human beings rather than analytical instruments. Arguably there are ethical reasons to create a good working environment and rewarding role for panellists, but ensuring panel motivation is also good practically and economically for the organisation engaging the sensory panel.

It can be helpful to consider what motivates sensory panellists given the particular context of their role. Reward theory suggests that this may be different for different individuals, but there are likely to be some commonalities to be aware of. For example, Lund et al. (2009) investigated the factors that affect and influence trained panellists' motivation. Surveys revealed that extra income and a general interest in food were the key drivers in inspiring people to become panellists, whilst enjoyment in being a panellist, interest in food and extra income were key drivers for people to remain panellists. This points to the importance of enjoyment and interest in addition to the expected financial motivations.

Motivation and role satisfaction can be measured and monitored in a range of ways, some more formal than others. For example, a structured 360-degree performance evaluation system for all sensory staff will enable each panellist to provide feedback to other members of the team and receive their own, in a safe and supervised manner. Informal chats between a panel leader and the panel and individual panellists will also help to ensure good communication and increased motivation. Regular process improvement sessions in which the sensory management, panel leaders and the panel all participate can help with continuous improvement and improve panel and panellist motivation, especially if suggestions are taken seriously and implemented where pragmatic and in line with business objectives.

Another formal approach is to run an anonymous feedback survey for panellists in conjunction with other forms of evaluation and communication. A survey can create a platform where grievances can be highlighted without compromising existing working relationships, especially if managed by a neutral third party. According to Kapparis et al. (2008) questionnaires can be used to set the correct framework for organisations to build relationships with employees. The authors suggest that the survey process needs to be understood by all and its aims clearly explained. Sharing final results with participants is also advised as a way of increasing organisational transparency, clarifying the importance of the original objectives and building trust.

11.4.3   Comparison with other panels, benchmarking and proficiency testing

When considering panel performance, benchmarking and proficiency methods can be very powerful in helping an organisation to understand if their panel and panellists are performing as well as other similar panels and/or delivering results that are within an externally defined target or industry standard. Reproducibility (the ability of the panel or individual panellists to provide the same output from the same samples on separate occasions) is sometimes also seen as an element of proficiency. Benchmarking and proficiency testing can be designed to validate general acuity and discrimination skills or performance in a specific methodology and for a specific sample type.

Resources and services exist that can help with benchmarking and proficiency testing. Commercial taster validation schemes such as that offered by Aroxa (www.aroxa.com/about-taster-validation) can be used to screen panellists and monitor and benchmark their abilities on an ongoing basis. Aroxa offers two types of sensory proficiency testing scheme for professional taste panels (http://www.aroxa.com/about-validate): The first assesses the ability of panellists to identify flavour attributes presented at low levels in samples. The second relates to scaling ability and involves ranking followed by rating nine samples for a single attribute. Aroxa is one of several possible suppliers offering similar third-party systems. Another potential supplier is FlavorActiV (www.flavoractiv.com). The fact that these offerings can include; samples, testing schemes, make up instructions, evaluation methods and benchmarking data; makes the task of proficiency testing easier for end users and provides a concrete outcome.

Benchmarking for panels and panellists can be applied at the methodology level and therefore be designed to compare specific performance indices. For example, SensoBase was created in 2006 for gathering sensory profiling data sets. The idea was to offer a free innovative statistical analysis of a data set that the owner deposited into a database. According to researchers managing and working with SensoBase, the concept has been well received, and as of 2016 SensoBase contained the data from 1084 studies conducted by 56 companies in 10 different countries. The database can be used to benchmark panel performances and place individual panellist and panel measures in context. For example, an analysis of the database found the standard deviation of scores over replicates measuring individual repeatability was 0.94 on average on a 0-to-10 scale (Peltier et al., 2016a).

Proficiency testing can also be applied to panels carrying out discrimination testing. For example, Sauvageot et al. (2012) compared 15 groups of assessors from 9 laboratories for performance on triangle tests according to the ISO standard on two pairs of soft drinks. The groups differed in practice level with triangle tests. The study found that two of these assessor groups had results in '*large disagreement*' with the others, although all laboratories were accredited by a French accreditation society for the triangle test. The authors noted a large diversity in instructions to assessors, and they suggest that differences in wording may have an unexpected impact. This research demonstrates the real value of proficiency testing for benchmarking performance and starting the journey towards correcting for any problems found.

11.4.4   *Communication and feedback of performance issues*

Performance issues should be identified from monitoring systems and corrected via an established communication, training, retraining and validation system. It can be helpful to take an action standard approach for panel performance and link the measuring, monitoring and visualisation of key performance statistics with training and retraining via feedback systems. Figure 11.10 shows the basic design of a typical panel monitoring and performance feedback system.

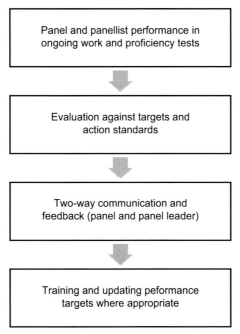

Figure 11.10 Typical panel monitoring and performance feedback system.

It should be noted that an output of the performance monitoring and feedback system is often retraining, but can also be updating of the performance targets themselves.

When correcting for performance issues preemptive training is the preferred route, rather than correction per se. For example, in the context of maintaining the effectiveness of a sensory quality control/assurance program, Everitt (2010b) advised refresher training on a formal basis at least once a year to realign and revalidate assessors' ability to identify and score specified levels of quality.

Training and/or retraining approach will vary depending on the type of sensory panel and the methodologies being used. For example, a core issue for many forms of descriptive analysis is training to improve panellist to panellist agreement. Tools exist for immediate feedback during the training phase in this area. For example, the Compusense Feedback Calibration Method (FCM) is described (Compusense, 2015) as a calibration technique that uses immediate feedback in line scales to train panellists rapidly and reliably. As they use line scales, the panellists receive this immediate feedback comparing their scores with established range values set for each product attribute. According to Compusense, users of FCM can see reductions in training time of up to 50%.

Tools such as FCM are undoubtedly useful in certain contexts, but defining the appropriate level of panellist agreement, and the way this should be obtained can be a challenging philosophical as well as a practical question. How much should a panellist be expected to alter their internal scaling to be in line with the rest of the panel or predefined targets? As mentioned in the introduction to this chapter, panellist agreement with respect to trends between samples is a prerequisite for good performance for most sensory methods, whereas the importance of quantitative scaling calibration will depend on type of method.

Retraining should be focused on both general panel work areas and specifically in problem areas. This is likely to be necessary for the panel as a whole and also include elements tailored to individual panellists. The practical implications of a panel leader carrying out personalised training programme elements for each assessor can considerable. Some panel leaders overcome this in part by using a mentoring approach where one experienced assessor works with a novice or poorer performing colleague, comparing and working on techniques for evaluation, describing sensations, vocabulary use, etc. The mentoring approach can be helpful to improve motivation and develop good team relationships within the panel even when there is no panel leader resource issue.

Feedback should always be given in a fair and considerate manner and using appropriate communication tools. For example, when training panels for profiling methods, to help visualise levels of agreement, the panel is often shown product by attribute line graphs indicating the scoring of both the panel on average and that for individual panellists; but each panellist may only be given enough information to identity of their own data. The concept of considered communication being an essential part of good performance is demonstrated by the fact that the ASTM is developing (at the time of writing) a *Standard Guide for Communication of Assessor and Panel Performance* (ASTM WK49780). This is intended for use with discrimination and profiling panels and will provide guidance on types of feedback and appropriate uses of feedback.

Linking panel performance monitoring, communication and action with an effective human resources system is also a challenge. This is where preagreed targets can help define the scope of evaluation. It's important in every context to work with human resources professionals and to understand employment law and local practices to ensure that panellist recruitment, monitoring and performance assessment practices are legal and ethical as well as effective.

11.5   Newer developments

Panel performance measurement is taking on a larger and more important role within sensory panel management. As this happens, existing measures and analyses are adapted, new measures and techniques are proposed for newer sensory methods, and statistical and visual tools are being offered to increase effectiveness and efficiency. This section introduces a selection of these developments.

11.5.1   Panel performance for newer and specialised methods

The types of sensory tests being used is expanding and changing. As well as the more standard profiling, discrimination and quality methods there is widespread use of methods such as time intensity (TI), temporal dominance of sensations (TDS) and a range of rapid profiling methods. This is necessitating different approaches and solutions for panel performance measurement.

For example, below are some suggestions for these newer and adapted methods:

- ASTM E1909-13 discusses panel performance monitoring and feedback in the context of TI evaluation. The standard suggests that performance should be monitored during training and evaluation, and a target level of individual panellist and total panel performance should be set at the start of a study. Within panellist consistency is a key performance measure for TI evaluations, as between panellist variation is expected for most TI curve parameters. The exception is potentially maximum intensity, especially if quantitative references for this parameter are used in panel training.
- Lepage et al. (2014) developed a dedicated panel performance approach for TDS to account for the specific nature of this methodology. This includes three panel behaviour indicators: (1) The maximum frequency of selection for each attribute at the panel level (to identify questionable attributes); (2) For each panellist, the average number of attributes selected per evaluation (to give some clues about the differences in individual behaviours with respect to use of the attribute list) and (3) The average evaluation duration for each panellist (to understand if some panellists need more time to evaluate the samples than others). In addition, the approach specifies four panel performance indicators: (1) Panel discrimination; (2) Panellist discrimination; (3) Panel disagreement and (4) Panellist disagreement; all of which are calculated at four time period and attribute combination levels.

- Hopfer and Heymann (2013) suggested using the people performance index (PPI) as a measure of individual assessor performance in projective mapping type tasks. A blind duplicate sample was presented within the sample set. The PPI is the ratio of the Euclidean distance between two replicated products and the maximum Euclidean distance between two different products in the projective mapping plot and can range between 0 and 1.

The above examples represent useful, and in some cases comprehensive and complex, frameworks for panel performance focused on specific methodologies but based on the core principles of sensory panel performance monitoring. This demonstrates how the key concepts can be applied and extended to any methodology that may be developed in the future.

11.5.2 Adaptations and new statistics applied to key performance measures

Panel performance is a developing area: refinements are often proposed for some of the most widely used panel performance measures. For example, for sensory profiling the Mixed Assessor Model (MAM) ANOVA approach breaks down the concept of agreement into two different areas: scaling differences and pure-disagreement/crossover. According to Brockhoff et al. (2015) the MAM model takes scaling differences into account by including product averages as a covariate in the modelling and allowing the covariate regression coefficients to depend on the panellist. This gives a more powerful analysis by removing the scaling difference from the error term. The authors report increased frequency of significant product differences using the model. Peltier et al. (2014) applied the MAM model into a unified system for monitoring panel performance resulting in the MAM Control of Assessor Performances (CAP) table. The MAM-CAP table summarises product, scaling and disagreement effects, as well as individual panellist performance for discrimination, scaling, disagreement and repeatability. The package for running MAM-CAP in the Free Software R is available from TimeSens (www.timesens.com).

A further refinement of the MAM model is to break the scaling coefficient into two components: an overall scaling coefficient independent of attribute that is related to the psychological approach of the panellist to scaling in general; and a corrected scaling coefficient for each attribute, indicating the specific sensitivity of the panellist for that attribute (Peltier et al., 2016b). Developments such as the MAM approach use statistical and psychological concepts to help to quantify important elements of panel performance. But the general usefulness of an extended model such as MAM may depend on profiling approach. For example, when using a system based on absolute scaling, it may not be appropriate to remove scaling differences from the error term in analyses.

A general approach for comparing panel performance across methods could help to understand their relative effectiveness and value and also could give insights into overall panel and panellists' abilities and performance trends. Some researchers are examining statistical formats for achieving this holistic approach to panel performance: For example, Bi and Kuesten (2012) proposed applying the intraclass correlation coefficient (ICC) as a type of general framework for monitoring and assessing panel performance from a diverse range of experiments that generate different types of data (e.g., continuous data, ordinal data, ranking data and various types of choice data). The authors point out that the ICC can measure similarity between assessors as well as sensitivity of panels and assessors. They suggest using the ICC as a quality index and limit, with larger values indicating better performance, and values below a specified limit indicating that either samples are not different or that the sensory data for the attribute may not be valid and reliable.

11.5.3   Tools available for panel performance monitoring

In recent years measuring and monitoring sensory panel performance has become easier due to specialised digital tools that can collect data, analyse data (often in real time), automatically calculate performance indices, monitor performance trends over time, and create graphics and visuals to easily inspect panel performance and suggest corrective action where necessary. Packages such as XLSTAT, SenPAQ, PanelCheck, RedJade, Fizz, Compusense, EyeOpenR, TimeSens, R and others, all have some functionality in this area. Outputs from a few of these have been shown in some of the examples above. Using these types of tools, sensory professionals should be able to spend substantially less time analysing data to be able to make informed decisions about quality of data collected and effectiveness of panel training. Each offering has specific features and functions of interest, so the choice of package or tool to use will depend on context and budget. It should be noted that the various packages also use different terminology, or the same terminology in different ways, as well as different statistical models and criteria, so it is necessary to become familiar with each tool and its approach.

One example is a comprehensive sensory application by RedJade (www.redjade.net). Included in the sensory software suite is a panel performance module which automatically produces a Panel Analysis and Panel Performance summary for every study. These are designed to be a quick overview of how cohesively the panel performed in determining product similarities and differences, and how well the panel used the language to encompass their perceptions of products and to suggest training actions (see example Panel Summary output sheet in Figure 11.11 using data featured in the Panel performance measures for profiling methods section above: Note that in this case, interaction refers to the fact that significance was lost due to interaction). In addition, the tool produces metric and graphical displays that summarise the performance of each panellist. This incorporates a summary table that gives suggestions for actions to focus future panel performance monitoring and/or training (see example of output for Panellist 10 in Figure 11.12). A useful feature of the panel performance tool is that the criteria for establishing performance thresholds can be customised depending on the panel/project context (the example outputs in Figures 11.11 and 11.12 use default settings). RedJade also has a feature allowing for panel performance evaluation across multiple studies which aggregates the data from many studies across a panel.

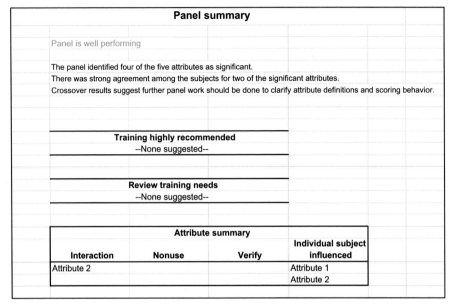

Figure 11.11 RedJade panel performance summary sheet output.

Judgments	Actions	Explanation
Well-performing discriminator		This subject failed to find significance on 20% of the attributes that had a $p < .05$; the mean percentage for the panel is 14%
Well-performing subject SD		This subject had a subject SD score that was above 16.7 on 0% of the attributes
Poor crossover	Monitor performance	This subject had a crossover score that was above 20 on 25% of the significant attributes
Well-performing range usage		This subject had product ranges that were either at least twice or less than half of the mean of all subject ranges on 0% of the attributes
Poor scale usage	Monitor performance	This subject had a scale center score that was either at least twice or less than half of the mean of all subjects on 40% of the attributes

Figure 11.12 Panel performance diagnostics summary table for Panellist 10 from RedJade.

11.6   Conclusions

Monitoring panel performance is now recognised as one of the key elements necessary for successful sensory panel management. Therefore, including panel performance procedures within normal operations has become a necessity. There are many tools and methods that allow sensory professionals to monitor, communicate and take action to improve both panel and panellist performance. Measures related to the core concepts of discrimination, repeatability and agreement are being developed for new methods and also being applied to a range of panel contexts. This is allowing sensory evaluation to move towards a culture of continuous improvement where effectiveness, reliability and accuracy are all important and are seen as a shared responsibility of both sensory professionals and sensory panellists.

Dealing with issues in sensory panels

12

12.1 Introduction

Trying to understand the behaviour of some people is like trying to smell the colour nine.

The statement above gives a nice sensory interpretation of how it feels sometimes to deal with difficult people, but this chapter should help you in your dealings with the panel in difficult situations and might even help with your interactions with other people as well.

It's useful to understand the causes of difficult behaviour first, as this can help us prevent people becoming difficult in the first place. We will start by considering attributes of a difficult person and why people are difficult, and then work our way through topics such as how to prevent recruiting difficult panellists, helping new recruits to not be difficult, panel motivation, how to deal with difficult people and finishing up by considering ourselves: are we causing the issues?

12.2 What is a difficult person?

How would you describe a difficult person or panellist? A panellist who always turns up late? Someone who is always talking over others? Someone who does not listen to the instructions or the other panellists? Someone who is always complaining, especially when you leave the room? Someone who rarely contributes to the discussions or who contributes too much? Or maybe you have a panellist who is often rude or belittles other panellists. Any of these descriptions might warrant the label 'difficult', but let's get back to basics – why are they being difficult?

Think back to the last time you might have been rude to someone, or spoke over someone or said less than polite things about someone behind their back. Why did you act that way? Maybe you were stressed or worried about something. A relative may have been ill or a recent bill may have been a lot higher than you expected. Or maybe the discussion you were involved in was worrying you to start off with. Maybe you were unsure about completing that report on time or taking part in a new work activity. Someone else may not understand why anyone would be worried about that; they will probably have forgotten that they were worried the first time they took on that duty as it was too long ago to recall. They do not see why it's such an issue for you. And, in disregarding your concern, they have shown you that they do not understand your beliefs. You both might be thinking that you want to win, and the other person to lose. You might be thinking that they might blame or criticise you, and they might be thinking that you might blame or criticise them! Because of all these emotions, one of

Sensory Panel Management. https://doi.org/10.1016/B978-0-08-101001-3.00012-4

you, or both of you, becomes 'difficult': speaking over the other person, not listening, maybe playing the silence card or just being generally rude.

But most likely, you did not start out with the intent of being 'difficult' – it was simply your emotions and feelings at the time. And this is often the case with other people, although of course there are some whose sole purpose in life is being difficult…

12.3   How to prevent recruiting difficult people

The first step is to make time in the screening for a personal interview with each potential panellist. You do not need to interview everyone who applies, just those who pass the sensory tests. During the interview, you could ask them some of the questions below:

- Why would you like to join [insert company name] panels?
- What interests you about the role?
- The panel meets [everyday] from [10–12]. How many sessions do you think you would be able to make?
- If your friends had to describe you in three words, what do you think they would say?
- If you were an animal, what type of animal would you be?
- What are your hobbies?

Involve your current panellists, if you have any, in the screening process. They will be great judges of personality and will quickly let you know if there is an interviewee who may be 'difficult' or who they may find difficult to work with. Ask them to help you out with:

- Directing the interviewees from reception
- Help answer interviewees' questions about what it's really like to be a sensory panellist
- Waiting with the interviewees between interview stages. During this 'informal' chatting period you will be amazed what the interviewees will disclose.

Asking the receptionist or staff on the security gate, about the interviewees can also be very useful. Someone who is rude to the company staff they first meet might indicate that they have the potential to be disruptive later – or it could just mean they were having a bad day or worried about the interview or worried about being late. You will need to consider all aspects of the person before writing someone off.

It's very useful to run a mock panel session with potential panellists (see Chapter 5, Section 5.4.5.12). You will get the opportunity to see who may be 'difficult' in the real work situation. If you hope to use your panel for quality control tests that involve an agreement about the final product score, for example, you could set up a test with a product that has passed, a borderline product and one that would fail and see how the potential panellists would work together. If you run profiling sessions, the mock panel session can be invaluable in deciding which panellists to recruit. However, do not write off a panellist who is a little quiet, as he or she may just be a little shy in speaking up amongst people they do not know: be more wary of the person who is very quick to contribute and seems to be running the session for you. If you are only recruiting a small number of people to make up numbers in an existing panel, running mock sessions with existing panellists can be helpful in identifying panellists who would work well with the team. If you are recruiting externally, it can be difficult to arrange a mock panel session as you may be interviewing up to 100 potential panellists to recruit your panel. One option is to have a two-phase interview and invite those who pass the sensory screening tests to attend on another occasion for the one-to-one interview and the mock panel session. If your panellists are not really going to be working together as a team, you may not need to run a mock panel session to assess personality, but it would be worth considering if you conduct panel training sessions that involve teamwork elements. If you are recruiting for internal panellists from the staff on site, a mock panel session is much easier to arrange as you can set up a mock panel session after the screening tests or on another occasion.

BS EN ISO 8586:2014 recommends a different approach to recruitment as shown in Figure 2.5, where training and validation comes before employment. The panellists are essentially recruited on a probationary period which ends once the training programme and validation is complete. There are many advantages to the BS/ISO approach. It allows you to fully assess the panellists, both in terms of their sensory abilities as well as their personalities. By having the panellists take part in many different types of activities, discussions and tests, you will be able to build up a much better impression of their sensory abilities than that just gathered from the sensory screening tests. You will also get a better understanding of how they might fit into the team, their potential to be difficult, how well they understand and follow instructions, and also their abilities at more in-depth skills such as the use of line scales. Some panellists may also find that they are not suited to the role. Maybe they thought they would be assessing products and telling the product developers what to do, or had another completely different impression of the role or they simply just do not seem to fit in with the team. Either way, having the probationary period makes it easier for the panellists to bow out gracefully or for you to tell them that they have been unsuccessful.

This makes this approach, of employment for a probationary period, seem ideal, however, there are some drawbacks. Firstly, if you are going to lose some panellists through this approach, you will need to recruit more people than you will need. This may not be an issue if you have the facilities and resources to cope with training more panellists. Secondly, people may make friends with the panellist who is not successful, and this can cause bad feelings among the remaining panellists who may think their friend was perfectly capable of doing the role. But I think the benefits outweigh the issues. If you think about it, you will need to have the best people you can get and you do not want to invest a lot of time training them to find that they are, after all, unsuitable or that they leave because it was not quite the role they imagined. In either case you will be in the situation where you have to go through the recruitment steps all over again: not the best outcome.

Do not recruit someone with excellent sensory skills if they seem to have the potential to be 'difficult': at some point you will regret it. They might cause disruption on a day-to-day basis or even cause other panellists to leave the panel.

12.4   Helping recruits not to be difficult

From the outset, it is a good idea to be clear about your expectations and how you will manage the panel. Give clear guidelines and rules for the panellists about expected behaviour. Explain why it's important to be punctual and to let someone know if they cannot attend a session. It might also be a good plan to explain up front that you probably will not be able to give the panellists much information about why the tests are being carried out. This will help prevent them complaining later on that they are treated like mushrooms (kept in the dark). One panel I worked with were successful in gaining a patent for a particular ingredient in a complex product and the project team manager treated them all to a lovely dinner one evening. It all worked really well until she told the panellists exactly what the patent was for, despite my asking her not to mention it. The next time we assessed that product, all eyes were on *that* ingredient and completely missed describing other elements of the product. It can be difficult to motivate panellists when they cannot see exactly what they have achieved, but in the example the panellists were delighted to have been treated to a dinner and told they had helped contribute to a company patent: no more information was necessary for them to be motivated.

You can also give the panellists a list of ideas that will make the panel sessions run smoothly and efficiently or better still ask them to generate the list themselves. Some examples are given in Figure 12.1 and the standard sensory rules are given in Table 12.1.

Be punctual for sessions

One person speaking at a time (although the panel leader may need to interrupt to help us achieve our session goals)

Listen when others are speaking – each person's suggestions are equally valued

Be considerate and polite – discussions and disagreement will be about the topic not about the person

Don't be stubborn – listen to others' opinions and be willing to compromise

Be keen to improve your skills and performance – feel free to ask for help from your colleagues or the sensory team

Speak up if you think you have the answer

Use your list of objectives and KPIs to become the best panellist ever

---

Figure 12.1 Helping the panel sessions run smoothly and efficiently.

It can be really useful to create a video of a sensory panel in action and to schedule this into one of your training sessions. Include examples of the panel behaving well so that there are clear examples of expected behaviour. For example, demonstrate panellists listening to each other, discussing ideas but not arguing and even panellists compromising where necessary with good grace. You could ask colleagues to help you if you do not already have a panel, or just record one of your standard panel sessions if you have a panel already. Remember to ask for permission from the panellists prior to the recording. You could also demonstrate the difficulties created when the panel behaves 'badly', to give clear examples of the behaviour to avoid. Show the outcomes when the panellists do not listen to instructions or each other, when they do not follow the standard rules and also demonstrate the impact of that panellist who is always late.

If you do decide to include the negative behaviours in your video, you might run the risk of actually creating the bad behaviours. This is because if someone says to you, for example, *don't* imagine a huge hairy spider sitting next to you, what happens? Suddenly you have a huge hairy spider sitting next to you. So even though I told you 'don't imagine', what did you do? Yes – you went ahead and imagined. This is why toddler books recommend that you tell small children *how* to behave and not *how not* to behave. As soon as you put the idea in the child's head about jumping on the bed at Grandma's house, by telling them not to, then they will want to do it. Same with dieting. You tell yourself that you are not going to eat that chocolate bar and what happens? You eat it (Or is that just me?). And while we are on the topic of bad behaviours, remember that you are the role model for the panel. If you get out your phone and look at it while someone is speaking, if you are rude and grumpy to the panel, if you belittle them and treat them like children, you give them the right to look at their phones, act like children and be rude and grumpy too! This is one of the best known ways to

Table 12.1 Panellist dos and don'ts

Do	Don't
Listen	Use powerful fragrances, soaps, deodorants, etc.
Maintain good hygiene	
Rest your senses between samples	Smoke before a session
Cleanse the palate or test site properly where applicable	Talk over other panellists
	Eat or drink within 30 minutes of a test
Take sensory testing seriously	Eat or drink any strong flavours within an hour of a test (e.g., mint, chilli)
Do not rush – take enough time when carrying out tests	
Switch off your phone	Do not participate in a test when you cannot smell
Respect and follow test protocols, procedures and instructions	Do not participate in a test if you are unwell
	Do not take part in a session if you have a strong dislike for the type of food/drink
Ask questions if you are unsure	
Tell us if you have any issues	Do not participate if you have too much prior knowledge
Consider others' feelings	

prevent behaviour issues – role model good behaviours. So think carefully about the way you run the panel sessions. Are you sometimes a bit snappy when you are running out of time or maybe you roll your eyes when Mildred suggests another attribute, or even enjoy a small smile with another panellist when Bill is rambling on as usual. We are all guilty, but by demonstrating these behaviours we reinforce the belief in our panellists that they can go ahead and do the same.

Another good way to prevent bad behaviours is to create a supportive and motivating environment for the panellists. Some elements of this are covered in Chapter 4 regarding becoming an excellent panel leader, but more information about motivating different types of panellists is given below. The first section is about motivating all panellists, the second includes some specifics for an internal panel and the third section includes ideas for motivating panellists during panel sessions. Motivation is important because it provides a reason for action and when people are motivated to do a good job they work enthusiastically and positively to get results. One of the dictionary definitions for motivation is 'to stimulate in a way that gets positive results' and that is certainly what we need. The ASTM *Guidelines for the selection and training of panel members* (ASTM, 1981) states 'A system for maintaining panel interest and morale is critical to continued participation and performance'.

12.4.1   Motivating all panellists and giving feedback

If you would like to find out what motivates your panellists – ask them. This is exactly what Lyon (2002) did by collecting questionnaires from various member companies' panels. They found that a number of different things motivated panellists and not just the pay. The panellists were motivated by the team spirit, the social aspect, unusual products to try and interest in the role itself. From feedback from some of my panellists, one of the things they enjoy most is describing their role to new people they meet. People are generally fascinated to meet a 'food taster', 'armpit assessor' or 'car door shutter' and this helps motivate the panellists as they realise their job is interesting to outsiders.

To help motivate your panellists you could advertise and promote the importance of their role by speaking in meetings about your work and sharing sensory results, but remember to tell the panellists about it afterwards! Sharing your passion in sensory science with people in your company and your panellists can be a great motivator for all concerned. You could also consider asking for space in the company blog (or even external magazines or local/national newspapers) to tell people about the importance of your panellists. Sensory science is an area that people tend to be very interested in and an article about the tasting of chocolate or smelling smelly armpits for a living can be quite compelling.

Always reward your panellists for participation: not results. For example, imagine an internal panellist who enthusiastically and regularly takes part in sensory discrimination tests but only detected the odd sample in the last 30% of triangle tests: no more than just chance. When getting this feedback on choosing the odd sample, he might be very disappointed and think that he is 'rubbish' and decide that there is really no point in coming along to more sessions. But wait, what if each of these tests was on products that were incredibly similar and the results from all panellists indicated that the two products could be used interchangeably? Doesn't that mean that this enthusiastic panellist was in fact right in his judgement? It seems wrong to make him feel bad in this situation. It might be better to wait for the results to come in for all panellists and give feedback then. Continuous monitoring of discrimination data, comparing each panellist's discrimination to the discrimination of the panel as a whole, can be very useful for recording panel performance and giving more realistic feedback. For panellists who do not do as well at the discrimination as others, remedial sessions where they reassess the same samples (without knowing that they are repeating the test) and are then allowed time to assess the samples with their results in front of them, can be very beneficial in helping them discriminate samples. If you also know from the previous results how other panellists described the difference, you can give this information to the panellist after the test or set the test up as an attribute-specific discrimination test instead.

Now imagine an external panellist who found a particular product set the panel was working on quite difficult. Maybe the rest of the panel found it easy to detect and describe a range of attributes in the samples, but this panellist, although she tried really hard, was unable to detect and rate some of the attributes. She contributed to the discussions about defining attributes and even brought in some references to demonstrate some of the products' other aspects, but she just did not have the ability to detect one or two attributes. In this case, there is not much either of you could do to improve her performance, apart from to agree that it's not a fault as such or that she should not be penalised for her inability to detect these notes. Of course, if these products are regularly assessed by the panellists and the attributes are critical to product quality, you may need to consider a different approach.

Whether an internal or external panel, it's a good plan to give each panellist feedback on their work and if possible make this part of the overall performance monitoring/appraisal system (see Section 4.2.4). This can work particularly well with an internal panel if the number of sessions voluntarily attended is part of the company's appraisal system. Make sure that this does not detrimentally affect people who are unable to take part because of their sensory abilities or because they are physically not on site enough to become a good panellist. Feedback helps embed the learning for a new panel, helps existing panellists build on their sensory experiences and, if done well, means that the sensory panel's results will continuously improve.

When giving feedback consider the different types of personalities and cultures in your panel. Imagine how the people you are giving feedback to will receive the feedback as this will give you the chance to consider the best approach for different outcomes. Think about how you would like the situation to be resolved as this can give you a clear idea of the behaviour you are expecting.

Never give negative feedback in front of the group. If you are running a feedback session that involves showing several panellists' data such as a quantitative descriptive profile, one way to share everyone's data is to code up each panellist so that only they know which is their data. One approach I have found useful in disseminating feedback from quantitative descriptive profiles, is to train the panellists to assess their own data. This has several benefits as I am sure you can imagine! Firstly, because the panellist has to assess their own data, you know that they have fully understood what their results mean. Secondly, it saves you having to actually tell a panellist if they have not performed as well as you had hoped. Thirdly, as the panellist is making their own assessments, they will see where they can improve more easily than by simply being told 'you need to improve your replicates' (because being told 'you need to improve your replicates' does not help them improve their replicates nor show them why they need to improve their replicates). Unfortunately, getting panellists to do their own panel performance assessments does not mean that you can take it easy, as you will need to do the checks yourself to plan the feedback session and choose which attributes, plots and data you are going to share.

One handy tip for giving feedback is in the use of the word 'even'. Consider you are on the receiving end of this feedback: 'You need to do better'. Think for a moment – how would you feel if someone said that to you? Now, how do things change if instead the sentence was, 'You need to do even better?' Notice the change? That word can make a huge amount of difference to the way the feedback is received. Another similar approach is to classify panellists in terms of 'good', 'very good' and 'excellent' rather than 'poor', 'OK' and 'good'.

There are some other ways to help motivate panellists for both internal and external panels:

- You might like to consider giving all your panellists ties, T-shirts, hats, staplers or something similar to show how important they are to the company. This can really work well for internal panellists who do not work as a team within sensory science, as it fosters team feeling and recognition for contributing to creating excellent products.
- Send each panellist a short newsletter two or three times a year with some news from the world of sensory science.
- Visits: see if there are other sensory panels locally that could visit you or that you and the panellists could visit – particularly useful if their product set is not so delicious as yours…
- Other products: arrange for some other product tastings – perhaps a range of new chocolate biscuits that is been heavily advertised or a wine tasting (if allowed). Do not advertise this in advance – let it be a surprise!
- Incentives: consider including incentives for regular attendance. For example, 'attend 20 sessions and receive a voucher for a cinema ticket (or 2)'. Where this is not allowed, 'thank you' – verbally or a simple certificate made in presentation software – can work wonders.
- Attendance: saying to a panellist 'thanks for turning up for the last 15 out of 20 sessions' can really boost their incentive to attend even more regularly – because they realise they are important to you.
- Optical illusion day: share some optical illusions for fun – particularly helpful if you have some waiting time between tests.
- Collage: consider asking panellists to create a collage for a difficult attribute such as fresh or creamy or soft – it might actually help!
- Have a regular celebratory panel party. This could be a Christmas or summer event where all the panellists get together. It need not be costly as panellists are generally more than willing to bring in an item to consume (and discuss it at great length!).

- Colours: bring in some colour charts and ask the panellists to create names for the colours – you could do food names or holiday location names, for example.
- Experts: invite an expert to come and give a talk to the panellists. Maybe a wine tasting or perfumer or chef...
- Do a smell walk – for more information see http://sensorymaps.com/about/.
- Run a quiz. This could be sensory orientated (for example, play different sounds and see if the panellists can recognise them) or just a 'normal' quiz with general knowledge questions.
- Ask the panellists to make lists of songs, books and films that all mention a particular sense or are related to sensory science in some way (for example, the book *Perfume* by Patrick Süskind).
- Bonus payments: when panellists have met certain criteria, or have been working on the panel for a certain number of months or years.
- Leavers: if a panellist decides to leave, and you value their opinion, chat to them – listen to their reasons. It might help you prevent losing more panellists.

12.4.2   Motivating internal staff

Taking part in tasting sessions is time-consuming for you running the sessions as well as for people attending, therefore it's not a good plan to invite internal staff to the initial recruitment and training sessions if you think they are not going to participate regularly. The same applies to interested and motivated staff where they might not be available for a sufficient proportion of time. Panellists need to flex and train their sensory 'muscles' regularly to be good panellists. Do not force people who are not interested enough: it will be very hard to improve their assessment skills and to motivate them. And unmotivated panellists produce poor data because they will not be interested enough to concentrate on the task at hand: their mind will be elsewhere thinking about how to word that difficult email or what to buy for lunch. And of course, as mentioned earlier, do not invite people if you think they might cause you issues. Where possible, have a sufficiently large number of panellists that you can call on so that you are not always relying on the same people.

A neat trick, particularly for internal panel sessions, is to set the session start time at 10 minutes past the hour as this encourages punctuality. For example, say you set your session for 10 past 10 in the morning. While people are at their current role, they see the time is approaching 10 and think, 'I have a panel session at 10'. They then get up from their desk (at 10!) to head to the panel session, but of course it takes 5 minutes to get to the sensory laboratory, and then they have to stop and chat to Bob who they pass in the corridor about that important report... And they arrive at 9 minutes past the hour – just on time. And while we are on the topic of timings, another great way to motivate internal panellists is to schedule sessions that are not too long: just long enough for you to get great data and not *so* long that the panellists dread attending.

Try to arrange sessions and tests that are enjoyable and interesting to help motivate your panellists to attend. If this is not possible, arrange it so that the occasional test is on some different products or try out a different type of test instead of the usual. For example, if you always run triangle tests, try running a duo-trio test with the same samples and compare the results next time you have enough sample. Most of all, be nice to the people attending your tests: be positive and encouraging, and remember to say thank you.

For quality control panels, involving the panellists in the troubleshooting exercises or sharing the results can be motivating. Sharing information from the results from in-trade purchasing[1] can be very beneficial in demonstrating the impact of the quality team's work on product quality. Publicising this information on the company intranet or via newsletters can also help increase panel motivation.

12.4.3   Motivating all panellists during sessions

The relationship you have with your panellists is important. As the panel leader, you may have management responsibilities for the panel (e.g., regular performance reviews) or you may not. Either way, the relationship needs to be friendly and based on respect. Remember to not be over friendly with a selection of panellists or have favourites, as this can cause issues if you need to take them in hand for any reason. Having favourites can also create bad feeling for the 'less than' favourites. One of the best ways to create a good relationship and to build up rapport, is to talk to each individual panellist on a regular basis. You could give them feedback about the sensory work, say 'thank you', tell them that their results were really valued by client X or simply ask them how they are. This may take some time if you have an internal panel but is less stressful than chasing around trying to encourage people to come to tasting sessions. By building rapport with the panellists and developing a good working relationship, you will be far less likely to have issues with bad behaviour.

Make sure that the area where the sensory testing is conducted is pleasant, clean, tidy and that the heating or air conditioning is working well. Have the samples ready for the panellists when they arrive and try to avoid slow or annoying data collection software. Try to avoid creating errors (in sample labelling, computers and printouts) as this can make the panellists wonder why they should try so hard if you do not know which sample is which… Make sure that all sensory staff are welcoming, polite and professional. Adopt the policy 'the panellist is always right' – it's generally the case!

It's a good idea to make a written plan before each panel session and tell the panellists what 'we are hoping to achieve in today's session'. If you ask them, 'Does the plan sound OK? Any suggestions/problems?' this makes the panellists feel like they have an important role and that they have a say in the plan for the sample assessments.

Facilitating is key for a successful panel session and to help you create and maintain a happy panel. You might find it beneficial to attend some facilitator training, such

[1] In-trade purchasing: samples are bought from sales locations and analysed to check quality through distribution and sales channels.

as the sessions run for focus group leaders. Some key ideas for facilitating a good panel session are listed below.

- Room layout: have a round or U-shaped seating arrangement if possible. Circulate where people sit session to session.
- Do not allow people to dominate and do not allow people to waste time. Look at your session plan and explain why 'we' need to move on. Explain again if necessary what the objectives are for the session, e.g., 'I think we need to move on and try the next sample if we're going to get everything done that we want to today'.
- Be respectful and polite – what goes around comes around: if you are off hand and negative, the panellists will be too. You might have to dust off your acting skills if you are having a bad day.
- Actively listen to what people are saying. This means concentrating on each word so that if necessary you might be able to paraphrase it back to them. This might take some practice because people tend to speak much slower than our brains can actually cope with, so the brain goes off to think about something else that it has the 'spare capacity' for. For example, to plan the next stage of the session or think about the experimental design for that awkward-to-cook product. To actively listen, you need to pay attention to the person, ask them questions and paraphrase back to them. Let's look at an example. Mildred is telling you about the smoky attribute in sample three. You might say, 'When have you experienced that smell before?' or 'What type of product also smells smoky?' or 'Can you think of a good reference we might try so that everyone understands the smoky term?' or 'So you think the smoky note you're picking up in sample three is like smoked cheese. Is there a particular brand we might try as a reference?'
- If you think that what the panellist said was probably not entirely helpful at this stage, thank the person for their comment, ask them another question if you think they might have the answer but have not quite got there, or turn to someone else and ask for their input, but be careful that your tone of voice does not sound negative.
- It can be useful to assign a different panellist each session as a timekeeper as this makes it feel like a team effort to keep to time, rather than just your role.
- If you find you are struggling to keep up with writing notes, ask the technician or panel assistant to help, or ask a panellist (remember to rotate who you ask session to session if you ask a panellist).
- Use the panellists' names to ask them questions. At the end of each session do a quick run through – did everyone offer their opinion? If someone looks as if they would like to make a comment but cannot get a word in edgeways, raise your hand, look directly at them and ask them if they would like to comment.
- If people are starting to get out of line or off topic, remind them about the panel rules or refer to the video describing good and less-than-good panel behaviours.
- A simple way to stop someone rambling on is to look away from them and explain why you need to interrupt them and move on. If you are lucky, your timekeeper may begin do this for you by the swift glance at the watch and back at the panel leader.
- If someone demonstrates difficult behaviour, talk to them about it after the panel. Not tomorrow. Not next week. But today.

- Humour can be useful to create a nice relaxed working atmosphere.
- It's a good idea to have a flipchart 'parking sheet' for issues that were not able to be 'solved' today, as this way panellists will feel that you listened to them and valued their questions and opinions. It also helps in case you actually forget an important point about that sample or that attribute, so it can be a good memory jogger too.

12.5   How to deal with difficult people

Let's start with a question. How do you feel and how do you react when you are faced with a difficult person? Take a minute to think about it.

We might think, 'They're being really difficult!' As a consequence, we might feel defensive, frustrated, confused, worried, upset, angry… and we might react by being loud, obnoxious, arrogant and rude. The 'difficult person' then becomes defensive, frustrated, confused, worried, upset, angry… and they might react by being loud, obnoxious, arrogant and rude. If you are having a discussion with someone who is angry and you simply try to placate them, this can come across as not listening to their point of view, and they become even more angry in their frustration to be heard. It's a vicious circle with both parties feeling that it's the other person who is being difficult. So even though we think we are lovely people, we have a lot of potential to be viewed by others as 'difficult!' And difficult people (people who are always difficult, I mean – not you) believe that the world is a tough place, they believe the world is unsafe and full of difficulties… and there we are being difficult and so reinforcing their beliefs! And when we reinforce their beliefs, we reinforce their difficult behaviour… and therefore the difficulty will occur again. What can we do?

Let's split it up into a few parts to start with. Let's imagine you are at home telling someone else about that difficult person. You might start by saying, 'That person made me so angry!' But hang on a moment… no one can actually control your emotions but you – no one can make you feel a certain way unless you let them. If you do let them change how you feel, for example, if you act hurt, they may well take advantage. If you let them get under your skin, they will have power over you. And remember, the most important thing is that you *cannot* change *them*.

But you can stop them affecting you. You can be indifferent, neutral and control your own emotions – decide how you would like to react. You can be polite, understanding and even help them! Stay calm – keep your head up and if applicable try a smile. Do not take it personally – it's their problem, not yours… Press the pause button. You do not respond badly out of choice… What outcome would you like? Put yourself in their shoes – have some empathy. Why are they behaving in this way? Recognise their emotions. Compassion helps us deal with them. Be receptive to what they are saying – listen 100%.

One time, when I was running a training session for a group of new panellists, I had to deal with a difficult situation. The new panellists had all attended the first session and learnt something about the company and how their senses worked. They had also had time to meet the other panellists and have a friendly chat. It was the following day and 15 of the panellists had arrived but not Belinda. I started the session by saying that Belinda could catch up when she arrived and would one of them make some notes so she would know what she missed. After five minutes, she exploded into the room, angry because she had not found anywhere to park and had to leave her car in the road outside. She was shouting and being very rude. Her face was red and she was clutching her bag to her body with her hands in fists. I put down what I was doing and said to her that I would help her find a space. I took her to the window and we looked out at the rows of cars packed on the forecourt. I said to her, 'Goodness me, it's packed today! No-one is going to be going anywhere fast!' I pointed to a couple of cars that she could pull in behind, as I knew the members of staff would not mind being blocked in: she would be heading home long before they would.

When she came back from parking her car, she had composed herself and slipped quietly into the empty seat left for her. She quickly caught up with the task at hand and made a couple of great contributions in the session.

Let's try to understand how she was feeling. Firstly, she knew she was late and was probably kicking herself for not leaving earlier. She had become angry *with herself*. Secondly, she was embarrassed. She had to walk in to a room full of almost-strangers she had only met yesterday, *and* her new boss, and admit she was late. She had started to develop a relationship with these people and now she was worried what they might be thinking about her – fancy being late on what was only her second day! And lastly, she was so frustrated about not being able to park when she arrived, that she could not think what to do and had become even more angry! And hence the explosion into the room. Imagine what would have happened if I had returned her anger with, 'That's not my problem! Please stop disturbing my session!'

Next time you are in a similar situation and you don't want to be saying later, 'That person made me SO angry!' try to put yourself in their shoes. Listen to what they are saying. Face them and maintain eye contact if you can. If it's appropriate smile or at least arrange your face so it's not a replica of their anger. Focus entirely on what they are saying. Note their body language. If they are complaining, let them have their say. You could maybe ask them a question. For example, 'How do you think we could solve this?' Do not immediately offer a solution – try to reach a solution together. In my example above I gave Belinda the solution. Perhaps I might have asked her where she might park, but I was worried she might think I was repeating the issue at hand and trying to wind her up further. The solution depends on the situation.

One thing that is important when dealing with people in this type of difficult situation is the words that you use and your tone of voice. Someone once told me that whatever someone said before the word 'but' in a sentence was untrue. The person listening hears the word 'but' and feels that the beginning of the sentence is a lie. For example, 'I'm really interested in helping you out on your project, but...' or 'I did hope to attend that panel session but...'

If you read those sentences out loud you will hear the ending that the listener hears: 'I'm really interested in helping you out on your project, *but I'd rather not*' or 'I did hope to attend that panel session *but I have far too much to do to waste my time doing things for you*'. The speaker may not have meant it, but by including the 'but' they have created that imaginary (possibly) negative sentence in the ears of the listener.

You might be thinking that 'therefore' or 'however' could replace the 'but' instead, but I am afraid they are simply posh replacements for the word 'but'. If you do not believe me, try out the sentences above. Do you still hear the imaginary negative ending? Probably.

You might be hoping that I am going to give you a solution to this but… Well, actually I am and it's very simple. Just replace the 'but' with 'and'. Let's try that old favourite, 'I hear what you're saying but…' That translates to 'I did not hear what you're saying and I do not care'. Whereas 'I hear what you're saying *and…*' That translates to 'I heard what you said and I really care'.

Now, let's think about your tone of voice. Imagine you are explaining to someone that you heard what they said and you understand. If you can hear that evil little gremlin in your head adding to the end of your sentence '… this person is so annoying/ stupid/rude', I'm afraid that has the same result as simply saying, 'I didn't hear what you said and I don't care'. If you successfully use a neutral tone you will have been successful in getting over the message of 'I heard what you said and I really care'. You might have to practise!

Let's look at an example. Let's say you have been asked to create some descriptive profiles of a range of products over their shelf life. Someone else had been doing this role for a while, but they are not available to finish the job. You are quite excited about the idea as you are very motivated to ensure that consumers only buy products that taste as good at the end of shelf life as at the beginning. But the first panel session does not go well: two panellists turn up late; trying to get contributions to the attribute definitions was like pulling teeth; and a couple of panellists seem to have a rather nasty tone when talking to the others. Generally the panel do not seem to be engaged in the work and did not even seem to want to be there! It would probably be pointless to continue the next session without some interventions, so you get together with another couple of panel leaders to devise a plan. This is what you decide:

1. You will start the next session by explaining to the panel what you expect from them and what the goals of the project are (not mentioning shelf life obviously). You will demonstrate your enthusiasm for the project and how you cannot deliver to senior management without their support. Select one or two items from the motivating panellists list (see Section 12.4) and implement or plan.
2. You will then ask the panel to create their own list of ground rules so that they can 'police' these themselves. This might look similar to Table 6.2: Panellist dos and don'ts, but as this panel have been working together for a while, simply giving them the list might well be counterproductive. They might think that you are belittling them or treating them like children.
3. While the panel are creating the list, stay in the room (but appear to do some other work) so that you can listen in to the discussions.
4. Make mental notes about the various panellists to determine if there are one or two that may have caused the panel inertia or poor motivation. You can also use any observations from the panel session. If you find this is the case, speak to each person individually about how you will be needing their help in creating a great working environment. Focus on their behaviour (for example, 'I noticed that you did not contribute to discussions about that difficult attribute') and not on their attitude or your feelings. Ask them how they might contribute more in the future. Let them know that you will check in with them after session 5, say, to let them know how things are going.

Part four

The future of sensory panels

Future of sensory panels **13**

As the saying goes, 'It is difficult to make predictions, especially about the future', but we can at least look through the literature and topics at conferences and make some suppositions about what this means for the future of sensory panels. For example, at the 11th Pangborn conference held in Gothenburg in Sweden, four speakers presented their ideas about the future of sensory and consumer science research. The audience were invited to vote on the importance of these ideas for the future of our field (Jaeger et al., 2017b). There were four main topics delivered by four experts:

- Understanding individual differences in sensory perception (Joanne Hort).
- The role of context and situation in future research (Christelle Porcherot).
- Consumers' decision-making processes (Gastón Ares).
- Future sensory and consumer research in industry (Suzanne Pecore).

Many of these topics are related to the recruitment, training and data collection from sensory panels and are discussed in more detail below in the relevant sections. There are also many publications and websites related to the future of sensory science and I have tried to distil from these what the impacts are for sensory panels. External factors will also impact the future of panels in sensory science. Factors such as employment laws, working habits and technical advancements are also covered.

13.1   The future of the recruitment of sensory panels

There are several things that may change the way we recruit sensory panels in the future: artificial intelligence (AI), consumers versus trained panellists, online recruitment, gender specifications, panellist age, the 'gig economy' and job sharing.

One of these is the use of AI in recruitment. The initial person assessment steps in the job application process (filling out the forms, online games and the creation of interview videos) can be performed by algorithms (Fox Business, 2017). The AI filters out potential unsuccessful employees by assessing the speed of question answering, vocabulary use and facial expressions. In the example described by Fox Business, even the application process created automatically completed forms from candidates' LinkedIn profiles. What is also interesting about this from a sensory scientist's point of view (apart from the possibility of time-saving) is that the candidates played 12 short online games which assessed various skills such as memory and concentration. That would be a really neat screener for sensory panellists! The European Sensory Network's ConsumerFacets test is a good example of this type of approach for assessing consumer behaviour (Hübener, 2017). The test was built because the researcher found that segmentation on demographics did not 'sufficiently explain consumer behaviour' and therefore they wanted to understand more about consumers' personalities and motivations. There are 15 scales that can be used to determine psychological drivers of liking and choice, and the questions can be used in addition to the usual recruitment questionnaire.

The type of person we might recruit for specific tests will also change. Many new analytical sensory methods are conducted using untrained consumer panels as opposed to trained panellists. For more information see Section 13.2.

Panellist recruitment via the use of online panels is a growing area, and panellists can be recruited from databases across the globe in a few clicks. One thing we must ensure when making use of this excellent resource, is data quality. There is no point collecting data from consumers if they are not the right consumers or if they are not motivated to give us the best data. If this is the umpteenth test that they have taken part in this month, their data might well not be worth having. BS ISO 11136 (2014) recommends a period of at least three months between tests on the same product type.

Another change might be in the use of gender-specific questions. At the moment the majority of tests ask whether the consumer is male or female. Some tests include an additional 'prefer not to say', and others include a long list of other options (for example, genderless, third gender, transgender). We often include gender questions by default, but perhaps we could simply miss out the question if the information was not actually required in the analysis. There is also more interest in running analytical sensory panels with different ages (see, for example, Methven et al., 2016).

With the increase in different working structures and balanced lifestyles, such as flexible employment and job sharing, we might find panel recruitment becoming easier. As people work in different ways, perhaps taking on several jobs to balance child care, family commitments and hobbies, less people will be working '9 to 5' and will be available for panel work. Obviously, they will need to be interested in the role and happy to commit to the company employing them.

13.2 Running sensory tests 1: who should we use for our sensory tests?

In sensory science we have become very used to the idea that trained panels are used for analytical tests (discrimination and descriptive tests, for example), and naïve or untrained panels are used for hedonic tests (tests to understand liking and preference) (Meiselman, 2013). However, since the 1980s, there has been some blurring around the edges of this 'right person for the right job' type of segregation. Of course, no one is suggesting that our trained panel can provide us with actionable data relating to hedonics (although there are some who still seem to pursue this): our analytical panellists are trained to leave aside their likings and preferences, and obviously there is no way that a handful of panellists could be representative of the target market. However, the use of consumers to give sensory scientists analytical information about products has grown, and this growth has been fuelled by the rise in rapid methods as these can be performed by naïve consumers as well as by semi-trained or even highly trained panellists (Valentin et al., 2012). Many publications and presentations have shown that consumers can produce reliable analytical-type results, in fact we have known this since free choice profiling first started being used, so what should we do in the future? Do we need to train the people who take part in our sensory experiments or just recruit naïve consumers for our analytical tests each time?

Ares and Varela (2017) raised the question again about the choice of person to take part in sensory tests and the paper, as well as the papers from the six authors who subsequently commented, is an interesting read. One of the reasons for using a trained panel in discrimination and descriptive work is the assumption that they are more sensitive than consumers; however, Ares and Varela (2017) suggest that this is not down to sensory acuity per se but due to test and task familiarity. For example, Jaeger et al. (2017a) found that when using consumers for the rather demanding task of Temporal Check-All-That-Apply (TCATA), a familiarisation task (in a between-subjects study) increased discrimination between the samples. Labbe et al. (2004) reviewed several publications that assessed the difference between trained panellists and naïve consumers in sensory profiling tasks, prior to their experimentation to assess the impact of training. They classified the collection of publications into three: those that showed that training had a significant impact on panel performance, those that showed that training had a low impact and those that showed no impact of panel training. Their experiment was split into three stages with the panel profiling samples after each stage so that the results could be compared. Their experimentation started with the recruitment and screening of a group of people who had not taken part in sensory profiling before. These people profiled eight coffee products in duplicate and then commenced their training. The first part of the training consisted of familiarising the panel with the vocabulary for coffee and then learning to rank and then rate a selection of very different samples from the original set. Additional training was then given to the panellists based on their performance in the second profile. They then profiled four samples and their data was assessed. The data indicated that no further panel training was necessary, and therefore the panellists went on to rate all eight samples again. The authors assessed panel performance through an assessment of discrimination and consistency with the other panellists. They found that training helps the panellist become familiar with the products and that this increases the ability to detect differences. Comparison of the data from the first (untrained) profile to the last (trained) profile showed that panel performance had increased in terms of consistency, scale use and discrimination. It's difficult to isolate the impact of test and task familiarity from the whole training programme in this experiment, as with many others, as we do not have a 'control' group as used in the TCATA experiment (Jaeger et al., 2017a). Therefore, it is not easy to say categorically whether the training received by a panel starting out in their profiling careers improves discrimination due to test/task familiarity or due to some other aspect of the training process (Köster, 2003). In my experience it is a mixture of both. Certainly, once panellists have taken part in a test a number of times they can stop paying attention to *how* to do the test and start paying more (of that) attention to the product differences and similarities.

Another part of the question about who to use for the tests relates to relevance. Ares and Varela suggest that the use of consumers gives an output that is closer to what a consumer might perceive. Guerrero (2017) states that any of the sensory tests we ask people to take part in are 'far-removed from the actual consumption situation, so none of them can accurately guarantee the sensory acceptance of the product in the market'. When did you last do a napping of 20 jams or 10 liver patés, at least in your home environment? And if you were assessing *a* jam at home you may well have it spread with butter as well, on some rather tasty farmhouse loaf.

There is also the question of whether consumers are able to give reliable descriptive information due to reasons such as different interpretations of the same attribute name, less readily available vocabulary, or the use of hedonic-related descriptions ('I do not like it' is not especially helpful as a descriptor). Symoneaux (2017) says in his comments on the Ares and Varela paper: 'I wonder if consumers can really describe something that they do not know how to name? How can you describe an odour or an off-flavour when you have never smelt it before or learnt its name?' and this neatly sums up the other side of the argument. But in deciding who to use for your test doesn't it really depend on your objective? Maybe you really need consumer language to describe your product to use in marketing campaigns (use consumers) or maybe you really need to understand how the changes in ingredients and processing, change your product in an experimental design approach (use a highly trained panel). If you are trying to determine which of several samples to take forward to the next stage of analysis perhaps you could use any type of panel and have the option of conducting some training where required. As Guerrero (2017) says, the different panel types and different methods all have their own advantages and disadvantages, so why don't we use them in a complementary way rather than argue one way or another?

13.3   Running sensory tests 2: how should we be presenting products?

There have been many studies recently regarding the impact on the sensory perception of the product as related to the shape, colour, feel and size of the receptacle (glass, plate, bowl) the product is delivered in (see for example Mirabito, 2017; Cavazzana et al., 2017; Spence and Wan, 2015 for an excellent review). These experimenters, and many others, found that the size, shape and colour of the receptacle was able to change consumers' sensory perceptions of the product, resulting in products with increased odour intensity or different flavours, for example. Some authors also tested products in different receptacles with wine experts and found a similar effect. Taking all this into account, and as many products nowadays are consumed directly from the packet or container, perhaps we should pay more attention to the receptacles we use to present products to our trained panellists. We might also conduct experiments to help decide which receptacles give the 'best' profile of the products to the consumers and gain more understanding about how the change in perception occurs.

13.4   Running sensory tests 3: where or how should we conduct our tests?

There is also the question of ecological validity which was one of the voting points in Pangborn 2015 (Jaeger et al., 2017b) and was also brought up in Ares and Varela's paper regarding trained versus consumer panels (Ares and Varela, 2017). In the Pangborn 2015 voting, the focus was on conducting consumer tests in real or natural situations, or by suggesting a context via immersion methods perhaps through immersive virtual reality (VR). VR is the use of computer technology that can create a three-dimensional environment that the user can control. For example, the three-dimensional environment might be a shop with several shelves stacked with products. The user can 'pick up' the items and 'put' them in a virtual shopping trolley or basket. Experiments can be designed to be as close as possible to the reality of a standard shopping trip. It is even possible to use odours in VR. Combined with eye-tracking software, these experiments can give some very useful information about product choice. These are all incredibly interesting areas, and an increase in ecological validity for consumer tests makes a lot of sense, whether this is a result of testing in the home or restaurant, for example, or through the use of technology.

In the Ares and Varela paper the focus was on conducting *analytical* sensory tests in a more ecological manner, again with reference to immersive VR. Although this might be a fun, motivating task to do with the panellists, I am not entirely convinced of its relevance. In an analytical situation, say a discrimination test or a detailed description of the texture of a product, how relevant is an ecological situation? Can you imagine setting up a paired comparison test to assess whether an ingredient has made a change to your current product, say an ice cream, and asking the panellists to imagine themselves walking along a beautiful Cornish beach first? As Labbe (2017) says in his comments about the Ares and Varela suggestions for ecological validity in analytical sensory tests: '… for analytical tests, even though training and/or tasting protocols should be adapted for a product experience close to "consumer reality", if necessitated by the study objective, we should accept that the outcomes only partly predict consumer perception and acceptance'.

The use of technology to understand emotions through the use of facial recognition programs also seems to be getting a lot of attention (see, for example, Winthrope, 2017). Recording the eating experience in the sensory laboratory by video with an infrared thermal camera in each booth, allowed the researchers at Melbourne University to study unconscious reactions from consumers who might be too polite to give their honest opinions. The consumers' biometrics, such as body temperature, heart rate and facial emotional responses, are recorded to give this information and are then analysed using AI algorithms.

Galmarini et al. (2016) used sensory software to train the panellists to take part in a progressive profile from their home and presented the panellists with interesting snippets between waiting times to increase attention and motivation. With technology and software advances it is now possible to converse with panellists via the Internet (for example, using Skype or Google Hangouts) and therefore we could conduct training sessions in this way (if bandwidth allows!).

13.5   New methods and new uses for old methods

There always seems to be new interesting methods arriving on the sensory science scene, and the last couple of years has not let us down. Temporal Drivers of Liking (TDL) allows the researcher to assess how liking changes over the consumption period alongside a recording of the most dominant sensation through Temporal Dominance of Sensations (Thomas et al., 2017). It can be quite easy to see this method being extended to home and personal care products during use. Imagine a consumer using a product for cleaning laundry. TDL might be used to record liking and related sensations on package storage in the cupboard, selection of the right product quantity for the wash, transfer to the machine, assessment of foaming during the wash, removal of the clothes, drying, ironing and storage. This would allow a full understanding of liking over the whole use of the product and the related dominant attributes.

There will of course be a whole host of new analytical sensory methods or the resurrection of older methods such as the ABX task (Greenaway, 2017). With the increase in interest in multisensory and holistic approaches to understand consumers' behaviour and the rise in social media, linking across all sources of information could also be developed to build more thorough insights and allow us to design better and better products. The use of combinations of sensory methods to dive deeper into sensory characteristics seems very likely. For example, using napping with consumers and highly trained panellists and linking the data to give greater insights. The use of rapid methods to profile items such as products with short shelf life or to include descriptive profiling in projects at an earlier stage will help improve new product development with this new flexibility (Delarue et al., 2015).

With the rise in the interest in ethnography, observing people's behaviour in particular situations, one could almost imagine a new discipline: 'Sensenography'. Building on the TDL approach, a consumer might be asked to describe their sensations in a real or VR, setting like a shopping mall, for example, whilst also reporting their liking for the environment. The researcher would also use the ethnography approach to deliver insights into the usual manner. The same approach could be used for the consumption or use of a product. This could even be combined with the biometric recording and analysis as mentioned in Section 13.4 to give a complete consumer output package to understand conscious and unconscious behaviours.

Masson et al. (2016) described the comparison of six qualitative consumer measures used to assess a product, and the methods used all seem to have potential in linking analytical sensory and consumer sensory data to give more insight into the holistic nature of consumer liking and purchase behaviours. Interest in these areas seems to have risen from the use of consumers in sensory methods such as napping and sorting because it has been found that consumers do not necessarily 'sort' products based on analytical sensory terms but on more subjective and complex aspects such as natural, artificial, refreshing and healthful (Masson et al., 2016). The methods assessed by Masson and colleagues were a sorting task with verbalization, repertory grid, projective methods such as word association, sentence completion and image association, and a ranking preference method, and they compared these to the findings from a focus group. All these methods have potential for use alongside methods which generate analytical sensory terms (sweetness level, intensity of cocoa aroma, greasiness in a hand cream, for example) to develop our understanding of consumer behaviours not just with products but with packaging and purchase scenarios.

13.6   New products to be tested

According to Mintel (2017) the key trend for 2017 for beauty and personal care products is 'Active Beauty'. This is basically products that will help consumers stay healthy and fit, such as products that protect the hair from pollution or sunlight, or clothes that communicate with phone apps. Other examples include skin creams that continue to deliver moisturisation days after application, lipsticks that condition the skin, and skin creams with zero oily feel after application. Further development of microencapsulation for slow release or active release of fragrances will increase the need for testing with temporal methods especially time-intensity using discrete time points (discrete or discontinuous time-intensity, see Hort et al., 2017 for some interesting applications of temporal methods). Products with active encapsulation such as fabric conditioners (the encapsulate is broken on wearing), or shampoo that releases fragrance on brushing, or deodorants that release additional perfume with the onset of activity, will certainly be a challenge for sensory both analytically and in consumer testing.

What will we be eating in the future? There seems to be a lot of interest in food sustainability and growing enough food to feed an increasing global population, and with this interest has come the assessment of novel foods such as the consumption of insects, or entomophagy. Although this might be a novel foodstuff for many people reading this chapter, according to The Guardian (2010) insects are already eaten in 80% of nations. The assessment of new foodstuffs originating from insects such as meat patties, energy bars or chocolate brownies is just around the corner.

New diets like gluten-free, dairy-free and eating less carbohydrates has certainly changed western eating behaviours, and companies, whose portfolios are based on the pre-free diets are innovating to increase sales (The Guardian, 2017). Items like fizzy milk, crunchy cheese and chewy yoghurts with mealworms might well change the way we approach sensory testing of these products.

The increase in new foods relating to health, such as functional foods and beverages, dietary supplements, engineered nanomaterials, as well as foods such as soft drinks containing vitamins, minerals and live bacteria, or alternative protein sources such as plant-based eggs or 'brewed milk', means we should keep a closer eye on new food regulations and what these might mean for us in sensory science.

As the benefits become more widely known, sensory science will be used in more and more different situations and across a wide range of products, such as the assessment of medical environments, the comfort of home furniture, the feel and sound of carpets, toys, children's clothing, spectacles and contact lenses, public transport, food for space travel…

13.7   Software and hardware advances

The ability to conduct sensory tests anywhere through the use of mobile devices or through easy connection to the Internet, will continue to impact on sensory testing. In quality control this will enable tests to be conducted across the whole process, from ingredient dispatch at the supplier, to testing in the factory and through to product delivery, resulting in a system such as 'Sensory Analysis Critical Controls' akin to Hazard Analysis and Critical Control Points or HACCP (Findlay and Findlay, 2017). With the option also to give panellists immediate feedback and the tester rapid results, the use of technology in this way can only increase.

The continued use of databases to store panellists' data will continue to be useful in the future. Panel performance information can be assessed quickly and easily through the use of queries, graphs and filtering mechanisms within the database. Information about the tests a panellist has completed, the samples assessed, the results and even the time taken for the tests, can be gathered or automated allowing panellists to be selected for the next test based on their performance. For consumer tests, information about demographics, previous purchases, favourite brands, etc. can be set as filters for selection of people very quickly and easily. Scheduling the consumer can also be done in an automated fashion with invitation and confirmation emails.

Questionnaires can also be created from templates or built from scratch and reused over and over again within the software, making tests easy to set up and run. Most software providers offer many different types of tests and will also set up tests for you if you have specific and complex requirements. Tests can be conducted anywhere without any need to install software, as the majority of sensory software providers offer web-based solutions. This means that a quality control test might be created in the United Kingdom and then staff at the factory in Singapore might assess the products five minutes later. A consumer test might be created in New York, and consumers from all over the globe might participate. And with the ability to show photos, videos, virtual shopping constructs, almost anything can be tested.

Consumer tests via mobile phones have allowed researchers to ask questions across all stages of the product purchase journey, from choosing products off the shelf, to experience at the till through to use at home. For fast-food purchases this can also now include a questionnaire about the eating experience of your burger and fries.

With very fast analysis and automatic report construction it appears that even analysed results will be ready at the drop of a hat. But beware; this analysis will still require your input in the checking of panel performance and robustness of the data. If, on querying the consumer data, you find that those people who had the control first gave a completely different result to those who had the test product first, you might need to delve a little deeper and not just rely on auto-reporting without your input.

Other types of software and data visualisation will continue to help sensory scientists assess and present their data. Free online software such as PanelCheck (http://www.panelcheck.com/) to help visualise panel performance data is a great example. Google's new data visualisation tool (Millar, 2017) could be really useful for demonstrating the impact of sensory results. Other online software data visualisation tools can be incredibly helpful in showing the detail behind sensory data. For an excellent overview of data visualisation and some very useful examples, see Andy Kirk's website 'Visualising data': http://www.visualisingdata.com/.

13.8   Instrumentation and robotics

There are 12 references to the electronic nose and four to the electronic tongue in the Journal of Sensory Studies, although not all of these papers are specifically about these devices, more a statement about human sensitivity being better than instrumentation. In Food Quality and Preference there were 10 papers which referenced electronic noses and one for the electronic tongue. It seems that the assessment of these devices to replace sensory panels is losing impetus, although there have been some fairly recent articles about the use of electronic noses for the detection of disease, dangerous gases and chemicals (for example, see Reuters, 2016). Previous issues with these devices had been the sensor sensitivity and stability, the lack of communication (the device might well recognise a wine as having an off-odour but would not be able to tell you what the off-odour smelled like) and the lack of temporal information (for example, in the assessment of pet foods, the device is unable to communicate that the initial aroma is overpowering and then the meaty notes become apparent). Interpreting the data from these devices to help explain the sensory properties of the food can be difficult: the fingerprint or profile generated by the electronic nose and tongue can be matched to other samples, but this does not tell you actually what the product smelled and tasted like (Koppell, 2014). However, Rodríguez-Méndez et al. (2016) state that electronic noses and tongues are a valuable tool for the wine industry particularly for quality control applications. A recent book entitled *Electronic Noses and Tongues in the Food Industry* (Rodríguez-Méndez, 2016) includes examples of the devices in use for many foodstuffs such as spices, rice and tea for the electronic nose, and beer, coffee and fruit juices for the electronic tongue.

When robots become commonplace, I guess we will have robots that can smell and taste, as well as see, hear and touch, and as they will be able to communicate, unlike the electronic nose and tongue, perhaps they will replace sensory panellists (Explain That Stuff, 2017)? The question is how will they be programmed? Will they be programmed to match the sensitivity and variability of humans (Service, 2017)? Or will we just need the one robot in our sensory 'panel'? Will we actually be designing products for robots to use in the future and hence have robot sensory and consumer panels? How will the sensors be able to best capture the chemical senses? For a really interesting review of the use of smell in human-computer interaction see Efe (2017). For a glimpse into the future of digitising the chemical senses read Spence et al., 2017.

13.9   Links with other disciplines and research groups

There are many different disciplines with links in sensory science: psychology, marketing, neuroscience, chemistry, statistics, neurobiology, etc. and working with all of these groups will help increase the impact of sensory science. Understanding more about how the sense of taste works or studying taste disorders can help identify new tastes and even help people with taste issues. Working with groups such as Fifth Sense (http://www.fifthsense.org.uk/; http://www.smelltraining.co.uk/) who support people affected by smell and taste disorders may well benefit both parties.

Maybe we will also have more interactions with game designers and entertainment providers as they incorporate multisensory aspects into their offerings. For example, see LOLLio, an interactive lollipop that 'dynamically changes its taste' (Murer et al., 2013) although there does not seem to be any recent publications about this device. And also innovations such as 9D television (Wired, 2016) and scent-emitting mobile phone attachments (CPL aromas, 2017). We should also probably be involved in more serious applications with mobile phones like the detection of body odour (The Guardian, 2017).

Sensory scientists are already involved in sensory branding and this looks like it will grow and grow as there is more interest in consumers' emotional attachments towards different brands, advertising and packaging. Interest in the area grew after the publication of *Brand Sense: build powerful brands through touch, taste, smell, sight and sound* (Lindstrom, 2005) and it is still very popular today (see, for example, McEachran, 2016 and Sandys, 2017).

13.10   The training of sensory scientists

It has been almost a quarter of a century since Lawless (1993) first published his paper regarding the teaching of sensory science and since that time things have moved on apace. There are many new courses in sensory science both online and in person. There has also been developments in statistics training and data interpretation for sensory science included in books and in the form of courses. Many of these are given in Chapter 14. There are still not enough well-qualified sensory scientists to fill all the vacancies, and as the use of sensory science is still growing, training in sensory science needs to increase. The Institute for Food Science and Technology offers a registered sensory science qualification in the United Kingdom, and an extension to offer this globally would help sensory scientists gain professional recognition. In the United Kingdom, apprenticeships in food science are being offered and it would be good to see a sensory science equivalent. Sensory scientists being trained formally in universities and colleges as well as gaining valuable experience on the job via an apprenticeship of this sort, may help increase the number of well-trained sensory scientists in the job market.

Where to find more information about sensory panels

Table 14.1 gives a list of sensory science textbooks that all have useful elements: it is difficult to recommend just one. Kemp et al. (2009) is a good book to get an overview of sensory as it is a little quicker to read than the others. Lawless and Heymann (2010) and Stone et al. (2012) include a wealth of information, particularly about quantitative sensory profiling, and some very useful checklists. Meilgaard et al. (2016) is particularly useful for information about the Spectrum profiling method. For information about quantitative sensory profiling both Hootman (1992) and Gacula (1997) are useful. Hootman (1992) dedicates a chapter to each of four methods: The Flavour Profile; Quantitative Descriptive Analysis (QDA); the Spectrum Descriptive Analysis Method and the Texture Profile. A useful page at the start of the book compares the four methods in a table format. Gacula (1997) is useful for its reprints of some inaccessible papers and also industrial applications of sensory science methods. Lawless (2013) is an excellent text for all things regarding quantitative sensory profiling and Chapter 9 (Using subjects as their own controls) is very informative and helpful.

An excellent text for consumer testing is Jaeger and MacFie (2010) as it includes many industrial applications and innovative approaches. An earlier similar publication by MacFie (2007) has an excellent readable chapter about preference mapping and partial least squares. Resurreccion (1998) is also a useful text for consumer testing.

For more information about the senses, Mather's text (2016) is excellent with some very useful diagrams and explanations.

Books on 'rapid' methods include Varela and Ares (2014) and Delarue et al. (2015). They both have chapters about specific methods and often include the history of the method and why it was created, which makes for an interesting background read. The temporal methods book by Hort et al. (2017) is very useful with chapters about each of the temporal methods and their applications in home and personal care as well as food and beverages.

And of course, you cannot have a list of sensory science texts without Amerine et al. (1965) which is well worth reading for the historical aspects as well as the authors' views on particular sensory science aspects. You can watch Rose Marie Pangborn lecturing at UC Davis on YouTube (https://www.youtube.com/watch?v=F_eo8fgL2Tc) which, despite the playback quality, is very interesting.

Many of the sensory textbooks include aspects about panellist health and safety and ethics, and another good resource is IFST (2017) and the various publications from the Market Research Society (for example, see MRS, 2014). Several other very useful guidelines and checklists are accessible from the MRS website (https://www.mrs.org.uk/).

For more information about sensory statistics Naes et al.'s (2010) book *Statistics for Sensory and Consumer Science* is an excellent resource. Lea et al.'s (1998) *Analysis*

Sensory Panel Management. https://doi.org/10.1016/B978-0-08-101001-3.00014-8

Table 14.1 List of sensory textbooks

Lawless, H.T., Heymann, H., (2010). Sensory Evaluation of Foods: Principles and Practices, second ed.	Stone, H., Bleibaum, R.N., Thomas, H.A., (2012). Sensory Evaluation Practices, fourth ed.
Meilgaard, M., Civille, G.V., Carr, B.T., (2016). Sensory Evaluation Techniques, fifth ed.	Kemp, S.E., Hollowood, T., Hort, J., (2009). Sensory Evaluation, a Practical Handbook.
Gacula, M.C., (1997). Descriptive Sensory Analysis in Practice.	Hootman, R.C, (1992). Manual on Descriptive Analysis Testing for Sensory Evaluation.
Jaeger, S.R., MacFie, H.J.H., (2010). Consumer-Driven Innovation in Food and Personal Care Products and MacFie, H., 2007. Consumer-Led Food Product Development.	Resurreccion, A.V.A., (1998). Consumer Testing for Product Development.
Mather, G., (2016). Foundations of Sensation and Perception.	Lawless, H.T., (2013). Quantitative Sensory Analysis. Psychophysics, Models and Intelligent Design.
Varela, P., Ares, G., (2014). Novel Techniques in Sensory Characterization and Consumer Profiling.	Delarue, J., Lawlor, J.B., Rogeaux, M., (2015). Rapid Sensory Profiling Techniques and Related Methods: Applications in New Product Development and Consumer Research.
Hort, J., Kemp, S.E., Hollowood, T., (2017). Time-Dependent Measures of Perception in Sensory Evaluation.	Amerine, M.A., Pangborn, R.M., Roessler, E.B., (1965). Principles of Sensory Evaluation of Food, Academic Press, New York.

of Variance for Sensory Data is a must read for anyone working with this analysis method. Although now rather outdated, O'Mahony's (1986) *Sensory Evaluation of Food, Statistical Methods and Procedures* is very readable and gives some excellent examples of statistics in action.

Journals are also a very useful source of information about sensory panels and new methods. For home and personal care applications, Household and Personal Care Today is a peer reviewed bimonthly journal and has some interesting articles relating to neuroscience and emotions as well as sensory science. The International Journal of Cosmetic Science often has many sensory-related articles relating to hand cream, hair, nail polish and body lotions as well as raw materials and ingredients. The Society of Cosmetic Chemists publish the Journal of Cosmetic Science for members and journal subscribers. Cosmetics is open access and occasionally has sensory articles but the other articles in the journal can often be very interesting.

For food-related products the journal Flavour often has interesting articles relating to sensory, perhaps not unsurprisingly given its title. The Journal of Texture Studies, The Journal of Food Science, Food Science and Technology and Chemosensory Perception also publish very relevant articles.

For sensory science, the main journals are The Journal of Sensory Studies and Food Quality and Preference. Both of these journals publish many articles every year on a wide range of topics and products.

The standards organisations are also another useful source of information. The ASTM used to be called the American Society for Testing and Materials and was formed in 1898 in the United States. They changed their name to ASTM in 2001. The ASTM publishes a wide range of standards across the globe in various 'volumes' covering, for example, construction, paints, textiles, water and energy. The sensory standards are curiously grouped with vacuum cleaners and homeland security under Volume 15.08. There are currently 37 standards relating to sensory science for home and personal care products and food and beverages. The complete list of standards can be found at the link below or by searching 'ASTM sensory' on the internet. https://www.astm.org/Standards/sensory-evaluation-standards.html.

The ASTM standards include guidance on the assessment of specific products such as those for measuring chilli heat, deodorants and shampoos, as well as fundamental topics such as serving protocols and threshold determination. For discrimination tests, there are standards for the triangle, same-different, directional difference, paired preference, duo-trio and tetrad tests. There are also standards on sensory theory and statistics such as time-intensity methods and estimating Thurstonian differences.

It is well worth being a member of the ASTM as there are opportunities to help develop world class standards, comment on existing standards as they come up for review, as well as work with a group of friendly technical experts. https://www.astm.org/MEMBERSHIP/MemTypes.htm.

BSI is the United Kingdom's National Standards Body and works with the ASTM and other standards groups such as the International Organisation for Standardisation (ISO), across the globe. Standards are developed by panels of experts chosen for each technical committee. The sensory committee is called AW/12 Sensory Analysis and has 35 current published standards relating to sensory science. These include fundamental standards such as guidance for the design of test rooms and the recruitment and training of staff, through to more specific sensory methods such as profiling, magnitude estimation and discrimination tests. The sensory discrimination standards include the triangle, ranking, paired comparison and duo-trio.

The full list of sensory standards can be found at the following link or by searching for AW/12 on the BSI homepage. https://standardsdevelopment.bsigroup.com/committees/50001678. The ISO lists all the international standards relating to sensory: https://www.iso.org/ics/67.240/x/ including those currently under development.

There are also various groups and committees that publish useful information as well as hold conferences and meetings where you can network and meet other people working in the field of sensory science. These are listed in Table 14.2 with the website link to find more information. The Institute of Food Science and Technology (IFST) Sensory Science Group hold a yearly sensory science conference (in the United Kingdom) and run several workshops every year. The also manage the IFST's sensory science training and professional recognition accreditation.

Table 14.2 Sensory science groups and committees

Group	Website
The Institute of Food Science and Technology Sensory Science Group.	https://www.ifst.org/communities-technical-networks/sensory-science
The Society of Sensory Professionals	http://www.sensorysociety.org/Pages/default.aspx
The Sensometric Society	http://www.sensometric.org/
ASTM E18 Group: Sensory Evaluation	https://www.astm.org/COMMITTEE/E18.htm
ESOMAR	https://www.esomar.org/
Market Research Society	https://www.mrs.org.uk/

Membership of The Society of Sensory Professionals gives you access to The Journal of Texture Studies and The Journal of Sensory Studies as well as useful meetings and publications regarding sensory science and professional development.

The Sensometric Society work on the link between sensory and statistics and hold a conference every two years which is well worth attending. The Society's aims are to

- increase the awareness of the fact that the field of sensory and consumer science needs its own special methodology and statistical methods;
- improve the communication and cooperation between persons interested in the scientific principles, methods and applications of sensometrics;
- act as the interdisciplinary institution, worldwide, to disseminate scientific knowledge on the field of sensometrics.

ESOMAR and MRS provide excellent resources for the sensory scientist in the form of checklists, guidelines, online conference attendance and courses to name a few. The MRS guidelines can be found at https://www.mrs.org.uk/standards/guidance.

References

ACNFP, 2017. Guidelines on the Conduct of Taste Trials Involving Novel Foods or Foods Produced by Novel Processes. [Online] Available at: https://acnfp.food.gov.uk/committee/acnfp/acnfppapers/inforelatass/guidetastehuman/guidetaste.

Adams, D.R., Wroblewski, K.E., Kern, D.W., Kozloski, M.J., Dale, W., McClintock, M.K., Pinto, J.M., 2017. Factors associated with inaccurate self-reporting of olfactory dysfunction in older US adults. Chemical Senses 42 (3), 223–231. https://doi.org/10.1093/chemse/bjw108.

Adhikari, K., Chambers IV, E., Miller, R., Vazquez-Araujo, L., Bhumiratana, N., Philip, C., 2011. Development of a lexicon for beef flavor in intact muscle. Journal of Sensory Studies 26, 413–420.

Albert, A., Varela, P., Salvador, A., Hough, G., Fiszman, S., 2011. Overcoming the issues in the sensory description of hot served food with a complex texture. Application of QDA, flash profiling and projective mapping using panels with different degrees of training. Food Quality and Preference 22, 463–473.

Albert, A., Salvador, A., Schlich, P., Fiszman, S., 2012. Comparison between temporal dominance of sensations (TDS) and key-attribute sensory profiling for evaluating solid food with contrasting textural layers: fish sticks. Food Quality and Preference. ISSN: 0950-3293, 24 (1), 111–118. https://doi.org/10.1016/j.foodqual.2011.10.003. http://www.sciencedirect.com/science/article/pii/S0950329311002163.

Amerine, M.A., Pangborn, R.M., Roessler, E.B., 1965. Principles of Sensory Evaluation of Food. In: Food Science and Technology Monographs. Academic Press, New York, pp. 338–339.

Ares, G., Jaeger, S.R., 2015. Check-all-that-apply (CATA) questions with consumers in practice: experimental considerations and impact on outcome. In: Delarue, Lawlor, Rogeaux (Eds.), Rapid Sensory Profiling Techniques and Related Methods. Applications in New Product Development and Consumer Research.

Ares, G., Varela, P., 2017. Trained vs. consumer panels for analytical testing: fuelling a long lasting debate in the field. Food Quality and Preference. ISSN: 0950-3293, 61, 79–86. https://doi.org/10.1016/j.foodqual.2016.10.006. http://www.sciencedirect.com/science/article/pii/S0950329316302117.

l'Association Française de Normalisation, 2013. AFNOR NF V 09-502: Analyse sensorielle — Directives générales pour un suivi, par approche sensorielle, de la qualité d'un produit au cours de sa fabrication. La Plaine Saint-Denis Cedex: AFNOR.

ASTM E1083 – 00, 2011. Standard Test Method for Sensory Evaluation of Red Pepper Heat. ASTM International, West Conshohocken, PA. www.astm.org.

ASTM E1871, 2010. Standard Guide for Serving Protocol for Sensory Evaluation of Foods and Beverages. ASTM International, West Conshohocken, PA. www.astm.org.

ASTM E1879, 2010. Standard Guide for Sensory Evaluation of Beverages Containing Alcohol. ASTM International, West Conshohocken, PA. www.astm.org.

ASTM E1909 – 13, 2013. Standard Guide for Time-Intensity Evaluation of Sensory Attributes. ASTM International, West Conshohocken, PA. www.astm.org.

ASTM E2299, 2011. Standard Guide for Sensory Evaluation of Products by Children.

ASTM E2943 – 14, 2014. Standard Guide for Two-Sample Acceptance and Preference Testing with Consumers. ASTM International, West Conshohocken, PA. www.astm.org.

ASTM E544 – 10, 2010. Standard Practices for Referencing Suprathreshold Odor Intensity. ASTM International, West Conshohocken, PA. www.astm.org.

ASTM International, 2012. E1810–12 Standard Practice for Evaluating Effects of Contaminants on Odor and Taste of Exposed Fish. ASTM International, West Conshohocken, PA. www.astm.org.

ASTM International, 2013. E1909–E1913 Standard Guide for Time-Intensity Evaluation of Sensory Attributes. ASTM International, West Conshohocken, PA. www.astm.org.

ASTM International, WK 49780 N. New Guide for Standard Guide for Communication of Assessor and Panel Performance. Online reference: https://www.astm.org/DATABASE. CART/WORKITEMS/WK49780.htm.

ASTM International, WK 8435. New Guide for Measuring and Tracking Sensory Descriptive Panel and Assessor Performance. Online reference: https://www.astm.org/DATABASE. CART/WORKITEMS/WK8435.htm.

ASTM Standard E253 – 16, 2016. Standard Terminology Relating to Sensory Evaluation of Materials and Products. ASTM International, West Conshohocken, PA. https://doi. org/10.1520/E0253-16. www.astm.org.

ASTM Standard E679 2004, 2011. Standard Practice for Determination of Odor and Taste Thresholds by a Forced Choice Ascending Concentration Series Method of Limits. ASTM International, West Conshohocken, PA. www.astm.org.

ASTM Stock #DS72, 2011. Lexicon for Sensory Evaluation: Aroma, Flavor, Texture, and Appearance. ASTM International, West Conshohocken, PA.

ASTM STP758, 1981. Committee E-18, Guidelines for the Selection and Training of Sensory Panel Members. ASTM Special Technical Publication 758, ASTM International, West Conshohocken, PA. www.astm.org.

ASTM-E2164 E2164–08, 2016. Standard Test Method for Directional Difference Test. ASTM International, West Conshohocken, PA.

Bartoshuk, L.M., 1978. The psychophysics of taste. American Journal of Clinical Nutrition 31 (6), 1068–1077.

Bartoshuk, L.M., 2000. Comparing sensory experience across individuals: recent psychophysical advances illuminate genetic variation in taste perception. Chemical Senses 25, 447–460.

Bartoshuk, et al., 2002. Labelled scales (e.g., category, Likert, VAS) and invalid across- group comparisons: what we have learned from genetic variation in taste. Food Quality and Preference 14, 125–138.

Bartoshuk, L.M., Duffy, V.B., Green, B.G., Hoffman, H.J., Ko, C.W., Lucchina, L.A., Marks, L.E., Snyder, D.J., Weiffenbach, J.M., 2004. Valid across-group comparisons with labeled scales: the gLMS versus magnitude matching. Physiology and Behaviour 82, 109–114.

Beeren, C., 2016. Sensory: itchy or scratchy? Lumpy or smooth? How to screen for texture sensitivity. In: IFST SSG Workshop Reading, 25th February 2016. Viewed 14th March 2016.

Bengtsson, K., Helm, E., 1946. Principles of taste testing. Wallerstein Laboratories Communications 9, 171.

Bi, J., Kuesten, C., 2012. Intraclass correlation coefficient (ICC): a framework for monitoring and assessing performance of trained sensory panels and panelists. Journal of Sensory Studies 27, 352–364.

Bianchi, G., Zerbini, P.E., Rizzolo, A., 2009. Short-term training and assessment for performance of a sensory descriptive panel for the olfactometric analysis of aroma extracts. Journal of Sensory Studies 24, 149–165. https://doi.org/10.1111/j.1745-459X.2008.00200.x.

Blakeslee, A.F., 1932. Genetics of sensory thresholds: taste for phenylthiocarbamide. Proceedings of the National Academy of Sciences of the United States of America 18, 120–130.

Bliss, C.J., Anderson, E.O., Marland, R.E., 1943. Technique for Testing Consumer Preferences, with Special Reference to the Constituents of Ice Cream. Storrs Agricultural Experiment Station.

Borg, G., 1982. A category scale with ratio properties for intermodal and interindividual comparisons. In: Geissler, H.-G., Petxold, P. (Eds.), Psychophysical Judgement and the Process of Perception. VEB Deutxcher Verlag der Wissenshaften, Berlin, pp. 25–34.

Boyar, M.M., Kilcast, D., 1986. Review food texture and dental science. Journal of Texture Studies 17, 221–252. https://doi.org/10.1111/j.1745-4603.1986.tb00550.x.

Brandt, M.A., Skinner, E., Coleman, J., 1963. Texture profile method. Journal of Food Science 28, 404–410.

Brockhoff, P.B., Schlich, P., Skovgaard, I., 2015. Taking individual scaling differences into account by analyzing profile data with the mixed assessor model. Food Quality and Preference 39, 156–166.

Brown, W.E., 1994. Method to investigate differences in chewing behaviour in humans. Journal of Texture Studies 25, 1–16. https://doi.org/10.1111/j.1745-4603.1994.tb00751.x.

Brown, W.E., Langley, K.R., Martin, A., MacFie, H.J.H., 1994. Characterisation of patterns of chewing behaviour in human subjects and their influence on texture perception. Journal of Texture Studies 25, 455–468. https://doi.org/10.1111/j.1745-4603.1994.tb00774.x.

BS EN ISO 13299: 2016. General Guidance for Establishing a Sensory Profile. International Organization for Standardization, Geneva, Switzerland.

BS EN ISO 5495: 2007+A1:2016. Sensory Analysis-Methodology – Paired Comparison Test. British Standards Institute, London.

BS EN ISO 8586:2014 Sensory Analysis – General Guidelines for the Selection, Training and Monitoring of Selected Assessors and Expert Sensory Assessors.

BS EN ISO 8589:2010+A1:2014 Sensory Analysis. General Guidance for the Design of Test Rooms.

BS ISO 3972:2011 Sensory Analysis. Methodology — Method of Investigating Sensitivity of Taste. International Organization for Standardization, Geneva, Switzerland.

BS ISO 4121: 2003. Sensory Analysis — Guidelines for the Use of Quantitative Response Scales. International Organization for Standardization, Geneva, Switzerland.

BS ISO 5496: 2006. Sensory Analysis. Methodology—initiation and Training of Assessors in the Detection and Recognition of Odours. International Organization for Standardization, Geneva, Switzerland.

BS ISO 11132:2012. Guidelines for Monitoring the Performance of a Quantitative Sensory Panel.

BS ISO 29842:2011+A1:2015. Sensory Analysis — Methodology — Balanced Incomplete Block Designs.

Byrne, D.V.O., Sullivan, M.G., Dijksterhuis, G.B., Bredie, W.L.P., Martens, M., 2001. Sensory panel consistency during development of a vocabulary for warmed-over flavour. Food Quality and Preference. ISSN: 0950-3293, 12 (3), 171–187. https://doi.org/10.1016/S0950-3293(00)00043-4. http://www.sciencedirect.com/science/article/pii/S0950329300000434.

Cairncross, S.E., Sjostrom, L.B., 1950. Flavor profiles – a new approach to flavour problems. In: Gacula Jr., M.C., (Ed.), Descriptive Sensory Analysis in Practice. Food Technology. 4, 3–8.

Campden, BRI, 2017. Sensory Training Aids. [Online] Available at: https://www.campdenbri.co.uk/training/sensory-training-aids.php.

Castura, J.C., Findlay, C.J., 2006. A system for classifying sensory attributes. In: Presented at "A Sense of Diversity: Second European Conference on Sensory Consumer Science of Food and Beverages" (Poster Abstract P33).

Cardello, A.V., Schutz, H.G., 2004. Research note – numerical scale point locations for constructing the LAM scale. Journal of Sensory Studies 19 (4), 341–346.

Carrington, D., 2010. Insects Could be the Key to Meeting Food Needs of Growing Global Population. The Guardian. [Online] Available at: https://www.theguardian.com/environment/2010/aug/01/insects-food-emissions.

Cavazzana, A., Larsson, M., Hoffmann, E., Hummel, T., Haeh, A., 2017. The vessel's shape influences the smell and taste of cola. Food Quality and Preference 59, 8–13.

Chambers, E., Sanchez, K., Phan, U.X.T., Miller, R., Civille, G.V., Di Donfrancesco, B., 2016. Development of a "living" lexicon for descriptive sensory analysis of brewed coffee. Journal of Sensory Studies 31, 465–480. https://doi.org/10.1111/joss.12237.

Chaya, C., 2017. Continuous time-intensity. In: Time-Dependent Measures of Perception in Sensory Evaluation. Wiley-Blackwell.

Chen, Y.P., Chung, H.Y., 2016. Development of a Lexicon for commercial plain sufu (fermented soybean curd). Journal of Sensory Studies 31, 22–33. https://doi.org/10.1111/joss.12187.

Civille, G.V., Dus, C., 1990. A development of terminology to describe the handfeel properties of paper and fabrics. Journal of Sensory Studies 5, 19–32.

Civille, G.V., Szczesniak, A.S., 1973. Guidelines to training a texture profile panel. Journal of Texture Studies 4, 204–223.

Civille, G.V., Lawless, H.T., 1986. The importance of language in describing perceptions. Journal of Sensory Studies 203–215.

Compusense, 2015. Feedback Calibration Methodology: Efficiently and Effectively Train Your Descriptive Analysis Panel. Online resource: https://compusense.com/en/test-better/white-papers/.

Corollaro, M.L., Endrizzi, I., Bertolini, A., Aprea, E., Dematte, M.L., Costa, F., Biasioli, F., Gasperi, F., 2013. Sensory profiling of apple: methodological aspects, cultivar characterisation and postharvest changes. Postharvest Biology and Technology 77, 111–120.

Cotton, D., 2014. Managing Difficult People in a Week. Teach Yourself Books. Hodder and Stoughton, London, UK.

Cover, S., 1936. A new subjective method of testing tenderness in meat—the paired-eating method. Journal of Food Science 1, 287–295. https://doi.org/10.1111/j.1365-2621.1936.tb17790.x.

Coulon-Leroy, C., Symoneaux, R., Lawrence, G., Mehinagic, E., Maitre, I., April 2017. Mixed profiling: a new tool of sensory analysis in a professional context. Application to wines. Food Quality and Preference. ISSN: 0950-3293, 57, 8–16. https://doi.org/10.1016/j.foodqual.2016.11.005.

Cover, S., 1940. Some modifications of the paired-eating method in meat cookery research. Journal of Food Science 5, 379–394. https://doi.org/10.1111/j.1365-2621.1940.tb17199.x.

CPL aromas, 2017. Tech Scent: Olfactory Ideas for the Modern World. [Online] Available at: https://www.cplaromas.com/fragrance-trends/tech-scent-olfactory-ideas-for-the-modern-world/.

Crocker, E.C., Platt, W., 1937. Food flavors–a critical review of recent literature. Journal of Food Science 2, 183–196. https://doi.org/10.1111/j.1365-2621.1937.tb16509.x.

Croy, I., Olgun, S., Mueller, L., Schmidt, A., Muench, M., Hummel, C., Gisselmann, G., Hatt, H., Hummel, T., December 2015. Peripheral adaptive filtering in human olfaction? Three studies on prevalence and effects of olfactory training in specific anosmia in more than 1600 participants. Cortex. ISSN: 0010-9452, 73, 180–187. https://doi.org/10.1016/j.cortex.2015.08.018.

Dawson, E.H., Harris, B.L., 1951. Sensory methods of measuring differences in food quality. In: Conference Proceedings, Bureau of Human Nutrition and Home Economics.

DBS Checks for Working with Children. 2017. https://www.gov.uk/disclosure-barring-service-check.

de Bono, 1998. Edward de Bono's Super Mind Pack. Expand Your Thinking Powers with Strategic Games & Mental Exercise. Dorling Kindersley, London.

de Bouillé, A.G., 2017. The A-not-A test. In: Rogers, L. (Ed.), Discrimination Testing in Sensory Science. A Practical Handbook.

da Silva, dos Santos Navarro, R.C., Minim, V.P.R., Carneiro, J.D.S., Nascimento, M., Lucia, S.M.D., Minim, L.A., 2013. Quantitative sensory description using the optimized descriptive profile: comparison with conventional and alternative methods for evaluation of chocolate. Food Quality and Preference. ISSN: 0950-3293, 30 (2), 169–179. https://doi.org/10.1016/j.foodqual.2013.05.011.

Dehlholm, C., 2012. Descriptive Sensory Evaluations: Comparison and Applicability of Novel Rapid Methodologies. University of Copenhagen (Ph.D. thesis).

Delarue, J., Lawlor, J.B., Rogeaux, M., 2015. Rapid Sensory Profiling Techniques and Related Methods: Applications in New Product Development and Consumer Research. http://www.sciencedirect.com/science/book/9781782422488.

Delahunty, C.M., Eyres, G., Dufour, J.P., 2006. Gas chromatography–olfactometry. Journal of Separation Science 29, 2107–2125.

Delwiche, J.F., Buletic, Z., Breslin, P.A.S., 2001. Covariation in individuals' sensitivities to bitter compounds: evidence supporting multiple receptor/transduction mechanisms. Perception and Psychophysics 63, 761–776.

DeRovira, D., 1996. The dynamic flavour profile method. Food Technology 50, 55–60.

Diamond, J., Lawless, H.T., 2001. Context effects and reference standards with magnitude estimation and the labelled magnitude scale. Journal of Sensory Studies 16, 1–10.

Dijksterhuis, G.B., Piggott, J.R., 2001. Dynamic methods of sensory analysis. Trends in Food Science and Technology 11, 284–290.

Di Donfrancesco, B., Koppel, K., Chambers, E., 2012. An initial lexicon for sensory properties of dry dog food. Journal of Sensory Studies 27 (6), 498–510.

Dove, W.F., 1947. Food acceptability—its determination and evaluation. Food Technology 1, 39–50.

Dove, W.F., 1953. A universal gustometric scale in D-units. Food Research 18, 427–453.

Drake, M.A., Civille, G.V., 2003. Flavor lexicons. Comprehensive Reviews in Food Science and Food Safety 2, 33–40. https://doi.org/10.1111/j.1541-4337.2003.tb00013.x.

Drake, M.A., Gerard, P.D., Wright, S., Cadwallader, K.R., Civille, G.V., 2002. Cross validation of a sensory language for cheddar cheese. Journal of Sensory Studies 17, 215–222.

Efe, A., 2017. Using olfactory displays as a non-traditional interface in human computer interaction. Journal of Learning and Teaching in Digital Age (JOLTIDA) 2 (2), 14–25. Retrieved from: http://joltida.org/index.php/joltida/article/view/30/94.

Eggert, J., Zook, K., 1986. Physical Requirement Guidelines for Sensory Evaluation Laboratories. ASTM Special Technical Publication 913, ASTM International, West Conshohocken, PA.

ESOMAR Data Protection Checklist. 2016. https://www.esomar.org/what-we-do/code-guidelines/esomar-data-protection-checklist.

EU General Data Protection Regulation. https://www.itgovernance.co.uk/data-protection-dpa-and-eu-data-protection-regulation.

Everitt, M., 2010a. Designing a sensory quality control program. In: Kilcast, D. (Ed.), Sensory Analysis for Food and Beverage Quality Control: A Practical Guide. Woodhead Publishing Limited, Cambridge (Chapter 1).

Everitt, M., 2010b. Going forward – implementing a sensory quality control program. In: Kilcast, D. (Ed.), Sensory Analysis for Food and Beverage Quality Control: A Practical Guide. Woodhead Publishing Limited, Cambridge (Appendix).

ExplainThatStuff, 2017. Robots. [Online] Available at: http://www.explainthatstuff.com/robots.html.

Fernández-Vázquez, R., Stinco, C.M., Hernanz, D., Heredia, F.J., Vicario, I.M., 2013. Colour training and colour differences thresholds in orange juice. Food Quality and Preference. ISSN: 0950-3293, 30 (2), 320–327. https://doi.org/10.1016/j.foodqual.2013.05.018. http://www.sciencedirect.com/science/article/pii/S0950329313001134.

Ferris, G.E., 1956. Taste Panels. https://ecommons.cornell.edu/xmlui/handle/1813/32710?show=full.

Findlay, M.P., Findlay, C., 2017. The future of sensory discrimination testing. In: Rogers, L. (Ed.), Discrimination Testing in Sensory Science: A Practical Handbook, first ed. Woodhead Publishing (Chapter 16).

Fjaeldstad, A., Petersen, M.A., Ovesen, 2017. Considering chemical resemblance: a possible confounder in olfactory identification tests. Chemosensory Perception 10, 42. https://doi.org/10.1007/s12078-017-9226-6.

Food Standards Agency, 2016. Importing and Testing Trade Samples. https://www.food.gov.uk/business-industry/imports/importers/trade-samples-testing.

Fox, A.L., 1932. The relationship between chemical constitution and taste. Proceedings of the National Academy of Sciences of the United States of America 18 (1), 115–120.

Fox Business, 2017. In Unilever's Radical Hiring Experiment, Resumes Are Out, Algorithms Are in. [Online] Available at: http://www.foxbusiness.com/features/2017/06/26/in-unilevers-radical-hiring-experiment-resumes-are-out-algorithms-are-in.html.

Gacula, M.C., 1997. Descriptive Sensory Analysis in Practice. Food and Nutrition Press, Trumball, CT.

Galmarini, M.V., Symoneaux, R., Visalli, M., Zamora, M.C., Schlich, P., 2016. Could time–intensity by a trained panel be replaced with a progressive profile done by consumers? A case on chewing-gum. Food Quality and Preference. ISSN: 0950-3293, 48, 274–282. https://doi.org/10.1016/j.foodqual.2015.10.006. http://www.sciencedirect.com/science/article/pii/S0950329315002621.

Galmarini, M.V., Visalli, M., Schlich, P., 2017. Advances in representation and analysis of mono and multi-intake temporal dominance of sensations data. Food Quality and Preference. ISSN: 0950-3293, 56, 247–255. https://doi.org/10.1016/j.foodqual.2016.01.011. http://www.sciencedirect.com/science/article/pii/S0950329316300118.

Green, B.G., Dalton, P., Cowart, B., Shaffer, G., Rankin, K., Higgins, J., 1996. Evaluating the 'labeled magnitude scale' for measuring sensations of taste and smell. Chemical Senses 21, 323–334.

Green, B.G., Shaffer, G.S., Gilmore, M.M., 1993. Derivation and evaluation of a semantic scale of oral sensation magnitude with apparent ratio properties. Chemical Senses 18, 683–702.

Greenaway, R., 2016. Overcoming challenges faced when screening for texture sensitivity in the personal care sector. In: IFST SSG Workshop Reading, 25th February 2016. Viewed 14th March 2016.

Greenaway, R., 2017. ABX discrimination task. In: Discrimination Testing in Sensory Science. A Practical Handbook. Woodhead Publishing.

Gridgeman, N.T., 1959. Sensory item sorting. Biometrics. 15 (2), 298–306. https://doi.org/10.2307/2527675.

Griffin, L.E., Dean, L.L., Drake, M.A., 2017. The development of a lexicon for cashew nuts. Journal of Sensory Studies 32, e12244. https://doi.org/10.1111/joss.12244.

Guerrero, L., 2017. Comments on Ares and Varela paper. Food Quality and Preference 61, 87–88.

Hall, B.A., Tarver, M.G., McDonald, J.G., 1959. A method for screening flavour panel members and its application to a two sample difference test. Food Technology 8, 699–703.

Hayes, J.E., Allen, A.L., Bennett, S.M., 2013. Direct comparison of the generalized visual analog scale (gVAS) and general labeled magnitude scale (gLMS). Food Quality and Preference 28 (1), 36–44.

Helm, E., Trolle, B., 1946. Selection of a taste panel. Wallerstein Laboratories Communications 9, 181.

Hoehl, K., Schoenberger, G.U., Busch-Stockfisch, M., March 2010. Water quality and taste sensitivity for basic tastes and metallic sensation. Food Quality and Preference. ISSN: 0950-3293, 21 (2), 243–249. https://doi.org/10.1016/j.foodqual.2009.06.007.

Holway, A.H., Hurvich, L.M., 1937. Differential gustatory sensitivity to salt. American Journal of Psychology 49, 37–48.

Hootman, R.C., 1992. Manual on Descriptive Analysis Testing for Sensory Evaluation. ASTM MNL 13, Philadelphia. https://doi.org/10.1520/MNL13-EB.

Hopfer, H., Heymann, H., 2013. A summary of projective mapping observations – the effect of replicates and shape, and individual performance measurements. Food Quality and Preference. ISSN: 0950-3293, 28 (1), 164–181. https://doi.org/10.1016/j.foodqual.2012.08.017. http://www.sciencedirect.com/science/article/pii/S095032931200184X.

Hort, J., Kemp, S.E., Hollowood, T., 2017. Time-Dependent Measures of Perception in Sensory Evaluation. Wiley-Blackwell.

Hort, J., Kemp, S.E., Hollowood, T. (Eds.), January 2018. Descriptive Analysis in Sensory Evaluation. Wiley-Blackwell. Expected publication. http://eu.wiley.com/WileyCDA/WileyTitle/productCd-0470671394.html.

Hough, G., 2005. Commentary: two types of sensory panels or are there more? Journal of Sensory Studies 20, 550–552.

Hough, G., et al., 2006a. Workshop summary: sensory shelf-life testing. Food Quality and Preference 17, 640–645. https://doi.org/10.1016/j.foodqual.2006.01.010.

Hough, G., Wakeling, I., Mucci, A., Chambers IV, E., Méndez Gallardo, I., Alves, L.R., 2006b. Number of consumers necessary for sensory acceptability tests. Food Quality and Preference 17, 522–526.

HSE, 2002. Control of Substances Hazardous to Health (COSHH). http://www.hse.gov.uk/coshh/index.htm.

HSE, 2017. Controlling the Risks in the Workplace. http://www.hse.gov.uk/risk/controlling-risks.htm.

Hübener, F., 2017. Tell Me What You Feel and I Tell You What You Like. [Online] Available at: http://www.esn-network.com/research/findings/view/articles/tell-me-what-you-feel-and-i-tell-you-what-you-like/.

Hurst, D., 2017. No Sweat: App Aims to Alert Office Workers when They Start to Stink. The Guardian. [Online] Available at: https://www.theguardian.com/technology/2017/jul/13/no-sweat-app-alert-office-workers-stink.

Hwang, S.-H., Hong, J.-H., 2015. Determining the most influential sensory attributes of nuttiness in soymilk: a trial with Korean consumers using model soymilk systems. Journal of Sensory Studies 30, 425–437. https://doi.org/10.1111/joss.12176.

IFST, 2017. IFST Guidelines for Ethical and Professional Practices for the Sensory Analysis of Foods. [Online] Available at: http://www.ifst.org/knowledge-centre-other-knowledge/ifst-guidelines-ethical-and-professional-practices-sensory.

IFT, 2017. Code of Professional Conduct for Members of the Institute of Food Technologists. [Online] Available at: http://www.ift.org/About-Us/Governance/Code-of-Professional-Conduct.aspx.

Imamura, M., 2016. Descriptive terminology for the sensory evaluation of soy sauce. Journal of Sensory Studies 31, 393–407. https://doi.org/10.1111/joss.12223.

International Standards Organisation, 2004. ISO 4120:2004: Sensory Analysis – Methodology – Triangle Test. Online reference: http://www.iso.org/iso/catalogue_detail?csnumber=33495.

International Standards Organisation, 2005. ISO 6658:2005: Sensory Analysis – Methodology – General Guidance. Online reference: https://www.iso.org/standard/36226.html.

International Standards Organisation, 2012. ISO 11132:2012: Sensory Analysis – Methodology – Guidelines for Monitoring the Performance of a Quantitative Sensory Panel. Online reference: http://www.iso.org/iso/catalogue_detail.htm?csnumber=50124.

International Standards Organisation, 2012. ISO 8586:2012: Sensory Analysis – General Guidelines for the Selection, Training and Monitoring of Selected Assessors and Expert Sensory Assessors. Online reference: https://www.iso.org/standard/45352.html.

Inventory and Common Nomenclature of Ingredients Employed in Cosmetic Products (INCI). 1996. http://eur-lex.europa.eu/legal-content/EN/TXT/?uri=CELEX:31996D0335.

Ishihara, S., 1994. Ishihara's Tests for Colour Blindness. Kanahara, Tokyo. 38 p.

ISO 13300-1: 2006 Sensory Analysis. General Guidance for the Staff of a Sensory Evaluation Laboratory. Part 1: Staff Responsibilities.

ISO 13300-2:2006 Sensory Analysis —General Guidance for the Staff of a Sensory Evaluation Laboratory —Part 2: Recruitment and Training of Panel Leaders.

ISO 16657:2006 Sensory Analysis – Apparatus – Olive Oil Tasting Glass.

ISO, 1994. ISO 11036: Sensory Analysis – Methodology – Texture Profile.

ISO, 2012. ISO 8586: Sensory Analysis — General Guidelines for the Selection, Training and Monitoring of Selected Assessors and Expert Sensory Assessors.

Jack, F.R., Piggott, J.R., Paterson, A., 1994. Analysis of textural changes in hard cheese during mastication by progressive profiling. Journal of Food Science 59 (3), 539–543.

Jaeger, S.R., MacFie, H.J.H., 2010. Consumer-Driven Innovation in Food and Personal Care Products. Woodhead Publishing.

Jaeger, S.R., Beresford, M.K., Hunter, D.C., Alcaire, F., Castura, J.C., Ares, G., 2017a. Does a familiarization step influence results from a TCATA task? Food Quality and Preference. ISSN: 0950-3293, 55, 91–97. https://doi.org/10.1016/j.foodqual.2016.09.001.

Jaeger, S.R., Hort, J., Porcherot, C., Ares, G., Pecore, S., MacFie, H.J.H., March 2017b. Future directions in sensory and consumer science: four perspectives and audience voting. Food Quality and Preference. ISSN: 0950-3293, 56 (Part B), 301–309. https://doi.org/10.1016/j.foodqual.2016.03.006.

Jellinek, G., 1985. Sensory Evaluation of Food: Theory and Practice. Ellis Horwood.

Jose-Coutinho, A., Avila, P., Ricardo-Da-Silva, J.M., 2015. Sensory profile of Portuguese white wines using long-term memory: a novel nationwide approach. Journal of Sensory Studies 30, 381–394. https://doi.org/10.1111/joss.12165.

Jung, R., 1984. Sensory research in historical perspective: some philosophical foundations of perception. In: Handbook of Physiology, The Nervous System. Comprehensive Physiology 2011, Suppl. 3, 1–74. https://doi.org/10.1002/cphy.cp010301.

Kapparis, E., Pfeiffer, J.C., Gilbert, C.C., 2008. Campden & Chorleywood Food Research Association Group. Guideline No. 57. Guidelines for the Motivation of Sensory Panels within the Workplace.

Kemp, S.E., Hollowood, T., Hort, J., 2009. Sensory Evaluation: A Practical Handbook. Wiley-Blackwell.

Kennedy, J., Heymann, H., 2008. Projective mapping and descriptive analysis of milk and dark chocolates. Journal of Sensory Studies 24, 220–233.

Kiehl, E., Rhodes, V., 1956. New techniques in consumer preference research. Journal of Farm Economics 38 (5), 1335–1345. Retrieved from: http://www.jstor.org/stable/1234552.

Koppell, K., 2014. Sensory analysis of pet foods. Journal of the Science of Food and Agriculture. https://doi.org/10.1002/jsfa.6597.

Köster, E.P., 2003. The psychology of food choice: some encountered fallacies. Food Quality and Preference 14, 359–373.

Koç, H., Vinyard, C.J., Essick, G.K., Foegeding, E.A., 2013. Food oral processing: conversion of food structure to textural perception. Annual Review of Food Science and Technology. 4, 237–266. https://doi.org/10.1146/annurev-food-030212-182637.

Labbe, D., Rytz, A., Hugi, A., 2004. Training is a critical step to obtain reliable product profiles in a real food industry context. Food Quality and Preference. ISSN: 0950-3293, 15 (4), 341–348. https://doi.org/10.1016/S0950-3293(03)00081-8.

Laffitte, A., Neiers, F., Briand, L., 2017. Characterisation of taste compounds: chemical structures and sensory properties. In: Flavour from Food to Perception. Wiley-Blackwell, Oxford, UK.

Larson-Powers, N.M., Pangborn, R.M., 1978. Descriptive analysis of the sensory properties of beverages and gelatine containing sucrose and synthetic sweeteners. Journal of Food Science 43 (11), 47–51.

Lawless, H., 1993. The education and training of sensory scientists. Food Quality and Preference 51–63.

Lawless, H.T., Heymann, H., 1999. Sensory Evaluation of Foods: Principles and Practices, first ed. Aspen, Maryland.

Lawless, H.T., 2013. Quantitative Sensory Analysis. Psychophysics, Models and Intelligent Design. Wiley-Blackwell.

Lawless, L.J.R., Civille, G.V., 2013. Developing lexicons: a review. Journal of Sensory Studies 28, 270–281. https://doi.org/10.1111/joss.12050.

Lawless, H.T., Heymann, H., 2010. Sensory Evaluation of Foods: Principles and Practices, second ed. Chapman & Hall, New York.

Lawless, H.T., Horne, J., Spiers, W., 2000. Contrast and range effects for category, magnitude and labeled magnitude scales in judgements of sweetness intensity. Chemical Senses 25, 85–92.

Lawless, L.J.R., Hottenstein, A., Ellingsworth, J., 2012. The McCormick spice wheel: a systematic and visual approach to sensory lexicon development. Journal of Sensory Studies 27, 37–47.

Lê, S., Lê, T.M., Cadoret, M., 2015. Napping and sorted napping as a sensory profiling technique. In: Rapid Sensory Profiling Techniques and Related Methods. Elsevier.

Lea, P., Naes, T., Rodbotten, M., 1998. Analysis of Variance for Sensory Data. Wiley.

Legarth, S.V., Zacharov, N., 2009. Assessor selection process for multisensory applications. In: Proceeedings of the 126th International Convention of the Audio Engineering Society, Munich, Germany, Paper 7788.

Lelièvre-Desmas, M., Valentin, D., Chollet, S., 2017. Pivot profile method: what is the influence of the pivot and product space? Food Quality and Preference 61, 6–14.

Lindstrom, M., 2005. Brand Sense: Build Powerful Brands through Touch, Taste, Smell, Sight and Sound. Free Press.

Lepage, M., Neville, T., Rytz, A., Schlich, P., Martin, N., Pineau, N., 2014. Panel performance for temporal dominance of sensations. Food Quality and Preference 38, 24–29.

Lillford, P., 2017. Texture and breakdown in the mouth: an industrial research approach. Journal of Texture Studies. 00, 1–6. https://doi.org/10.1111/jtxs.12279.

Lim, J., 2011. Hedonic scaling: a review of methods and theory. Food Quality and Preference 22, 733–747.

Liu, J., Grønbeck, M.S., Di Monaco, R., Giacalone, D., Bredie, W.L.P., 2016. Performance of flash profile and napping with and without training for describing small sensory differences in a model wine. Food Quality and Preference. ISSN: 0950-3293, 48, 41–49. https:// doi.org/10.1016/j.foodqual.2015.08.008.

Louw, L., Malherbe, S., Naes, T., Lambrechts, M., van Rensburg, P., Nieuwoudt, H., 2013. Validation of two napping techniques as rapid sensory screening tools for high alcohol products. Quality and Preference 30, 192–201.

Lucak, C.L., Delwiche, J.F., 2009. Efficacy of various palate cleansers with representative foods. Chemosensory Perception 2, 32. https://doi.org/10.1007/s12078-009-9036-6.

Lund, C., Jones, V., Spanitz, S., 2009. Effects and influences of motivation on trained panellists. Food Quality and Preference 20, 295–303.

Lyon, D.H., 2002. Guidelines for the Selection and Training of Assessors for Descriptive Sensory Analysis. Guideline No. 37. Campden and Chorleywood Food Research Association Group.

MacFie, H., 2007. Consumer-Led Food Product Development. Woodhead Publishing.

September 2014. Market Research Society Code of Conduct. https://www.mrs.org.uk/standards/ code_of_conduct/.

Market Research Society MRS, 2017. MRS Guidance. [Online] Available at: https://www.mrs. org.uk/standards/guidance.

Martens, M., Risvik, E., Martens, H., 2013. Matching sensory and instrumental analyses. In: Understanding Natural Flavours. Springer.

Masson, M., Delarue, J., Bouillot, S., Sieffermann, J.-M., Blumenthal, D., 2016. Beyond sensory characteristics, how can we identify subjective dimensions? A comparison of six qualitative methods relative to a case study on coffee cups. Food Quality and Preference 47, 156–165.

Mather, G., 2016. Foundations of Sensation and Perception. Psychology Press, Oxford.

McEachran, R., 2016. Multisensory Branding: Immersing All Five Senses. Virgin. [Online] Available at: https://www.virgin.com/entrepreneur/multisensory-branding-immersing-all-five-senses.

McEwan, J.A., Hunter, E.A., van Gemert, L.J., Lea, P., 2002. Proficiency testing for sensory profile panels: measuring panel performance. Food Quality and Preference 13, 181–190.

McEwan, J.A., Heinio, R.-L., Hunter, E.A., Lea, P., 2003. Proficiency testing for sensory ranking panels: measuring panel performance. Food Quality and Preference 14, 247–256.

McMahon, K.M., Culver, C., Castura, J.C., Ross, C.F., 2017. Perception of carbonation in sparkling wines using descriptive analysis (DA) and temporal check-all-that-apply (TCATA). Food Quality and Preference 59, 14–26.

Meilgaard, M., Civille, G., Carr, B., 2016. Sensory Evaluation Techniques. Taylor & Francis, Boca Raton, Florida.

Meiselman, H.L., 2013. The future in sensory/consumer research: ....evolving to a better science. Food Quality and Preference. ISSN: 0950-3293, 27 (2), 208–214. https://doi. org/10.1016/j.foodqual.2012.03.002.

Meilgaard, M., Civille, G.V., Carr, B.T., 2006. Sensory Evaluation Techniques, fourth ed. CRC Press, Boca Raton, FL.

Methven, L., Jiménez-Pranteda, M.L., Lawlor, J.B., 2016. Sensory and consumer science methods used with older adults: a review of current methods and recommendations for the future. Food Quality and Preference. ISSN: 0950-3293, 48, 333–344. https://doi.org/10.1016/j. foodqual.2015.07.001.

Methven, L., Rahelu, K., Economou, N., Kinneavy, L., Ladbrooke-Davis, L., Kennedy, O.B., Mottram, D.S., Gosney, M.A., 2010. The effect of consumption volume on profile and liking of oral nutritional supplements of varied sweetness: sequential profiling and boredom tests. Food Quality and Preference. ISSN: 0950-3293, 21 (8), 948–955. https://doi.org/10.1016/j.foodqual.2010.04.009.

Meyners, M., 2016. Temporal liking and CATA analysis of TDS data on flavored fresh cheese. Food Quality and Preference. ISSN: 0950-3293, 47, 101–108. https://doi.org/10.1016/j.foodqual.2015.02.005.

Meyners, M., Castura, J., 2014. Check-all-that-apply questions. In: Novel Techniques in Sensory Characterization and Consumer Profiling. CRC Press.

Mhra Retention of Trial Records. 2015. http://forums.mhra.gov.uk/showthread.php?1917-Retention-of-Trial-Records.

Millar, M., 2017. Google's New Viz Tool Makes Snappy GIFs Out of Your Data. Fastco Design. [Online] Available at: https://www.fastcodesign.com/90127277/googles-new-data-viz-tool-makes-snappy-gifs-out-of-your-data.

Miller, A.E., Chambers IV, E., Jenkins, A., Lee, J., Chambers, D.H., 2013. Defining and characterizing the "nutty" attribute across food categories. Food Quality and Preference. ISSN: 0950-3293, 27 (1), 1–7. https://doi.org/10.1016/j.foodqual.2012.04.017.

Minoza-Gatchalian, M., de Leo, S.Y., Yano, T., 1990. Quantified approach to sensory panellist selection. Food Quality and Preference (4), 233–241.

Mintel, 2017. [Online] Available at: http://www.mintel.com/press-centre/beauty-and-personal-care/beauty-and-personal-care-trend-2017.

Mirabito, A., 2017. Glass shape influences the flavour of beer. Food Quality and Preference. https://doi.org/10.1016/j.foodqual.2017.05.009.

Mojet, J., Christ-Hazelhof, E., Heidema, J., July 2005. Taste perception with age: pleasantness and its relationships with threshold sensitivity and supra-threshold intensity of five taste qualities. Food Quality and Preference. ISSN: 0950-3293, 16 (5), 413–423. https://doi.org/10.1016/j.foodqual.2004.08.001.

Morse, R.L.D., 1942. Egg Grading and Consumers' Preferences with Special Reference to Iowa Egg Marketing. Iowa State University.

MRS, 2014. MRS Guidelines for Research with Children and Young People. https://www.mrs.org.uk/standards/guidance.

MRS Guidelines for Research with Children and Young People. January 2012. Updated September 2014. https://www.mrs.org.uk/pdf/2014-09-01Children%20and%20Young%20People%20Research%20Guidelines.pdf.

Muñoz, A.M., Civille, G.V., 1998. Universal, product and attribute specific scaling and the development of common lexicons in descriptive analysis. Journal of Sensory Studies 13, 57–75. https://doi.org/10.1111/j.1745-459X.1998.tb00075.x.

Munsell, 2017. Munsell Colour. [Online] Available at: http://munsell.com/.

Munoz, A.M., 1992. Sensory Evaluation in Quality Control. Springer.

Muñoz, M., Civille, G.V., Carr, B.T., 1992. Sensory Evaluation in Quality Control. Van Nostrand Reinhold.

Murer, M., Aslan, I., Tscheligi, M., 2013. LOLLio: exploring taste as playful modality. Proc. of TEI, 299–302.

Nachtsheim, R., Schlich, E., 2013. The influence of 6-n-propylthiouracil bitterness, fungiform papilla count and saliva flow on the perception of pressure and fat. Food Quality and Preference. ISSN: 0950-3293, 29 (2), 137–145. https://doi.org/10.1016/j.foodqual.2013.03.011.

Naes, T., Brockhoff, P.B., Tomic, O., 2010. Statistics for Sensory and Consumer Science. Wiley.

Ng, M., Lawlor, J.B., Chandra, S., Chaya, C., Hewson, L., Hort, J., 2012. Using quantitative descriptive analysis and temporal dominance of sensations analysis as complementary methods for profiling commercial blackcurrant squashes. Food Quality and Preference 25 (2), 121–134.

Olabi, A., Lawless, H.T., 2008. Persistence of context effects after training and with intensity references. Journal of Food Science 73, S185–S189. https://doi.org/10.1111/j.1750-3841.2008.00732.x.

O'Mahony, M., 1986. Sensory Evaluation of Food, Statistical Methods and Procedures. Wiley.

O'Mahony, M., Masuoka, S., Ishii, R., 1994. A theoretical note on difference tests: models, paradoxes and cognitive strategies. Journal of Sensory Studies 9, 247–272.

Peltier, C., Brockhoff, P.B., Visalli, M., Schlich, P., 2014. The MAM-CAP table: a new tool for monitoring panel performances. Food Quality and Preference 32, 24–27.

Peltier, C., Visalli, M., Schlich, P., 2016a. Benchmarking panel performances and sensometric techniques thanks to the SensoBase. In: Workshop Presented at Eurosense, Dijon, 12–14 September.

Peltier, C., Visalli, M., Schlich, P., 2016b. Multiplicative decomposition of the scaling effect in the mixed assessor model into a descriptor-specific and an overall coefficients. Food Quality and Preference 48, 268–273.

Pereira, J.A., Dionísio, L., Matos, T.J.S., Patarata, L., 2015. Sensory Lexicon development for a Portuguese cooked blood sausage – Morcela de Arroz de Monchique – to predict its usefulness for a geographical certification. Journal of Sensory Studies 30, 56–67. https://doi.org/10.1111/joss.12136.

Peryam, D.R., Swartz, V.W., 1950. Measurement of sensory differences. Food Technology IV (10).

Petit, C., Vanzeveren, E., 2015. Adoption and use of flash profiling in daily new product development: a testimonial. In: Delarue, J., Lawlor, B., Rogeaux, M. (Eds.), Rapid Sensory Profiling Techniques and Related Methods. Applications in New Product Development and Consumer Research.

Peyvieux, C., Dijksterhuis, G., 2001. Training a sensory panel for TI: a case study. Food Quality and Preference 12, 19–28.

Pickering, G.J., 2009. Optimizing the sensory characteristics and acceptance of canned cat food: use of a human taste panel. Journal of Animal Physiology and Animal Nutrition 93, 52–60.

Pineau, N., Schilch, P., 2015. Temporal dominance of sensations (TDS) as a sensory profiling technique. In: Delarue, Lawlor, Rogeaux (Eds.), Rapid Sensory Profiling Techniques and Related Methods. Applications in New Product Development and Consumer Research.

Pineau, N., Cordelle, S., Schlich, P., 2003. Temporal dominance of sensations: A new technique to record several sensory attributes simultaneously over time. In: Fifth Pangborn, Symposium, 20–24 July.

Platt, W., 1937. Some fundamental assumptions pertaining to the judgement of food flavours. Food Science 2 (3), 237–249.

Prinz, J.F., 1999. Quantitative evaluation of the effect of bolus size and number of chewing strokes on the intra-oral mixing of a two-colour chewing gum. Journal of Oral Rehabilitation 26, 243–247. https://doi.org/10.1046/j.1365-2842.1999.00362.x.

Purcell, S., 2017. Duo-trio. In: Rogers, L. (Ed.), Discrimination Testing in Sensory Science. A Practical Handbook.

Rainey, B.A., 1986. Importance of reference standards in training panelists. Journal of Sensory Studies 1, 149–154. https://doi.org/10.1111/j.1745-459X.1986.tb00167.x.

Regulation of the European Parliament and of the Council on the Provision of Food Information to Consumers, 2011. Regulation of the European Parliament and of the Council on the Provision of Food Information to Consumers. http://eur-lex.europa.eu/legal-content/EN/TXT/?qid=1506709426871&uri=CELEX:32011R1169.

Regulation of the European Parliament and of the Council on the protection of natural persons with regard to the processing of personal data and on the free movement of such data, 2016. Regulation of the European Parliament and of the Council on the Protection of Natural Persons with Regard to the Processing of Personal Data and on the Free Movement of Such Data. http://eur-lex.europa.eu/legal-content/EN/TXT/PDF/?uri=CELEX:32016R0679&from=EN.

Remblay, F., Backman, A., Cuenco, A., Vant, K., Wassef, M.A., 2000. Assessment of spatial acuity at the fingertip with grating (JVP) domes: validity for use in an elderly population. Somatosensory and Motor Research 17 (1), 61–66.

Resurreccion, A.V.A., 1998. Consumer Testing for Product Development. Aspen.

Reuters, 2016. 'Electronic Nose' Could Sniff Out Dangerous Chemicals. [Online] Available at: http://www.reuters.com/video/2016/11/28/electronic-nose-could-sniff-out-dangerou?videoId=370572630.

Risvik, E., McEwan, J.A., Colwill, J.S., Rogers, R., Lyon, D.H., 1994. Projective mapping: a tool for sensory analysis and consumer research. Food Quality and Preference 5, 263–269.

Rodríguez-Méndez, M., 2016. Electronic Noses and Tongues in Food Science. Academic Press.

Rodríguez-Méndez, M.L., De Saja, J.A., González-Antón, R., García-Hernández, C., Medina-Plaza, C., García-Cabezón, C., Martín-Pedrosa, F., 2016. Electronic noses and tongues in wine industry. Frontiers in Bioengineering and Biotechnology. 4, 81. https://doi.org/10.3389/fbioe.2016.00081.

Rogers, L., 2010. Sensory methods for quality control. In: Kilcast, D. (Ed.), Sensory Analysis for Food and Beverage Quality Control: A Practical Guide. Woodhead Publishing Limited, Cambridge(Chapter 4).

Rogers, L., Raithatha, C., 2012. International Panel Performance Survey Practices and Priority. Poster Presented at 2012 IFST Professional Food Sensory Group Conference. Available at: http://www.laurenlrogers.com/panel-performance.html.

Rogers, L.L., 2017. History of sensory discrimination testing. In: Rogers, L. (Ed.), Discrimination Testing in Sensory Science. A Practical Handbook.

Rousseau, B., Ennis, J.M., 2013. Importance of correct instructions in the tetrad test. Journal of Sensory Studies 28, 264–269.

Rousseau, B., Meyer, A., O'Mahony, M., 1998. Power and sensitivity of the same-different test: comparison with triangle and duo-trio methods. Journal of Sensory Studies 13, 149–173.

Rutledge, K.P., Hudson, J.M., 1990. Sensory evaluation: method for establishing and training a descriptive flavor analysis panel. Food Technology 44 (12), 78–84.

Salles, C., 2017. Release of tastants during in-mouth processing. In: Flavour from Food to Perception. Wiley-Blackwell.

Sandys, A., 2017. Diving into Sensory Experience. [Online] Available at: http://www.trans-formmagazine.net/articles/2017/diving-into-sensory-experience/.

Sauvageot, F., Herbreteau, V., Berger, M., Dacremont, C., 2012. A comparison between nine laboratories performing triangle tests. Food Quality and Preference 24, 1–7.

Schifferstein, H.N., 2012. Labelled magnitude scales: a critical review. Food Quality and Preference 26 (2), 151–158.

Schimmel, M., Christou, P., Herrmann, F., Müller, F., 2007. A two-colour chewing gum test for masticatory efficiency: development of different assessment methods. Journal of Oral Rehabilitation 34, 671–678. https://doi.org/10.1111/j.1365-2842.2007.01773.x.

Schutz, H.G., Cardello, A.V., 2001. A labelled affective magnitude (LAM scale) for assessing food liking/disliking. Journal of Sensory Studies 16 (2), 117–159.

Schwartz, N.O., 1975. Adaptation of the sensory texture profile method to skin care products. Journal of Texture Studies 6, 33–42. https://doi.org/10.1111/j.1745-4603.1975.tb01116.x.

Sensory Dimensions, 2017. What is Free Choice Profiling? [Online] Available at: https://www.google.co.uk/url?sa=t&rct=j&q=&esrc=s&source=web&cd=4&ved=0ahUKEw-jv3KPlm5_VAhVMAcAKHSZrB7kQFgg2MAM&url=http%3A%2F%2Fwww.sensorydimensions.com%2Ffiles%2F1514%2F1227%2F7779%2FWhat_is_Free_Choice_Profiling.pdf&usg=AFQjCNE1SZd36bgdZj64fgeR7wY4ijH6uA&cad=rjt.

Service, R.F., 2017. Artificial Intelligence Grows a Nose. Science. [Online] Available at: http://www.sciencemag.org/news/2017/02/artificial-intelligence-grows-nose.

Sheehan, M., Marti, V., Roberts, T., 2014. Ethical review of research on human subjects at Unilever: reflections on governance. Biothethics 28 (6).

Sieffermann, J.-M., 2000. Le profil Flash. Un outil rapide et innovant d'évaluation sensorielle descriptive. In: Agoral 2000, XIIèmes rencontres L'innovation: de l'idée au succès. Tec. & Doc., Montpellier, France, pp. 335–340.

Spence, C., Wan, X., 2015. Beverage perception and consumption: the influence of the container on the perception of the contents. Food Quality and Preference. ISSN: 0950-3293, 39, 131–140. https://doi.org/10.1016/j.foodqual.2014.07.007.

Spence, C., Obrist, M., Velasco, C., Ranasinghe, N., 2017. Digitizing the chemical senses: possibilities & pitfalls. International Journal of Human-Computer Studies. ISSN: 1071-5819. https://doi.org/10.1016/j.ijhcs.2017.06.003.

Stampanoni, C.R., 1994. The use of standardized flavor languages and quantitative flavor profiling technique for flavored dairy products. Journal of Sensory Studies 9, 383–400. https://doi.org/10.1111/j.1745-459X.1994.tb00255.x.

Stevens, S.S., Galanter, E.H., 1957. Ratio scales and category scales for a dozen perceptual continua. Journal of Experimental Psychology 54, 377–411.

Stevens, J.C., Marks, L.E., 1980. Cross-modality matching functions generated by magnitude estimation. Perception and Psychophysics 27 (5), 379–389.

Stewart, J.E., Keast, R.S.J., 2012. Recent fat intake modulates fat taste sensitivity in lean and overweight subjects. International Journal of Obesity 36, 834–842. https://doi.org/10.1038/ijo.2011.155.

Stone, H., 2015. Alternative methods of sensory testing: advantages and disadvantages. In: Rapid Sensory Profiling Techniques. Woodhead Publishing Series in Food Science, Technology and Nutrition, Woodhead Publishing. ISBN: 9781782422488, pp. 27–51. https://doi.org/10.1533/9781782422587.1.27.

Stone, H., Bleibaum, R.N., Thomas, H.A., 2012. Sensory Evaluation Practices, fourth ed. Academic Press, San Diego. ISBN: 9780123820860. https://doi.org/10.1016/B978-0-12-382086-0.00009-1.

Stone, H., Sidel, J., Oliver, S., Woolsey, A., Singleton, R.C., 1974. Sensory evaluation by quantitative descriptive analysis. Food Technology 28 (11) 24–34.

Symoneaux, R., 2017. Trained panelists versus consumers for sensory description: comments on the opinion paper of Ares and Varela. Food Quality and Preference. ISSN: 0950-3293, 61, 96–97. https://doi.org/10.1016/j.foodqual.2017.01.011.

Szczepanski, E., Peltier, C., Visalli, M., Schich, P., 2016. Longitudinal Tracking of Sensory Panel Performances and Diagnoses of Their Evolutions. Poster 242 Presented at Eurosense, Dijon, 12–14 September.

Szczesniak, A.S., 1963. Classification of textural characteristics. Journal of Food Science 28, 385–389.

Szczesniak, A.S., Brandt, M.A., Friedman, H., 1963. Development of standard rating scales for mechanical parameters of texture and correlation between the objective and sensory methods of texture evaluation. Journal of Food Science 211, 397.

Teillet, E., 2015. Polarized sensory positioning (PSP) as a sensory profiling technique. In: Delarue, Lawlor, Rogeaux (Eds.), Rapid Sensory Profiling Techniques and Related Methods. Applications in New Product Development and Consumer Research.

Teillet, E., Schlich, P., Urbano, C., Cordelle, S., Guichard, E., 2010. Sensory methodologies and the taste of water. Food Quality and Preference 21, 967–976.

Thomas, A., Chambault, M., Dreyfuss, L., Gilbert, C.C., Hegyi, A., Henneberg, S., Knippertz, A., Kostyra, E., Kremer, S., Silva, A.P., Schlich, P., 2017. Measuring Temporal Liking Simultaneously to Temporal Dominance of Sensations in Several Intakes. An Application to Gouda Cheeses in 6 Europeans Countries. ISSN: 0963-9969. Food Research International. https://doi.org/10.1016/j.foodres.2017.05.035.

Thuillier, B., Valentin, D., Marchal, R., Dacremont, C., 2015. Pivot profile: a new descriptive method based on free description. Food Quality and Preference 42, 66–77.

Thurstone, L.L., 1927. The method of paired comparisons for social values. Journal of Abnormal and Social Psychology 21, 384–400.

Ting, V.J., Romano, A., Silcock, P., Bremer, P.L., Corollaro, M.L., Soukoulis, C., Biasiloi, F., 2015. Apple flavor: linking sensory perception to volatile release and textural properties. Journal of Sensory Studies 30, 195–210.

Torgerson, W.S., 1958. Theory and Methods of Scaling. Wiley.

Valentin, D., Chollet, S., Lelièvre, M., Abdi, H., 2012. Quick and dirty but still pretty good: a review of new descriptive methods in food science. International Journal of Food Science and Technology 47, 1563–1578. https://doi.org/10.1111/j.1365-2621.2012.03022.x.

Van Boven, R.W., Johnson, K.O., 1994. The limit of tactile spatial resolution in humans: grating orientation discrimination at the lip, tongue and finger. Neurology 44 (12), 2361–2366.

van Hout, D.H.A., 2014. Measuring Meaningful Differences Sensory Testing Based Decision Making in an Industrial Context; Applications of Signal Detection Theory and Thurstonian Modelling (thesis). Erasmus University, Rotterdam.

Varela, P., Ares, G., 2014. Novel Techniques in Sensory Characterization and Consumer Profiling. CRC Press, Taylor and Francis Group, FL, USA.

Vene, K., Rosenvald, S., Koppel, K., Paalme, T., 2013. A method for GC–olfactometry panel training. Chemosensory Perception 6 (4).

Verriele, M., Plaisance, H., Vandenbilcke, V., Locoge, N., Jaubert, J.N., Meunier, G., 2012. Odor evaluation and discrimination of car cabin and its components: application of the "field of odors" approach in a sensory descriptive analysis. Journal of Sensory Studies 27, 102–110.

Vidal, L., Antúnez, L., Giménez, A., Ares, G., 2016. Evaluation of palate cleansers for astringency evaluation of red wines. Journal of Sensory Studies 31, 93–100. https://doi.org/10.1111/joss.12194.

Wallop, H., 2017. Fizzy Milk or Crunchy Cheese, Anyone? The Food of the Future. The Guardian. [Online] Available at: https://www.theguardian.com/science/2017/jul/08/food-of-the-future-fizzy-milk-crunchy-cheese-mealworms-pasta-science.

Whelan, V.J., 2017a. Ranking test. In: Rogers, L. (Ed.), Discrimination Testing in Sensory Science. A Practical Handbook.

Whelan, V.J., 2017b. Difference from control (DFC) test. In: Rogers, L. (Ed.), Discrimination Testing in Sensory Science. A Practical Handbook.

Williams, A.A., Langron, S.P., 1984. The use of free choice profiling for the examination of commercial ports. Journal of the Science of Food and Agriculture 35, 558–568.

Winthrope, S., 2017. Measuring Deliciousness. University of Melbourne. [Online] Available at: https://pursuit.unimelb.edu.au/articles/measuring-deliciousness.

Wired, 2016. We Got Sprayed in the Face by a 9D Television. [Online] Available at: http://www.wired.co.uk/article/9d-television-touch-smell-taste.

Worch, T., Lê, S., Punter, P., Pagès, J., 2013. Ideal profile method (IPM): the ins and outs. Food Quality and Preference. ISSN: 0950-3293, 28 (1), 45–59. https://doi.org/10.1016/j.foodqual.2012.08.001.

Worch, T., Crine, A., Gruel, A., Lê, S., 2014. Analysis and validation of the ideal profile method: application to a skin cream study. Food Quality and Preference. ISSN: 0950-3293, 32, 132–144. https://doi.org/10.1016/j.foodqual.2013.08.005.

World Medical Association Declaration of Helsinki, 1964. http://www.who.int/bulletin/archives/79(4)373.pdf.

Yang, Q., Ng, M.L., 2017. Paired comparison/directional difference test/2-alternative forced choice (2-AFC) test, simple difference test/same-different test. In: Rogers, L. (Ed.), Discrimination Testing in Sensory Science. A Practical Handbook.

Zook, K., Wessman, C., 1977. The selection and use of judges for descriptive panels. Food Technology 11, 56–61.

Index

Note: 'Page numbers followed by "f" indicate figures, "t" indicate tables and "b" indicate boxes.'

Printed in the United States
By Bookmasters